Pleistocene and Palaeolithic Investigations in the Soan Valley, Northern Pakistan

H. M. Rendell, R. W. Dennell
and M. A. Halim

British Archaeological Mission to Pakistan Series 2

Editors: F. R. Allchin, B. Allchin, S. M. Ashfaque

BAR International Series 544
1989

B.A.R.

122 Banbury Rd, Oxford, OX2 7BP, UK

BAR -S544, 1989: 'Pleistocene and Palaeolithic Investigations in the Soan Valley, Northern Pakistan'

ISBN 9780860546917 paperback
ISBN 9781407348391 e-book
DOI https://doi.org/10.30861/9780860546917
A catalogue record for this book is available from the British Library

This book is available at www.barpublishing.com

Contents

List of Figures

List of Tables

Preface

This is the first monograph of the British Archaeological Mission to Pakistan's Potwar Project, and the second in the Mission's monograph series. It therefore seems appropriate that it should include a brief account of the history of the project and a few words about that of the Mission as a whole. The Cambridge Archaeological Mission to Pakistan, as it was then called, was initiated by two of the editors (F.R.A. and B.A.) in 1977, in response to an invitation from the then Director of Archaeology in Pakistan, Mr Ishtiaq Khan. The first two seasons (1977-78 and 1978-79) were devoted to working on projects in the North West Frontier Province, in collaboration with the Department of Archaeology of the University of Peshawar, the first at Lewan and the second at Taraqai Qila. The results of the Lewan Survey and excavation were the subject of the first monograph in this series.

From the outset it was clear that in order to gain a comprehensive understanding of the early cultures of northern Pakistan it was necessary to know about their environment. This applied equally to early settlements of the Holocene with which we were primarily concerned in Bannu, and to the Palaeolithic sites of the Potwar Plateau. Throughout the region environmental change, particularly in relation to the shifting channels of rivers and seasonal torrents, and widespread erosion and gullying are taking place rapidly today. The writings of early European travellers in the region and other records leave no doubt that changes of this kind, sometimes on a massive scale, took place during recent centuries, and there are clear indications in the landscape itself of similar changes taking place during early historic and later prehistoric times. Clearly some of these erosional processes are and were accentuated by human activities, but fundamental changes, brought about by natural causes, have been going on continually and are part of ongoing processes that extend back into Quaternary times and beyond. In order to understand all this, and to relate the remains of past cultures accurately to changes in the environment, we needed the collaboration of a geologist. This we found in Dr Helen Rendell (now senior lecturer in the Geography Laboratory, University of Sussex) who has worked with the Mission since its early days in the North West Frontier Province.

In the course of the second season's work in Bannu, Ishtiaque Khan invited me to undertake a revision of the work of De Terra and Paterson in the Soan Valley (De Terra and Paterson, 1939). This invitation was particularly welcome, as I had had considerable doubts about some of their basic assumptions and observations after I first visited the area on a brief survey with my husband and Professor A.H.Dani in 1963. My doubts regarding these aspects of their work caused me to have serious reservations about their conclusions, which were drawn after only a few weeks spent in what is obviously a geologically highly complex region. At that time, in the environment of the Potwar Plateau, which then seemed as strange as the landscape of the moon, I could not define exactly what was wrong, but certain things certainly did not bear out De Terra and Paterson's descriptions.

For one thing, the features that appeared to be what De Terra had called terraces, and upon which their relative chronology depended, did not seem to bear much resemblance to any terraces I had seen, nor to the definitions of terraces which I had read. Furthermore the relationship of the concentrations of stone artefacts to these 'terraces' was puzzling. De Terra and Paterson claimed that palaeolithic artefacts were associated with these terraces, although they were often unclear whether the artefacts were in the make-up of a terrace, or on its surface. These observations, combined with De Terra and Paterson's assumption that models drawn from the glacial valleys and associated river terrace systems of northern Europe could be applied uncritically to the northern Punjab, left me uneasy and unconvinced. Ishtiaq Khan's invitation, coming as it did in 1978, some fifteen years after my first visit to the Potwar, was a challenge in every sense, not least because by then De Terra and Paterson's monograph had been regarded as one of the pillars of South Asian Prehistory for almost forty years.

In a real sense the principal investigators of this project have taken up the work where De Terra and Paterson left it in 1935. After this long haitus there have been many changes in approach. However palaeolithic archaeology remains the same in one important respect: the development of early man and his hominid ancestors in all its aspects is still the central focus of research, and archaeologists are still dependent upon geologists, and now on physicists also, to date their material and to provide a secure environmental framework. When – as here – the primary objective is to calibrate the palaeolithic record, the geologist bears an even greater responsibility. Therefore it is appropriate that this monograph follows De Terra and Paterson's in that the geologist's name appears first, although the project was one of equal collaboration between geology and archaeology and the overall objective archaeological. The monograph begins with an account of the earlier work on the geology of the Siwaliks and the Potwar Plateau, of which the Soan Valley is part. There follows a reassessment of De Terra's geological framework, with the reasons for rejecting it, and then a detailed account of Dr Rendell's revised framework for the period from the Late Pliocene to the Late Pleistocene. This provides the basis for understanding the recent geological history and the palaeolithic material of the area. Dennell's critique of Paterson's work reaches similar conclusions to Rendell's of De Terra's work. It is suggested that the Soan Flake Tradition is not a valid concept and that it should be abandoned. In the remainder of the volume the authors begin the task of filling the vacuum with the evidence they themselves have brought together.

It will be shown in the following pages that our new perception of the situation is based upon a clearer understanding of the processes by which the Siwaliks and the Potwar Plateau were formed. This has been made possible by advances in geology and dating techniques since De Terra and Paterson worked there in the 1930's; and by the depth of understanding that has more recently been reached of the nature and timing of climatic, geomorphological and geological changes during the last few million years. The region with which we are concerned lies close to the collision zone between India and the main land mass of Asia, and undoubtedly it is this continuing collision between the plates of India and central asia that provides a most fundamental force behind the creation of the landscape we see today.

These new developments within the field of geology enabled Dr Rendell to carry out her revision of the Quaternary geology of the region that has provided the basis of our research. Some key sections in the Potwar had already been dated by palaeomagnetism as a result of work done by other teams in the 1970's, and more has been done by her, in collaboration with Dr Ernie Hailwood (Department of Oceanography, Southampton University) as part of this project. A major part of Dr Rendell's work has involved the mapping and analysis of fluvial and aeolian deposits, as these at present provide the most important sources of information on the environment of the Soan valley in terms of changes in drainage patterns and the timing of phases of erosion and deposition. Another important aspect has been her investigations of the Upper Pleistocene loess deposits which cover much of this region. One area where she has made an original contribution has been in the development of thermoluminescence (TL) as a way of dating the previously undatable. It is unfortunate that the Soan valley is very poor in fossil animal remains, as these might have provided much useful additional information. However, the Mission's palaeolithic project has been highly successful in obtaining a long and rich sequence of these from the Pabbi Hills and elsewhere in the Jhelum and Gujrat districts, some 80km east of the area covered by this volume, and these will be the subject of the next monograph. There are also indications that the future work of the Mission will involve the study of chemical and magnetic properties of palaeosols, and these might prove rather more rewarding than the study of pollen, which seems to be poorly preserved.

Collaborating with experts in so many other disciplines, and depending heavily upon their work to give meaning to what is often slender and sporadic evidence of the presence and activities of our early ancestors, the prehistorian can feel almost redundant. His role seems to be more managerial than that of an expert in his own field: gone are the days of the great individualist who could claim a new discovery as his and his alone, if indeed this was ever really the case. In one sense the managerial image is the right one, for the goal of all this scientific investigation is to gain an understanding of the cultural and physical development of early man and his immediate, hominid, ancestors: the prehistorian who is directing a multidisciplinary operation has constantly to bear this in mind, and it is often difficult to keep the balance right.

Once we had embarked upon the Potwar project it was clear that another active, younger prehistorian would be needed to help carry it through, and in 1980 we invited Dr Robin Dennell (then lecturer in the Department of Archaeology and Prehistory, University of Sheffield) to join us. This proved particularly opportune as in the following year I suffered an accident which reduced the extent to which I could participate. Dr Dennell collaborated with Dr Rendell in carrying out the initial survey of the Soan Valley, and the Riwat area in particular, and in focussing attention upon finding palaeolithic artefacts and debitage in secure contexts that could then be dated. It was this work that has broken new ground in providing the first well-dated palaeolithic evidence from the Potwar Plateau. In 1981 and 1982, he also conducted the excavation of a palaeolithic site near Riwat that was buried under loess and dated by Dr Rendell to ca. 45,000 years old. The account of this excavation occupies four chapters of the monograph, and was undertaken in collaboration with Mr M. Halim, who acted as the Department of Archaeology's representative before becoming Director of the National Museum in Karachi. The excavation of an open-air palaeolithic site of this kind called for all the normal skills of an experienced excavator, plus the imagination and ingenuity to adapt methods and techniques developed in Europe to the radically different conditions of northern Pakistan. As this was the first excavation of its kind to be attempted in Pakistan, it was necessary to improvise and to import some of the equipment, and to train Afghan labourers into the necessary skills of careful trowel- and brush-work.

Since completing this excavation and survey Dr Dennell has been the Field Director of the Potwar project. The focus of our activities has shifted to the Pabbi Hills, south of the Jhelum river, where we have been investigating an area very rich in fossils between 1.0 and 2.4 million years old. It will clearly be of great interest if some of the stone artefacts found in the course of collecting these fossils turn out to be of similar antiquity to those from Riwat that are dated to ca. two million years old.

The project has depended throughout upon the goodwill and the skills of all those associated with it, both Pakistani and British. We owe much to the then Director of Archaeology, Government of Pakistan, Mr Ishtiaq Khan for his initial invitation to tackle the revision of the early prehistory of the Potwar Plateau, and for his support and encouragement during the period when the Riwat survey and excavation were underway, and to the present Director General, Dr Ahmed Nabi Khan, for his continuing support during the time this work was being completed and the next phase of research was beginning in the Pabbi Hills. The third editor of this series (M.A.) and currently Deputy Director-General of the Department of Archaeology and Museums, Government of Pakistan, was assigned to the Potwar Project as the representative of the Department of Archaeology on the occasion of our preliminary survey in the Soan Valley in 1979, and has remained ever since as an advisor and colleague. The representatives of the Department who have accompanied us in the field have always offered friendly advice, and help with our problems, both archaeological and logistical.

We may not always have agreed, but our discussions have always been stimulating, and it is to their credit, and ours, that they have led to greater friendship and understanding. During the time that the work described in this monograph was in progress the Department's representatives have included Dr M. Ashfaque, Mr Khan Mohamed, Mr M. Halim and Mr Salim ul-Haq. During the later stages of the project we have also enjoyed the collaboration of the Geological Survey of Pakistan which has extended the scope of our work, and been mutually beneficial in a number of ways. We greatly appreciate the support of the Director General and all the officers of the Geological Survey who have helped us in the field, in making arrangements and in discussion. The contribution of all those from Britain who have taken part in the project is beyond estimation. Those taking part in the field work have done so without remuneration, giving their services and skills in return for the interest of participating in the research and the opportunity to go to Pakistan. The senior members of the Mission have all had full-time teaching appointments, or the equivalents, and most of their work has been done during vacations or sabbatical leave - and on occasions has only been made possible by their taking unpaid leave. The same has applied to much laboratory work, data processing, drawing and writing of reports, papers and monographs carried out in Britain. Our Dating Programs which have been in large part financed by grants from the SERC-SBAC, Southampton University, and the South Asian Society, also owe much to the work of Dr Rendell and Dr Hailwood.

All this research, including both fieldwork in Pakistan and laboratory work in Britain has needed funds to keep it going. Throughout, our principal source of finance has been the British Academy, through the Afghan and South Asian Societies, and on a lesser scale the Ancient India and Iran Trust. We have also had grants from the Royal Society, SERC-SBAC, the Royal Geographical Society, the Wenner-Gren Foundation, the Creswell Crags Visitor Centre, Sheffield University, Southampton University, Sussex University, Churchill College Cambridge, and the Smuts and Crowther Benyon Funds (University of Cambridge). To all of these we would like to express our gratitude.

Although the fieldwork presented in this volume has been undertaken by a team that was sometimes as small as two, and with resources far smaller than enjoyed by many comparable projects, it is hoped that the results are of interest to Pleistocene and Palaeolithic research in general. The subcontinent has been greatly overshadowed in recent years by the wealth of discoveries made in East Africa relating to very early toolmaking, and by European and Near Eastern research into the origins of modern humans. Hopefully, this monograph will begin to redress the balance.

Bridget Allchin,
Cambridge, September 1989.

Acknowledgments

The authors would like to thank the many people who have helped us both in Pakistan and Britain during the course of this work. In Pakistan, we gratefully acknowledge the encouragement and support from Mr. Ishtiaq Khan, Director-General of Archaeology and Museums, Government of Pakistan in the early stages of this project; and Dr. Ahmed Nabi Khan, his successor. Dr S.M. Ashfaque represented the Department of Archaeology during the 1979/80 season, and Dr M. Sharif provided assistance during the spring 1982 season. Valuable help was given by Mr. Saleem ul-Haq (National Museum of Karachi) and Mr. Bahadur Khan (Taxila Museum) whilst representing the Archaeology Dept. in the field. Mr. M. Gulzar, Curator of Taxila Museum, is also thanked for providing storage facilities, and facilitating accommodation at the Nikra Bungalow, Taxila. After 1985, we also received valuable field assistance from the Geological Survey of Pakistan, and we are indebted to its then Director-General, Dr. Wahuddin Ahmad, and to Dr. Ibrahim Shah, Deputy Director-General, Dr. Zaki Ahmad (Deputy-Director) and Dr. Mahmoud Raza (Deputy Director). Mr. Munsif Khan is also thanked for his patience and care when driving on our behalf. The excavation of site 55 would not have been possible without the assistance of the Afghan workmen at Riwat, who were a privilege and pleasure to work amongst, and Rehat, the chowkidar.

We also gratefully acknowledge the interest and advice from the successive representatives of the British Council, Islamabad: Mr. and Mrs. David Latter, Mr. Michael Eltington, Mr. Roger Budd and Mr. and Mrs.Adam Ladbury.

Members of the British Mission team who participated in the work at Riwat included Dr. P. Carter, Dr. L. Hurcombe, Ms. H. Jones, Dr. P.M. McKean, Mr. E. Moth, Dr. S. Sutherland as well as Mr. S. ul-Haq, from the Department of Archaeology, and Mr. M. Sikander, from the Geological Survey of Pakistan. To all these we are very grateful. Mr. Moth is especially thanked for the part he played in the excavation of site 55 in the autumn of 1982; and in drawing the plans, sections and artefacts from that site.

In Britain, we wish to thank Drs. B. and F.R. Allchin, the Directors of the British Archaeological Mission to Pakistan for their continual support. We are particularly grateful for the financial support of SERC-SBAC to fund the TL dating work at Sussex University, and for the support of the Ancient India and Iran Trust and the University of Southampton for the palaeomagnetic work carried out in the Oceanography Department at the University of Southampton. The TL dating work would have been impossible to undertake without the collaboration of Prof. P.D. Townsend, and we are similarly indebted to Dr. E.A. Hailwood of the University of Southampton for his collaboration on the palaeomagnetism work. We thank Mr. D. Steer, Mr. D. Crawford, Mr T. Gibbs, Dr. D. Gordon and Ms. B. Grealy for their assistance in various aspects of the laboratory work.

Finally, we must express our gratitude to other colleagues for their advice and comments on many different aspects of the work presented in this volume. We are particularly grateful to Prof. M.J. Aitken, Prof. L. Binford, Dr. S.G.E. Bowman, Prof. J.D. Clark, Prof. J.G.D. Clark, Dr. N.C. Debenham, Dr. P.F. Friend, Dr. R.G.H. Raynolds and Dr. A.G. Wintle.

In the preparation of this volume, the text-setting assistance of Kevin Crisp and the drawing skills of Sue Rowland and Eddie Moth are gratefully acknowledged.

HMR/RWD Sussex & Sheffield, October 1989

Authors affiliations:

Dr H M Rendell,
Geography Laboratory,
University of Sussex,
Falmer,
Brighton BN1 9QN

Dr R W Dennell,
Department of Archaeology and Prehistory,
University of Sheffield,
Sheffield S10 2TN

currently on secondment until 1992 to:

Ancient India and Iran Trust,
Brooklands House,
23 Brooklands Avenue,
Cambridge CB2 2VG

M A Halim,
National Museum of Pakistan,
Burns Gardens,
Karachi 8,
Pakistan.

Introduction

This study presents the results of eight season's investigations into the Pleistocene and palaeolithic sequence of the Soan Valley, northern Pakistan (Fig 0.1). Such have been the changes over the last forty years in our understanding of human origins that few students of the palaeolithic outside India and Pakistan will have heard of this tributary of the Indus, or can appreciate why it was once considered so important to studies of the Asian palaeolithic. Yet the Soan occupies a major place in the archives of 20th century research into human origins. In 1935, De Terra and Paterson conducted a major campaign of fieldwork into the Pleistocene succession and palaeolithic occupation of the Soan Valley, and their monograph (De Terra and Paterson, 1939) formed the basis of our knowledge of early man in the sub-continent for the next forty years.

Few surveys have been so brief, and yet had so long-lasting a significance. As De Terra (1939: 4) recalls in his account, the survey that he directed in the Soan Valley was accomplished in little more than a month. He arrived in Rawalpindi in April, 1935, and was joined by Paterson on the 18th of that month. By May, Paterson had moved on to Kashmir to join De Terra shortly before the latter contracted typhus and was laid up until late June. All in all, they could not have spent more than six weeks at the very most in the Soan Valley, and more probably, they spent only a month there - roughly the length of our first season in the same area in 1981.

This point seems even more incredible to us now than it did eight years ago, when we began what eventually amounted to over a year's observation between us in just a small part of the Soan Valley, in the Riwat area. For those outside Britain, it is often difficult to comprehend the scale of the Indian landscape, and the contrasts between it and ones more familiar in western Europe. At the end of the first season, we felt we had only begun to appreciate the geological complexity of this area; to identify the main problems that faced us; and the need to undertake several seasons of fieldwork to obtain even a preliminary assessment. That De Terra and Paterson managed to write a comprehensive synthesis of this region on the basis of only a month or so's work might seem surprising; that their conclusions were accepted for the next forty years seems even more so.

So much has happened in early man research since De Terra and Paterson's fieldwork that their results cannot be appreciated without reference to some of the issues that dominated research into human origins in the late 1930's. One that was particularly relevant to De Terra and Paterson's work in the Soan Valley was the apparent conflict between the palaeontological and archaelogical evidence for where man evolved.

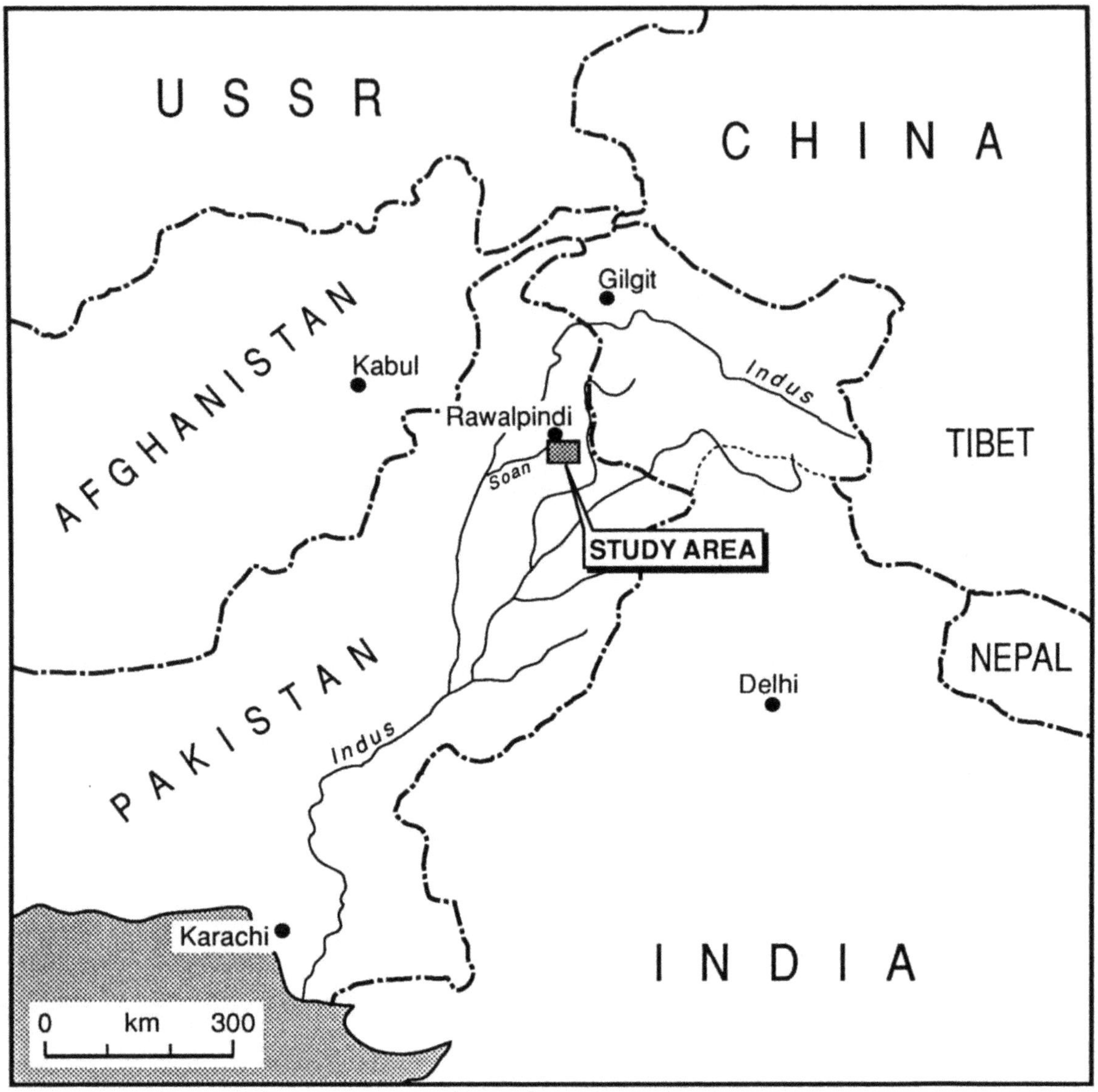

Figure 0.1: Location of study area

0.1 The palaeontological evidence for human origins in the 1930's: the EurAsian argument

In the 1930's, debates over where (and when) man evolved were even more intense than today, and many palaeontologists favoured Eurasia rather than Africa as the "cradle of mankind". Strange though it may seem by the standards of today, many palaeoanthropologists presented powerful theoretical arguments why the earliest evidence for man should have been found in the Siwaliks and adjoining regions of Asia, rather than in Africa. Matthew's (1914) *Climate and Evolution* had been one of the earliest attempts to apply the principles of biogeography and palaeoecology to evolutionary studies. He argued that the most primitive members of animal lineages would be found on the peripheries of their original distribution, and that the most advanced types would be near their centre of origin. According his theory, many animal lineages evolved in Central Asia: as it became higher, drier, more seasonal, and less forested, primitive types of animals would have migrated southwards into S.E. Asia and Africa without having to change. Those remaining would have had to evolve as a way of adapting to new circumstances, and in turn would later have colonised those areas occupied by their primitive fore-runners. In the 1920's, many realised the implications of this argument for studies of human evolution. Just as many today argue that our ancestors evolved in Africa because that is where we find our closest relatives - the gorilla and chimpanzee - so in the 1920's and 1930's, many argued the opposite, that they represent the descendants of primitive types of hominoids that had migrated from Asia before recognisable types of hominids existed. Osborn and Gregory of the American Museum of Natural History in New York were enthusiastic adherents of this view, which they modified to include the Siwaliks, arguing that the uplift of the Himalayas would have created the ideal conditions for early hominids (see e.g. Bowler, 1986: 173-185). It was with this model in mind that they supported Lewis's Yale-based expedition to the Siwaliks in 1934, in which De Terra participated, and which produced the remains of the Miocene hominoid, *Ramapithecus punjabicus.* Black (1925) also championed Matthew's views, and used them to argue for the importance of Chou Kou Tien (now Zhoukoudian) near Peking as an ideal place to find evidence of early man's earliest migrations out of Central Asia. Shortly before his death in 1934, he also encouraged De Terra to plan a similar campaign in the Siwaliks to cover the southern flank of this hominid diaspora (see De Terra and Paterson 1939: 2).

Such fossil evidence for human origins as then existed seemed to support the argument that hominids first evolved in EurAsia. After all, the oldest hominid remains, of *Eoanthropus dawsoni*, had been found at Piltdown in England (by a team that included none other than one of De Terra's group, the Jesuit Teilard de Chardin); and the next oldest remains of man found at that time were those of *Sinanthropus pekinensis* from Chou Kou Tien in China (Black 1926, 1934), and Dubois's finds of *Pithecanthropus erectus* in Java (Dubois, 1898). In addition, when De Terra and Paterson were conducting their fieldwork in 1935, Lewis (1934) had just found and reported on his discoveries in northern India of Miocene specimens of *Ramapithecus punjabicus*, which he described as "man-apes."

In Africa, by contrast, there were no comparable fossil human remains, except for those of the australopithecines, that most regarded (except for a few heretics such as Broome (1925) and Dart (1925)) as late in date, and indicative only of an extinct type of ape. The oldest acceptable human remains from East Africa were those found by Leakey (1936) at

Kanam and Kanjera, but their dating and context elicited considerable scepticism, and in any case neither seemed as ancient as the oldest from Asia. In addition, the human skelton found by Reck at Olduvai in 1913 seemed to be a comparatively recent burial.

To summarise, where better to look in the mid-1930's for our earliest ancestors than the Siwaliks of northern India? First, ever since Falconer's (1868) monumental study of Siwalik fossils, it was known to be immensely rich in fossil remains; secondly, as we have just seen, Lewis had recently found evidence there of a Miocene "man-ape"; thirdly, Todd (1932) had recently reported palaeolithic artefacts in the Bombay area (which De Terra and Paterson included on their itinerary); and fourthly, Asian research into human origins was encouraged and well-funded from American institutions such as Yale and the Carnegie Institute of Washington. It was thus entirely appropriate that De Terra was then based at Yale, and received sponsorship from the Carnegie Institute, which was shortly to fund also von Koenigswald's search for early hominids in Java (von Koenigswald, 1956: 92).

0.2 The archaeological evidence for human origins

Here, there was a notable discrepancy, in that Asia had produced hominid remains older than those from Africa, but the latter seemed to have earlier artefacts. Not surprisingly perhaps, Europe - the birthplace of palaeolithic archaeology - claimed the world's oldest artefacts for itself. Whilst many were doubting the so-called *"eoliths"* from southern England that supposedly belonged to the *"pre-glacial"* part of the Pleistocene, others did not, and people as respected as Reid Moir could command respect for his views on *"the culture of Pliocene man"* in his presidential address to the foremost prehistoric society of Britain (Reid Moir, 1932). Even without these *"eoliths"*, there seemed good stratigraphic evidence that the earliest bifacial implements from the Thames and Somme valleys dated to the first (*"Günz-Mindel"*) interglacial. Africa also appeared to have equally old, if not older, stone industries. In Uganda, Wayland (1934) had identified a Kafuan industry that was allegedly associated with the first (Kageran) *"pluvial"* of East Africa, and thus thought to be equivalent to the first Alpine glaciation, the Günz.

Asia had yet to produce artefacts as ancient as the oldest ones from Europe or Africa: despite the fossil hominid evidence, there were then no indications of Early Pleistocene tool-making in Asia, with the possible exception of those that von Koenigswald (1936) claimed from lower pleistocene contexts in Java shortly after De Terra and Paterson finished their fieldwork.

0.3 De Terra and Paterson's work

As we have seen, De Terra, based then in Yale, was encouraged by such palaeontologists as Black, Gregory, and Osborn to search in the Siwaliks for evidence of early man, which they believed would strengthen their arguments in favour of an Asian origin of man. In the end, the outcome of the De Terra and Paterson survey was, ironically, to support those archaeologists who claimed that tool-making began later in Asia than in Europe and Africa.

To put this in greater detail, they claimed that the earliest evidence for tool-making in the Soan Valley came from deposits that De Terra correlated with the **second** of the four glaciations that were supposed to have affected the Himalayas in the Pleistocene (De Terra and Patterson, 1939: 224-225). On the supposition that these glaciations had been synchronous world-wide, this conclusion meant that the earliest tools found by De Terra and Paterson were of the same age as those assigned to the second - or Mindel - glaciation in Europe. (see e.g. Leakey, 1953: 90 or Zeuner, 1946: 277). The further supposition that these four glaciations of Europe and the Himalayas corresponded to four *"pluvials"* in East Africa brought in the East African evidence. There, the second *"pluvial"* - the Kamasian - was thought to be represented at Bed I at Olduvai Gorge, where Leakey (1951) had already identified by 1935 an Oldowan stool-tool industry.

On these grounds, the earliest industries in the Soan were seen as being of early Middle Pleistocene age, and as old as those from Olduvai Bed I in East Africa, and the *"Mindel"*-age industries of western Europe. However, apart from a tentative *"pre-Soan"* industry (see chap.2), there was no indication of any early Pleistocene stone tool assemblages in the Indian subcontinent; as we have already seen, this did not appear to be the case in either Europe or Africa.

De Terra and Paterson thus strengthened the hand of those who maintained that Asia was settled later than Africa, the *"cradle of mankind"*. At the same time, they emphasised the distinctiveness of the Soan record, as being unlike the bifacial "hand-axe" tradition of Europe. Paterson in particular argued that the Soan Valley showed the local, gradual evolution of a flake-tool tradition that differed substantially from the bifacial industries of Africa, but which nevertheless displayed contact between people of different lithic traditions. Shortly after their work, their results became incorporated into Movius's (1948) grand synthesis of an Asian core and chopping tool tradition that co-existed in the Middle Pleistocene with the biface-and-flake traditions of Europe and Africa (see chap. 3).

These events are largely forgotten today, and the ideas of Matthew, Osborn, Gregory, Black and De Terra are treated little more than historical curios. As the evidence that hominids and tool-making first developed in Africa increased, so the Asian evidence declined in importance. For many reasons, few palaeoanthropological discoveries of importance were made in Asia after 1939, and to the present generation, the primacy of Asia in the search for human origins must now seem a quaint episode in the history of palaeoanthropology, and not the seemingly secure and immensely productive view it once was.

0.4 The longevity of De Terra and Paterson's work

The longevity of the results obtained by De Terra and Paterson affords a good - if somewhat rare - example of how a piece of academic research can survive through isolation. After 1945, their monograph remained dinosaur-like in an age of mammals. To put this point in a different way, it is as though no work took place at Olduvai (or elsewhere in East Africa) between 1939 and 1980. For people like us, faced with the task of resuming fieldwork in an area where no comparable work had taken place for forty-odd years, there are some substantial problems in following a trail that has grown so cold.

Since Terra and Paterson's work, there has been a world war that ended (for Europeans, the Americans and the Japanese) in 1945; the part of British India where they worked is now divided between India and Pakistan; compared with the 1930's, little fieldwork into human origins has taken place in Asia over the last thirty years; Piltdown Man and *"eoliths"* have been discredited; potassium-argon dating has pushed Leakey's Oldowan culture from Bed I at Olduvai back from the Middle Pleistocene to the late Pliocene; over the last twenty years, deep-sea cores have revolutionised our views on Pleistocene climatic change; no-one now believes in a four-fold glacial sequence for the Pleistocene, and fewer still can remember the plethora of names that marked each local sequence of glaciations or pluvials; and the last thirty years work into human origins has been dominated by evidence from Africa. Not surprisingly, De Terra and Paterson's work seems anachronistic.

Given this peculiar instance of academic survival in a subject noted for the high casualty-rate of ideas and claims, it is inevitable that much of our work is highly critical of our predecessors, and different from it in emphasis, outlook and results. Because their monograph is now almost old enough to be submitted to a dating laboratory, we have found it necessary to investigate the Soan Valley in two ways. One is in terms of what we have observed and found ourselves in our own fieldwork, and the other is by assessing their work as one would with a set of historical documents. For this latter reason, it has been necessary to voyage back into Pleistocene and palaeolithic work in the 1930's to see why they saw the past the way they did; and whether their observations are still valid even though the framework in which those observations were made has changed beyond all recognition. Put more precisely, three questions need to be answered: i) was De Terra basically correct in concluding that the earliest stone tool industries were Middle Pleistocene in date ? and ii) was Paterson correct in his views on lithic evolution in the Soan ? The third - and most difficult - question is what should be added to, or placed instead of, their results ?

This two-fold approach is reflected in the organisation of this volume. It begins where De Terra and Paterson ended. De Terra's geological conclusions are discussed in the next chapter, and Paterson's work in chapter 3. The rest of the volume presents our own material. Chapters 4 - 7 discuss the stratigraphy of the Soan valley, and the dating of palaeolithic artefacts that we have found. The results of excavating an open-air locality ca. 40-45,000 years old are presented in chapters 8-11. Finally, an overview of our results is presented in chapter 12.

Part I

The Background to the Present Study

Chapter 1

The Soan valley and the Pleistocene of northern Pakistan

"Nowhere else in the Potwar is the Pleistocene history so well recorded as along the Soan River and its tributaries" (De Terra and Paterson, 1939:279)

"The (Soan) area is thus a key to Siwalik stratigraphy and to the history of North-West India in the Pleistocene Period, and has been recognised as such by all previous workers" (Gill, 1952:380)

1.1 Introduction

The potential of the northwestern part of the indian subcontinent for studies of early human prehistory and Pleistocene geology was clearly recognised by De Terra (De Terra and Teilhard de Chardin, 1936; De Terra and Paterson, 1939). The combination of recent tectonic movements, Pleistocene glaciations, palaeolithic artefacts and pre-Pleistocene fossil hominoids made the Siwaliks as irresistible in the 1930's as they are today. De Terra's aim was to provide a chronology for the geological changes, the climatic fluctuations and the palaeolithic industries of what was then northern India. He did so armed with the tools available to a geologist in the 1930's. Inevitably he relied heavily upon the then standard techniques of stratigraphic and morphological correlation when working in the field, and related these observations to the growing body of evidence for four major worldwide glaciations during the Pleistocene, as first deduced by Penck and Bruckner (1909) for the Alpine foreland. Whenever major advances occur in a field of study, such as the development of radiometric and other dating techniques, the judgement of work predating such developments needs to be undertaken with care. Hindsight is often uncharitable, and is especially unfair if researchers are blamed for not using techniques that were then unavailable. In this chapter, we will focus on the geological sequence developed by De Terra, and on the basic tenets which underpinned his chronology, and we will end by demonstrating why his approach and results were ultimately flawed.

In undertaking fieldwork in the northern part of the subcontinent De Terra had at his disposal the work of previous researchers on both the Siwalik sequence (eg. Pilgrim, Wadia) and on the glacial history of the Himalayas (eg. Dainelli, Norin). In addition to this large body of secondary information, which included one inch to the mile geological sheets, he had the evidence of his own eyes. His perception of what he saw would have been conditioned by the way in which Pleistocene geology was practised in the 1930's. He would undoubtedly have been aware of the general consensus that the Pleistocene had featured four major worldwide glaciations, and that each glacial event was composite in nature (e.g. Eberl, 1930). He would also have been aware of Wayland's arguments for "pluvials" in East Africa, and debates as to whether these corresponded to northern latitude glacials or interglacials. De Terra undoubtedly saw the Indian subcontinent and southeast Asia as key areas for linking glacial and pluvial sequences. Like many other Pleistocene geologists of the time, he saw river valleys as providing the main key to unravelling the climatic history of the Pleistocene. The main assumptions were that major river valleys would contain sequences of terraces, of which the highest would have been the oldest, and the lowest the youngest, with each terrace corresponding to a continental-scale glacial or interglacial or, in lower latitudes, pluvial or interpluvial. One particular example of this type of study which certainly influenced Paterson (see chap.2), and possibly also De Terra, was that conducted in the Somme valley by Breuil and Koslowski (1931), not least because De Chardin had worked with both Breuil and De Terra. This type of study rested on the assumption that the terrace sequences were climatically-controlled and therefore that the terrace sequence was imposed on a tectonically stable landscape. This assumption of stability, although reasonably valid in northern France, is simply inapplicable in the Soan area, where the Siwaliks exhibit one of the most tectonically-eventful Pleistocene histories in the world.

Studies on the Ice Age in India (De Terra and Paterson, 1939) is a fascinating, and often puzzling text. Few geologists have managed to write 500 pages based on only six month's fieldwork that included a bout of typhoid and encompassed four areas as diverse as the Soan, the Kashmir and Narmada valleys, and Bombay. There are numerous minor errors (east substituted for west, map numbers wrongly transcribed). The text is full of descriptions and conclusions, but remarkably short of both analysis and explicit statements of underlying assumptions. Certain key sections in the first part of the volume, dealing with the Pleistocene geology of Kashmir, were written by Paterson. This change of authorship, presumably a result of De Terra's illness, reveals distinct differences in approach to Pleistocene geology. Paterson was clearly concerned with decoding morphology, and piecing together the history of a particular landscape using methods originally developed in the United States by W.M. Davis and other geomorphologists. De Terra, by contrast, put far more emphasis on stratigraphy and stratigraphic correlation, as practised in Europe.

1.2 De Terra's Pleistocene sequence

De Terra's aims appear to have been: first, the identification of deposits either directly (e.g. moraines) or indirectly (e.g. terraces, loess or erratics) related to Quaternary climatic change; second, the development of local sequences and, therefore, relative chronologies for the glaciated and unglaciated areas of northern India; and third, the regional correlation of these sequences. De Terra's sequence is summarised in Table 1.1.

De Terra's concern for stratigraphic correlation is reflected in a particularly revealing passage summarising the sequence in Kashmir:

> *"The reconnaissance of the Pleistocene history of the Kashmir Valley can be properly evaluated only if we can show that the geologic records are not peculiar to this region but that they are found also in neighbouring areas.*

Depositional Unit		Correlation with Alpine glacial sequence
Pink loam/silt/gravel	Terrace 4	4th glaciation (Würm)
Thin loam	Terrace 3	3rd interglacial
Potwar Loessic Silt	Terrace 2	3rd glaciation (Riss)
Upper Terrace Gravel	Terrace 1	2nd interglacial
- - - - - - - - - - - - - - - erosion/tilting - - - - - - - - - - - - - - -		
Boulder Conglomerate		2nd glaciation (Mindel)
- - - - - - - - - - - - - - - erosion/tilting - - - - - - - - - - - - - - -		
Pinjor Beds		1st interglaciation
Tatrot Beds		1st glaciation (Günz)
- - - - - - - - - - - - - UNCONFORMITY - - - - - - - - - - - - -		
Dhok Pathan		

Table 1.1: Relative Chronology of Pleistocene Deposits of the Potwar according to De Terra and Paterson (1939)

In the foregoing chapter, an attempt was made to correlate the stratigraphic sequence of Kashmir with that found in the foothills of the Pir Panjal. In this we were greatly aided by the fact that the glaciated tract merges into the nonglaciated region of the plains, in which fossiliferous Upper Siwalik beds are present. Indeed, the fossil and stratigraphic records of the Siwaliks, as worked out mainly by various members of the Geological Survey of India, provided a certain standard by which we could measure the correctness of our interpretation. This is especially apparent in the correlation of the Boulder conglomerate zone with the second glaciation, the conglomerate lying unconformably on tilted beds that contain an early Pleistocene fauna." (De Terra and Paterson, 1939:212).

The key to this chronology lies first in the recognition of a sequence of different deposits (moraines, outwash, fluvioglacial, alluvial) or landforms in the Kashmir Valley, its tributary valleys, in the valleys of the Indus, Jhelum and Chenab, and on the plains, and secondly, in the basic assumption that similar sequences, found in different areas, were synchronous.

In identifying the glacial deposits of northern India, De Terra was aided by previous work by Middlemiss (1896), Dainelli (1922) and Norin (1925). Evidence of glaciation was first recognised by Middlemiss (1896) in the Hazara area, where he noted the presence of moraines at 6,000 ft (1830m). Subsequently Norin (1925) identified terminal moraines at 6,800 ft (2070m) in Kashmir and at 8,000 ft (2440m) in Baltistan. Like Dainelli (1922), De Terra recognised four main phases of glaciation, but disagreed with Dainelli's dating of the sequence, preferring an early Pleistocene age for the earliest glacial phase. De Terra and Paterson argued for an early Pleistocene date ("first interglacial") for the Lower Karewas on the basis of palaeontological evidence. They continued

"The folding of these beds ... was simultaneously recorded by the disturbance of the older Upper Siwalik beds, which are unconformably overlain by Boulder Conglomerate. This conglomerate, merging into morainic outwash represents the great fan stage which Dainelli assigned to his second glaciation ... Moreover, the fauna of the Boulder Conglomerate and of its equivalents in central India is of Middle Pleistocene type, which argues against a later age, as do also the archaeologic records ... The great dividing line is the fan or Boulder conglomerate stage, which separates lower and upper Pleistocene, and this corresponds to the maximum glaciation of the Pir Panjal." (1939:214-215).

In Kashmir the "Boulder Conglomerate Stage" was associated by De Terra with moraines of the second glacial age. He appeared to have reasoned that all Boulder Conglomerate was the same age ie. second glacial and thus equivalent to the Mindel glaciation in Europe. The possibility that these deposits might be either of different ages or time-transgressive does not seem to have occurred to him.

The correlation of the "Boulder Conglomerate Stage" with the second glaciation lies at the heart of De Terra's chronology. By equating the Boulder Conglomerate with a glacial phase, De Terra was the first to identify a climatically-related deposit within the Siwalik sequence. This apparent breakthrough enabled him to link the Upper Siwalik sequence with the relative climatic chronology deduced by Penck and Bruckner (1909) for the Alps. Paterson's summary of the glaciation of the Sind and Liddar valleys (1939:78-80, Fig.58) even presented a chronology deduced on the basis of the archaeological record to determine

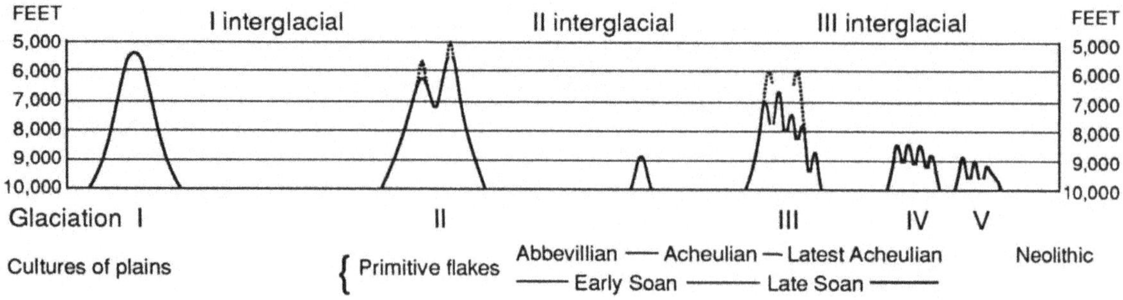

Figure 1.1: Paterson's diagram to illustrate the relative durations of glacial and interglacial periods (Himalayan slope of Kashmir)

the relative durations of glacial and interglacial phases. Paterson's figure illustrating the relative durations of glacials and interglacials is shown above (Figure 1.1)

In the absence of glacial moraines, De Terra's sequence for the Himalayan foothills rested on the identification of deposits that can be correlated with the glacial sequence established in Kashmir, namely; erratic blocks, loess deposits and, in particular, terrace sequences. The first two can be considered briefly before we turn to the terrace sequences, which require a much lengthier discussion.

Erratic blocks

The so-called erratic blocks of the northern Punjab attracted the attention of a number of geologists. Wynne (1879,1881), and later Cotter (1933), sought to explain the presence of a number of erratic blocks found in the vicinity of Attock in terms of catastrophic flooding by the Indus. On the other hand, Coulson (1938) favoured ice-rafting rather than catastrophic flooding as an explanation while De Terra and Teilhard de Chardin (1936:794) unequivocally identified a gravel deposit in the Attock area as fluvioglacial outwash containing *"large erratic blocks and glacially facetted boulders ... which can be traced back to the mountain front"*. De Terra and Teilhard de Chardin related this deposit to the Boulder Conglomerate, and, therefore, to the second glacial.

Loess

De Terra recognised loess as a glacially related deposit in the Potwar almost by default. His reasoning is rooted in a chronological argument which runs as follows:

(i) *"...the Potwar silt in the Soan area is cut into by a terrace, which is the third in a uniformly observed system of five terraces ..."*

(ii) *"...In ...Kashmir we found terrace 3 cut into the moraines of the third glaciation"*

(iii) *"It is our suggestion that the Potwar silt originated during the third glaciation".*

(iv) *"This relationship with the glacial cycle would speak in favour of the windblown origin of the Potwar silt ..."* (De Terra and Teilhard de Chardin, 1936:814).

Thus the loess of the Soan valley was thought to be of the same age as some of the glacial moraines of Kashmir: once again, similarity of sequence denoted synchroneity of age.

In this particular case, the alleged contemporaneity of the loess and a glacial moraine enabled De Terra to link a particular deposit to a specific type of climate.

1.3 De Terra's sequence in the context of the 1930's

De Terra's reliance on terrace sequences for glacial and interglacial correlations was very much in keeping with Pleistocene geology from the 1920's to the early 1950's. River terraces were commonly used to correlate the Quaternary sequences of areas that were not glaciated, and palaeontological evidence, where available, was used to confirm the general validity of regional correlations. These faunal data were rarely sufficiently precise to refine or refute a regional chronology; indeed in many cases the faunal assemblages were dated by the geological sequences rather than *vice versa.*

During the interwar years, the general validity of the four-fold glacial sequence that Penck and Bruckner (1909) had identified along the river valleys of the northern Alps seemed confirmed by data from Britain and North America. During the 1930's, the hegemony of northern European glacial sequence was reinforced by discovery of a apparently matching sequence of wet or *"pluvial"* phases that was identified in the terrace sequence along the Kagera and Vaal rivers in eastern and southern Africa (Söhnge *et al*, 1937; Wayland, 1934).

The use of terrace sequences as indicators of climatic change tended to be rather indiscriminate in two important respects. The first was the issue of tectonic instability, mentioned already, and the second concerns the assumption that glacials and pluvials were synchronous. Both of these issues were highlighted by Zeuner (1946, 1959), one of the most thoughtful critics of his time. He cautioned Pleistocene geologists as follows: *"Successions of climatic terraces are found in many countries. Where local conditions, or tectonic movements or changes in sea-level, have interfered with their development, it is often difficult, if not impossible to reconstruct the succession of climatic events ... In all cases a careful study of the system of terraces in question is needed in order to establish their climatic origin, or otherwise."* (1959:47).

Zeuner (1950) also questioned the rather simplistic equation of glacials and pluvials in a monograph based on fieldwork in peninsular India:

> *"The theory holds that pluvials occur all over the world during the phases of glaciation in temperate latitudes. Whilst it can be shown that this theory is justified in the strict chronological sense for the Mediterranean i.e. north of the Dry Belt, it is by no means certain whether the pluvial phases in the equatorial and monsoonal belts were contemporary with northern glaciations or not. The assumption has often been made that they were, but it has never been possible to prove this."* (1950:42)

Despite De Terra's recognition of the fact that the sequences and landforms he was studying reflected the interaction of climatic and tectonic agencies, his identification and interpretation of the terrace sequence in the Soan valley is entirely climatic. Again, his work has to be judged in the context of its time. He apparently opted for studying climatic change by assuming that the imprint of climatic events on a landscape would obscure effects due to local tectonic events.

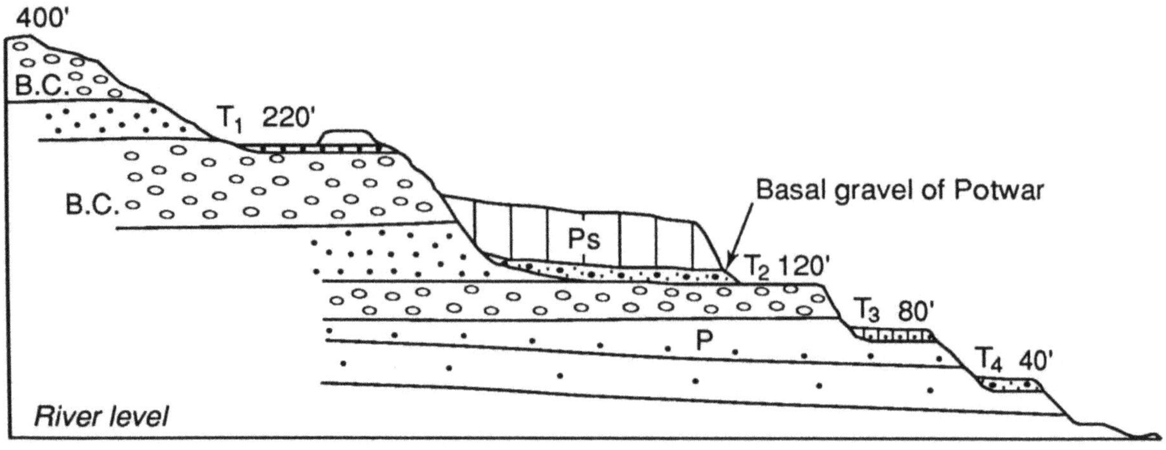

B.C., Boulder conglomerate; Ps, Potwar silt; P, Pinjor clay.

Figure 1.2: Idealised section through terrace sequence at Chauntra (De Terra and Paterson, 1939)

With this background in mind, we can now consider his terrace sequence in greater detail.

1.4 The Soan terrace sequence and associated palaeolithic sites

The Pleistocene chronology of the middle Soan valley was developed on the basis of the identification of four terraces, which postdate the deposit which De Terra mapped as "Boulder Conglomerate", and which were therefore younger than the second glacial. The terraces are correlated in turn with the second interglacial (T1), third glacial (T2), third interglacial (T3) and fourth glacial (T4). This sequence provided a relative chronological framework for the palaeolithic assemblages studied by Paterson. Close examination of De Terra and Paterson's text reveals that evidence for the existence of the terraces is far more elusive than one might imagine. Although a generalised section is presented of the terraces above Chauntra (1939:Fig.177), De Terra and Paterson (1939:282) admitted that no terraces were seen in the key area of the Soan valley to the southeast of Rawalpindi. In addition, terraces are marked specifically on only four of the 14 stratigraphic sections (1939:Figs.162-175) which accompany the description of the terraces and palaeolithic sites. The idealised figure of the terrace sequence at Chauntra (Figure 1.2) is literally too good to be true.

The Soan Valley Sites

De Terra and Paterson identified nine main sites or areas of palaeolithic artefacts within the middle Soan Valley, namely: Malakpur, Adiala, Hassan Khan Dhok, Milestone 163 (GT Road), "Section 11" (right bank of Soan, north of Chauntra), Chak Singhu, Balawal, Gila Kalan and Milestone 23 (Chakri-Sihala Road) (see Figure 1.3). The evidence for the age of these sites as presented by De Terra and Paterson will now be reviewed in some detail. In each case, details of the stone artefacts refer to Paterson's classification of tool typologies (De Terra and Paterson, 1939:301-312); these are discussed more fully in the following chapter.

At Malakpur evidence of a *"Late Soan Industry"* was found at the base of a section of loess at the junction with planed-off beds of tilted Dhok Pathan conglomerates. De Terra and Paterson commented *"... undoubtedly these (artefacts) were manufactured on a terrace (?T3) which at that time was less dissected than now* " (1939:284). Their reasoning appears to have been as follows: Late Soan industries require a younger terrace; therefore, if Late Soan industries are present then a younger terrace (ie.T3) must also have been present. There is, in fact, no evidence of a terrace at Malakpur, and all that this section indicates is that the assemblage was older than the loess and younger than the (Miocene) Dhok Pathan strata.

The same is true of the sites north of Adiala and near Hassan Khan Dhok, where artefacts were again found at the junction between the planed-off surface of Middle Siwalik Conglomerates and the overlying loess. These two sites yielded *"Late Soan"*, and rolled *"Early Soan"* artefacts and one *"Abbevillian-like handaxe"*. Although they speculated that the pebbles and gravel represent a fluvial deposit *"... the implements are somewhat rolled but not so much so that we must necessarily assume river transport"*, De Terra and Paterson commented at some length on the relationship between the underlying conglomerates and the artefacts, remarking that:

> *"This deposit provided the Soan people with ideal raw material (quartzite pebbles) for the manufacture of tools; in fact, all these sites are so closely associated with outcrops of the Middle Pliocene conglomerates that one can almost predict sites wherever this formation underlies the peneplain."* (1939:286).

These remarks echo Wadia's earlier comment that struck stones are found wherever suitable raw material is encountered (Wadia, 1928:346). They also highlight De Terra and Paterson's dilemma: were they studying deposits that were exposed when artefacts were made and discarded, or fluvial deposits in which stone tools were incorporated? It is this relationship between palaeolithic sites and sources of raw material that gives us the key to understanding the distribution of palaeolithic sites in the Soan Valley, and to realising the difficulties of dating them.

These uncertainties are clear in De Terra and Paterson's evaluation of the Milestone 163 site (or complex of sites). This site yielded three *"cultural layers"* related to the "Boulder Conglomerate" and the overlying "Potwar Gravel" and "Potwar Loessic Silt". The upper few inches of the "Boulder Conglomerate" contained *"the oldest industry represented by rolled crude hand axes of Acheulian type and "Pre-Soan" flakes."* De Terra and Paterson continued: *"... very probably the conglomerate at this horizon is an old surface or terrace remnant of T1."* (1939:282). The basal gravel associated with the loess contained *"Early Soan"* tools while fresh flakes of *"Late Soan B"* were found in the base of the loess. The relationship between the Milestone 163 complex and the terrace sequence remains unclear since *"... the Soan ... has here denuded the terraces, some of which reappear a few miles downstream"* (1939:282), with only the oldest horizon tentatively associated with T1. Their assertion is that if a terrace had existed here it would have had a very early date, thus confirming the apparent age of the artefacts.

In contrast, at the Section 11 site, across the Soan to the north of Chauntra, worked pebbles of *"Early Soan"* type were found on a conglomerate ledge. De Terra and Paterson comment that *"... the high altitude of this terrace ledge suggests that this is T2 of Third Glacial age"* (1939:287). This remark contradicts their earlier observation that: *"the right*

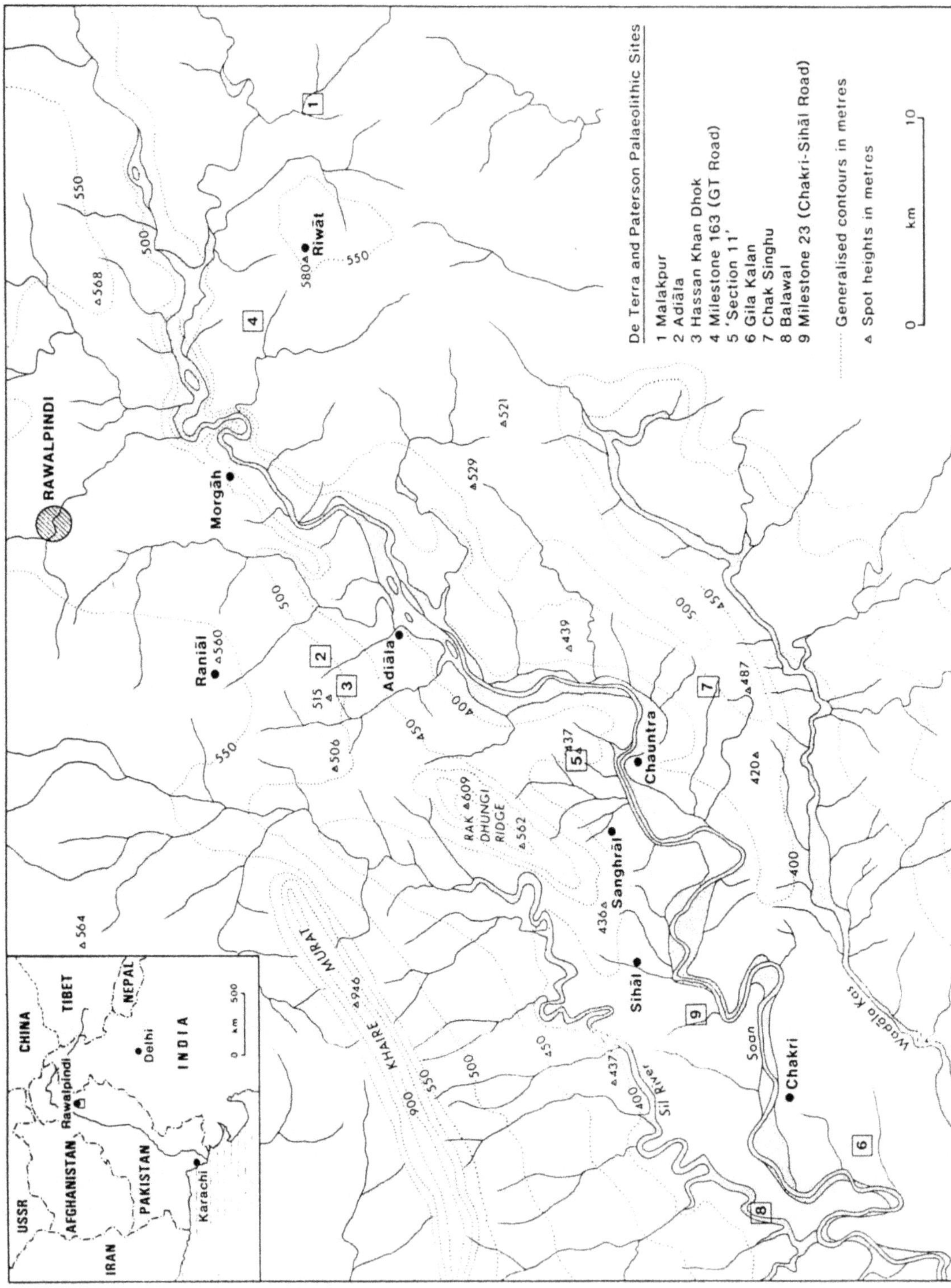

Figure 1.3: location of De Terra and Paterson's palaeolithic sites within the Soan Valley

bank (of the Soan) above Chauntra carries what looks like dissected remnants of a uniform terrace level, but there is no way of telling whether this is not merely a ledge of resistant gravel above Pinjor clays" (1939:276). Given their ambivalence, there is no reason to accept their dating of this locality. Their observations simply indicate the difficulties encountered in distinguishing between cut and fill terraces in the field.

At Gila Kalan, a *"Soan Workshop"* was discovered on the surface of the "Boulder Conglomerate", and covered by "Potwar loessic silt". De Terra and Paterson comment that: *"... the surface of this conglomerate is planed and may well represent T2"* (1939:292) There seems no reason to accept this assertion. The only conclusion allowed by their observations is that this locality pre-dates the loess and lies on a conglomerate of indeterminate age.

The Chak Singhu locality proved more complex since *"no terrace flats are preserved in the neighbourhood (of the gravel ridge) ... adjacent to this gravel ridge lies a pinkish sandy silt in which were found some 100 artefacts of Abbevillian and Acheulian type"* (1939:290). As far as the chronology of this site in concerned, De Terra and Paterson (1939:290) reasoned as follows

> *"... the ridge gravel can belong only to a fill stage during which the then less dissected slope was buried by ancient Soan drift..it would represent a terrace gravel of Third glacial age..while the adjoining sandy silt must be Third Interglacial, representing a thin deposit of T3. Accordingly, the early palaeolithic handaxes were embedded in a Third Glacial deposit and redeposited in the following interglacial stage, together with Late Soan tools".*

This reasoning is a good example of the kind of circular argument used to construct the terrace chronology: typologically early implements must have been derived from an early deposit, which in turn has to be ancient because it contains early types of artefacts.

The Balawal site comprises a cemented gravel patch resting unconformably on Middle Siwalik strata (including conglomerates) and overlain by "Potwar loessic silt". The site is related to T2 in a figure (Fig.173A) but not in the main body of the text. As at Malakpur, Adiala and Hassan Khan Dhok the only statement that can be made with any confidence about the age of this locality is that it is older than the loess but younger than the Miocene.

The Milestone 23 (Chakri-Sihala road) site yielded *"Pre-Soan Flakes"* in association with the conglomerate *"clearly proving the stratigraphic value of these artefacts in differentiating between terrace gravel, Boulder conglomerate and Pliocene conglomerates"* (1939:292). However, the type of association is not defined, and it is unclear whether these artefacts were on or in this conglomerate. One suspects the former, as it was a surface locality and there is no mention of any excavation. Once again, it seems that stone artefacts were found on a surface of unknown age. Also, as will be pointed out in the next section, there is little reason to accept *"Pre-Soan flakes"* as humanly-struck.

Thus four sites, if Balawal is included, appear to be related to outcrops of (quartzite-bearing) Middle Siwalik Conglomerates; a further four sites are related to De Terra and Paterson's "Boulder Conglomerate" and are tentatively linked to Terrace 1, in one case, and Terrace 2 in the others. Finally, the Chak Singhu site is related to Terrace 3.

The evidence put forward by De Terra and Paterson fails to sustain the existence of any link whatsoever between Pleistocene river terraces and palaeolithic sites. Terrace sequences in the middle Soan valley appear to be highly fragmented and, for the most part, erosional features. All the archaeological evidence is based on surface collections and there are no means of associating artefacts with secure stratified contexts. Although the attribution of particular sets of artefacts to particular terraces enabled De Terra and Paterson to construct their palaeolithic chronology, we have demonstrated that there is no basis for these attributions and therefore that the palaeolithic sequence devised by Paterson is wrong. On the evidence presented by De Terra and Paterson, the only point that is clear is that **palaeolithic artefacts are associated with outcrops of quartzite-bearing Middle and Upper Siwalik conglomerates, or with spreads of quartzite pebbles and cobbles derived from these conglomerates.**

1.5 The Boulder Conglomerate

De Terra correlated the "Boulder Conglomerate" which he mapped within the Soan valley with the "Boulder Conglomerate" of Kashmir and with the Upper Siwalik Boulder Conglomerate Zone of Pilgrim (1913). Unfortunately, De Terra made a serious error in his mapping of the Soan synclinal fold structure. He followed Wadia (1928: Plate IX, Fig.1) in interpreting the fold structure as symmetrical, and with steeply dipping limbs. The fold is, however, strongly assymmetrical, with a steeply-dipping northern limb and a gently dipping southern one. Only later was this structure mapped correctly by Gill (1952:386), who remarked on the subsequent confusion over the "Boulder Conglomerate":

> *"De Terra's restriction of the term 'Boulder Conglomerate' to a formation which has a maximum thickness of about 300 ft (91m) is somewhat confusing in the field, since the term could perhaps be more appropriately applied ... to the 2-3000 ft (610-910m) of Siwalik conglomerate below".*

Gill pointed out that confusion as to the structure of the syncline and the relationship between Siwalik and post-Siwalik strata is heightened if sections are drawn with vertical exaggeration. To emphasise this, Gill's sections through the syncline are compared with De Terra and Paterson's section in Figure 1.4.

De Terra's mapping error was compounded by his association of the post-Siwalik limestone-rich conglomerates on the northern limb of the synclinal fold with Siwalik quartzite conglomerates on the southern limb of the fold. He speculated that this difference in composition could be explained by different source areas (De Terra and Teilhard de Chardin, 1936:799). In fact, as Gill (1952) pointed out, there is no need to postulate the existence of two ancient rivers in the Soan, each carrying a distinctly different sediment type, as the two conglomerates are of different ages.

Neither De Terra nor Wadia (nor us) were helped by the virtual absence of diagnostic fossils in Upper Siwalik deposits exposed the middle Soan valley. Wadia (1928:344) noted that the Boulder Conglomerate stage in the Soan *"yield(ed) only indeterminable proboscidean bones"*. Therefore, until the development of radiometric and palaeomagnetic dating techniques, the stratigraphic correlation of the Soan sequence with other areas depended totally on the colour, texture and associations of deposits, as revealed in sections, and their similarity, or otherwise, to sequences elsewhere. Lithological correlations of this kind are fatal in sequences like the Siwaliks for several reasons: deposits are often highly variable laterally; many Siwalik sequences are repetitive so that deposits are not time-specific; and, for tectonic reasons, many of the deposits are time-transgressive

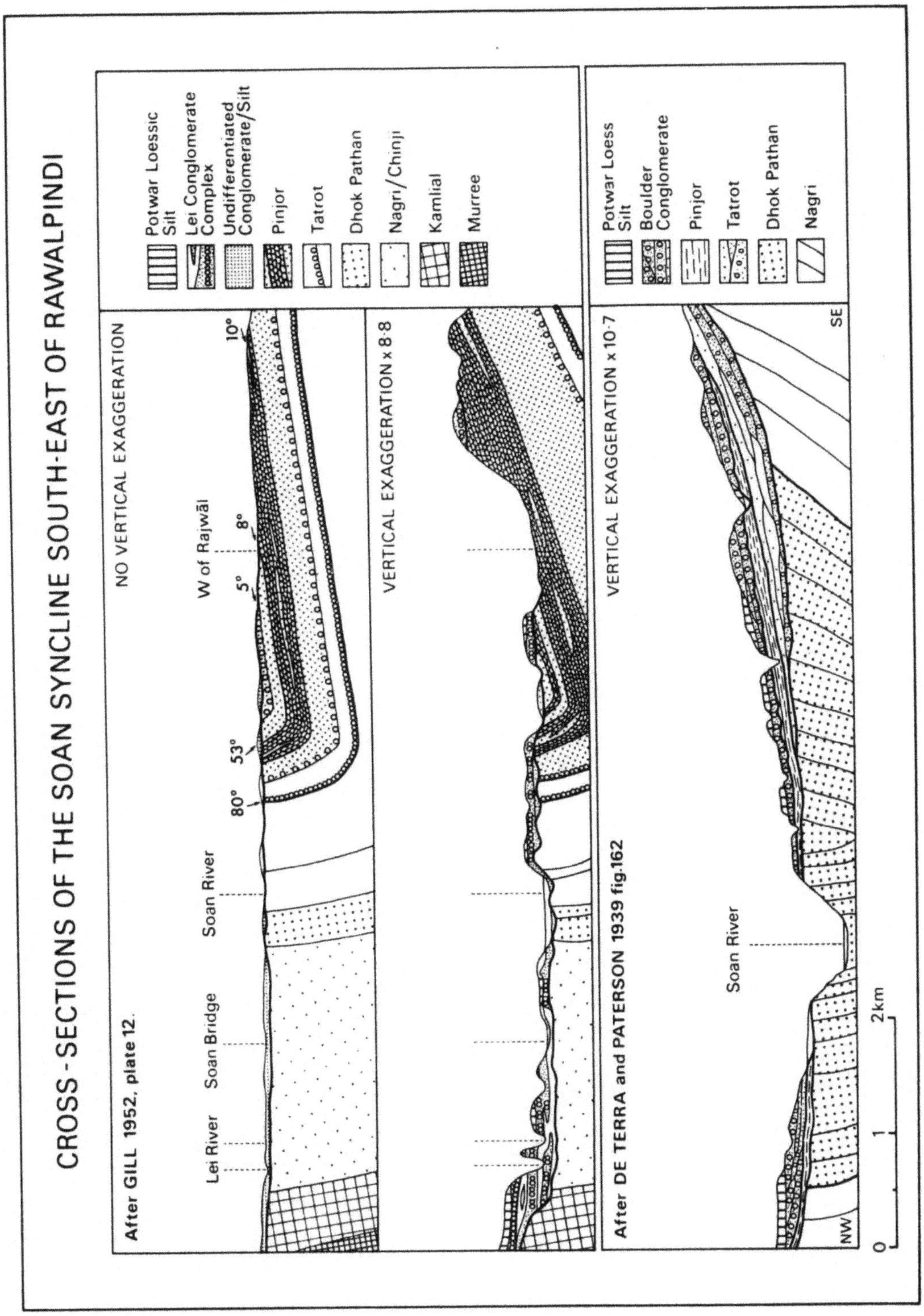

Figure 1.4: Cross-sections through the Soan Syncline after Gill (1952) and De Terra and Paterson (1939).

(see chapter 3).

It is important to note that, in the absence of palaeontological evidence, De Terra used the following criteria to recognise the Boulder Conglomerate:

(i) its relation to the (overlying) younger beds, especially Potwar Silt

(ii) its *"tilted status and general unconformity above older Siwalik beds"* (1939:265)

and:

(iii) *"it was important to observe that specimens of Pre-Soan industry, coarse flakes or flaked pebbles occur in the upper layers of the fan and pebble gravels but nowhere in younger formations"* (1939:265).

As Stiles (1978) points out, the inconsistency of the third criterion is that since a conglomerate containing *"Pre-Soan Industry"* is, by definition, the "Boulder Conglomerate", then this industry cannot be found in younger formations. Conversely, the "Boulder Conglomerate" was held to be early because it contained what seemed to be the earliest-looking artefacts. As we have already seen, the Soan terrace and palaeolithic sequences are also riddled with this type of circular reasoning.

1.6 Summary

1) De Terra's fieldwork was very much a product of its time. His view of the Pleistocene was derived from a large body of European work which appeared to confirm the existence of four major world-wide glaciations during the Pleistocene. He inherited a tradition of fieldwork which had been developed in tectonically-stable areas, and which stressed the importance of establishing (climatically-related) river terrace sequences as the key to the Pleistocene record of unglaciated areas.

2) With this background in mind, it is unsurprising that he correlated the boulder fans and moraines of Kashmir with the "Boulder Conglomerate" of the Soan and correlated both with the second (Mindel) glaciation in Europe. It is also unsurprising that he looked for and, found river, terraces in the Soan. He also assumed that the Pleistocene record within the Soan valley was essentially complete, thus allowing a convenient correlation of each "terrace" with either a glacial or an interglacial episode.

3) In the field, he was forced into making a major error by assuming that Wadia had mapped the structure of the Soan syncline correctly. He therefore mapped a 100m thick post-Siwalik conglomerate as an Upper Siwalik conglomerate and called it the "Boulder Conglomerate". This misnomer has proved to be a major source of confusion for subsequent workers.

4) In our opinion, both De Terra and Paterson's geological and archaeological sequences have to be rejected. We see no evidence of a major series of river terraces within the Soan Valley; it is also our view that none of the palaeolithic sites described in the 1939 monograph can be dated relative to one another or absolutely; and none of the archaeological material can be firmly related to particular geological contexts. In the course of this monograph we will attempt to put another sequence in its place.

Chapter 2

The Soan Valley and the Palaeolithic in the Northern Sub-Continent

"Just as in the Abbevillian-Acheulean cultures there is an evolutionary trend toward finer and neater forms of hand axes, so in the Soan pebble tools there can be traced a development toward smaller and more finely made types" (De Terra and Paterson 1939: 312).

"The Barnham sequence of the Clactonian shows progressive and continuous development of technique" (Paterson and Fagg 1940: 28)

2.1 Introduction

As noted already, De Terra and Paterson's (1939) monograph had an immensely long-lasting effect on palaeolithic research in the northern sub-continent as well as in Asia as a whole. Indeed, if longevity is taken as an index of success, theirs was one of the most important palaeolithic monographs published this century. Their results became an integral part of Movius's (1944, 1948) synthesis of the Asian lower palaeolithic, in which he grouped the Soan industries of the Potwar, the Patjitanian of Java, the Anyathian of Burma, the Choutoukien material from China, and the Kota Tampan industries of Malaya into an Asian *"chopper-chopping-tool"* tradition that was largely distinct from the African and European hand-axe and flake tradition. There were two areas where these two traditions appeared to overlap. One was southern England, where the Clactonian flake industry overlapped in time and space with Acheulean hand-axe assemblages, and the other was the Punjab, where Acheulean-type hand-axes were found in some Soan flake assemblages.

De Terra and Paterson's results have also been included in virtually every regional survey of the palaeolithic of the sub-continent. Good recent syntheses of this material include Agrawal (1982); Allchin and Allchin (1968, 1982); Gupta (1979); and Sankalia (1974). Various papers have also attempted to compare the Soan Valley assemblages with those of, for example, India (Sankalia, 1978) and Soviet Central Asia (Boriskovsky, 1978).

Post-war palaeolithic fieldwork in northern India and Pakistan adhered remarkably faithfully to De Terra and Paterson's results, as did Asian research in general to Movius's views. In Pakistan, further collecting took place by Graziosi (1964), Johnson (1972) and Krantz (1972), but these aimed at (or at least resulted in) supplementing rather than challenging the findings of De Terra and Paterson. Allchin (1963) tried to synthesise the evidence from India and Pakistan whilst pointing out the stratigraphic problems that first needed resolving, and laid the foundations for this present study (Allchin, 1981). Stiles (1978) was one of the few to have attempted a dispassionate and critical look at this work, and his paper is one of the most useful critiques of these data in recent years. Similarly, Davis (1987) has made several useful and perceptive points about the material from the Soan Valley, and its alleged relationships with Central Asia.

The longevity of Movius's synthesis on the Asian palaeolithic is clearly shown by a major conference held in Montreal in 1973 simply because – in Movius's own words : *"scarcely anything substantial has been added to our general understanding of the fundamental problems of the core and flake assemblages of South and East Asia"* (Movius 1978: 354) in the thirty five years since his synthesis. He further added: *"the past three decades of research have produced relatively little other than a few thousand more artifacts, some from new localities. When one realises that the same old methods have been employed in most instances in gathering and analysing the data, perhaps this is not surprising"* (ibid. 354).

We should thus follow Movius's call for new methods, and begin with a re-appraisal of the pre-war work of De Terra and Paterson. To do so, it is worth recapitulating their main archaeological conclusions. These were, first, that the earliest stone tools in the Punjab dated from the base of the Middle Pleistocene; secondly, that the Soan assemblages demonstrated *"an excellent example of the evolution of a flake culture in a small area"* (De Terra and Paterson, 1939: 311); and thirdly, that during part of this time, there were two *"traditions"* represented in the Punjab: a dominant and longer-lived one represented by the Soan flake culture, and a shorter-lived and intrusive one that used Abbevillian-type hand-axes. How valid are these conclusions fifty years after they were drawn?

2.2 The antiquity of man in the Potwar: dating

The first of their conclusions - that the antiquity of man in the Punjab extends to the base of the Middle Pleistocene - is essentially a geological problem. As Rendell (1984 and the previous chapter) has indicated, De Terra's chronological framework has to be abandoned if only for the simple reasons that his work does not even provide a secure relative chronology, and the number of Pleistocene climatic fluctuations now recognised greatly exceeds the number of river terraces that he identified.

The archaeological implications of the above are obvious and far-reaching: if the Soan Valley assemblages cannot be dated relative to each other, there are no grounds for inferring diachronic change in the stone tool assemblages from the Potwar. On these grounds alone, Paterson's typological divisions of the assemblages into *"pre-Soan"*, *"Early Soan"* and *"Late Soan"* cannot be maintained; nor can even a division between Lower and Middle Palaeolithic (see below also, this chapter). On a larger scale, suggestions that there were migrations, diffusion or other forms of contact between the Potwar and adjacent regions are of little value if it is not known whether the assemblages in a donor area are earlier or later than, or contemporary with, those in a recipient zone. Thus suggestions by Paterson and Drummond (1962; manuscript prepared 1940) that there were migrations between the Punjab and S.E. England or southern Africa by Soanians,

Clactonians, and Fauresmith peoples merely on similarities in some tool types are - to say the least - unwarrantable, as is Gupta's (1979) attempt to suggest that people migrated from the Soan into Central Asia. Nor can we talk, as Johnson (1972) did, of "Levallois-Mousterian influences" somehow infiltrating their lithic techniques into the Potwar until it is established that these "foreign" traits are indeed later than their first apearance further west.

2.2.1 The "meaning" of the early palaeolithic record: typological refinement and unilinear change.

As indicated in the previous section, the bulk of De Terra and Paterson's conclusions have to be rejected because of failings in their geological framework. Even if this were not the case, however, there would still be serious grounds for questioning Paterson's conclusion that: *"The Soan flake industries ... provide an excellent example of the evolution of a flake culture in a small area"*. Before going any further, however, it is necessary to examine the background to his work in the Soan Valley. There are two related issues here that need considering. The first is the influence of French scholarship upon British palaeolithic research in the 1930's, whilst the second is Paterson's own post-graduate work on Clactonian assemblages of southern England.

2.3 The French connection: the Somme and the Soan Valleys

Paterson's work is no exception to the unremarkable point that we are all the victims of our times. He worked in an age where it was commonly assumed that typological progress was an integral part of the palaeolithic record. This view has deep roots, and can be traced back to 18th century views of the inevitability of social, economic and technological progress that found it most optimistic expression in the French and Scottish Enlightenment of the 1760-1780's. That in turn, early in the last century, underpinned Thompson's Three Age system of Stone, Bronze and Iron. With the recognition of the antiquity of man by the 1850's, and the growing amount of palaeolithic material from western Europe, notions of technological progress were extended to the palaeolithic, which was soon sub-divided into a Lower, Middle and Upper. By the 1880's De Mortillet had vitually enshrined the belief in technological progress to a law-like status, with his axiom of *loi de progrès universal*. De Mortillet's radical political views encountered sustained opposition from the Church and State which eventually resulted in the establishment of the *Institut de Paléontologie Humaine* as a rival to his *Musée D' Histoire Naturelle* (Hammond, 1982). Evolutionary views were voiced more cautiously by those in the I.P.H., notably by the palaeontologists Gaudry, and later Boule and his student, Teilhard de Chardin. As both palaeontologists and devout Catholics, Boule and De Chardin collaborated with a talented group of clerically-led prehistorians, notably the Bouyssonie brothers, and the Abbé Breuil, the doyen of French palaeolithic studies. The significance of this collaboration on studies of Neanderthals in particular and human evolution in general has been thoughtfully addressed by Hammond *(ibid.)*. What has received less attention is the significance of Breuil's work on lower palaeolithic studies, and particularly that on the Somme Valley in northern France (Breuil and Koslowski, 1931; Breuil, 1939). What Breuil showed was that large numbers of stone assemblages from an intensively-studied series of river terraces could be used to devise a series of 11 lithic *"stages"* of progressive refinement in lithic technology, defined largely by the extent of rolling and patination, and evidence of refinement in terms of artefact-thinning and retouch. Ideally, each assemblage was stratigraphically distinct from the others, and typological developments could be cross-checked against their relative stratigraphic position. However, if this could not be done, the artefacts themselves could be grouped into *"early"* and *"late"* types on the basis of their shape, wear and amount of retouch. In general, flaked stone that was rolled, chunky and plain implied an early date; fresh, thin and extensively retouched ones implied lateness, and the cumulative experience of prior trial and error.

This work profoundly affected two studies undertaken a few years later. One was Louis Leakey's monograph on Olduvai Gorge (Leakey, 1951), which mirrored this approach very carefully, and ended with 11 stages of Acheulean industries. This is not surprising, as Louis Leakey worked in the Thames area at Swanscombe, whilst his then fiancée worked at Clacton with Oakley, and Leakey would have been well aware of this work. (Although his study was not published until 1951, it is worth noting that this work was based on survey and excavation undertaken at the same time as De Terra and Paterson, and would have been published in 1939 had not war broken out, see Leakey (1951:xv)).

The other study that showed Breuil's influence was Paterson's in the Soan Valley. As with Leakey, it is inconceivable that Paterson would not have known of Breuil's work, and extremely unlikely that he was anything but deeply influenced by it. As way of reinforcing the "French connection" between the Somme and the Soan, it should also be noted that Breuil was president of the Prehistoric Society in 1934, whilst Paterson was beginning his post-graduate research under Miles Burkitt at Cambridge, and also that De Terra's colleague, Teilhard de Chardin, knew both Boule and Breuil well.

2.4 Paterson and the Clactonian

At the time of the Soan survey, Paterson was the holder of a Cambridge research studentship, and thus a contemporary there of Louis Leakey. Although Paterson's is best remembered for the one season of fieldwork that he did with De Terra in the Soan Valley in 1935, his main research concerned the Clactonian flake assemblages of Barnham St. Gregory in southern England. The relevance of this research to the Soan Flake Industry has been consistently - and regrettably - overlooked. One reason is that he does not refer to it in his monograph with De Terra, and one can suppose only that he wrote his contribution before publishing his own papers on the Clactonian. A second is that in Europe, his work on the Clactonian was soon superceded by that of post-war scholars. It is, however, of fundamental importance in understanding the way that he perceived the flake assemblages of the Soan Valley in 1935.

Barnham St. Gregory

Here, Paterson was examining an assemblage of flake tools from the top two feet (60 cm) of a gravel horizon. Bearing in mind the type of geological deposit and the extent to which artefacts and other clasts can be mixed within it, many (this author included) might feel that it contained at most one assemblage. Paterson, however, distinguished no less than four: B and C came from the top of the gravel, and were distinguished by their typology, and also by their state of wear and patination; industries D and E came from the top two feet of the same gravel, and were distinguished the same way.

After examining each group, Paterson (1937: 135) proposed that this site showed *"a series of industries, showing a progressive development along indigenous lines, uncontaminated by contact with other culture techniques. The technique evolves continuously in the production of finer flakes, prepared with increasing skill. More shapely tools are produced and finished with progressively finer retouch. At the same time the angle of striking platforms decreases, the bulb increases in depth, and simple facetting of platforms appears. ...New forms of tools appear towards the end of the series but their origin is seen to lie in previous types".*

Paterson adopted exactly the same procedure in the Soan:

"The early Soan tools differ in patination and state of wear and have therefore been divided into three groups, here termed A, B and C...The earliest, A, is heavily patinated....and thoroughly worn...Group B is patinated like A but not worn at all, and Group C is less patinated and fairly fresh" (De Terra and Paterson 1939: 305). He similarly divided the *"late Soan"* industries: in late Soan A, *"some of the specimens are worn"*, whereas those of late Soan B were *"exceedingly fresh and unworn"* (De Terra and Paterson 1939: 309-310).

Not surprisingly, he came to exactly the same results:

"In the early stages the flakes are crude and simple, often rather Clactonian-like in appearance, with primary flaking of a very haphazard nature. In the late Soan, alongside the simple forms, there are other flakes, showing a development in technique, with much more regular primary flaking and often with faceted platforms, denoting careful preparation of the core in a manner reminiscent of the Levalloisian. Retouch, when necessary for the type of flake implement required, is much finer, but with the improvement in technique flakes and blades were produced with so fine and regular an edge that retouch was rarely needed." (De Terra and Paterson 1939:311-312).

His assumption that assemblages from the same deposit could be separated on the basis of patination and wear was not equally shared at the time. Oakley and Leakey (1937) detected no change in their Clactonian assemblages from a similar riverine context to that from Barnham St. Gregory, and suggested, albeit with great diffidence:

"If it is permissible to make a generalisation, it might be said that there is often little justification for taking the degree of rolling as a prima facie criterion of relative age in Drift deposits. Tools deriving from contemporary workshop sites located by the riverside at irregular intervals down a valley would tend to be transported downstream in course of time to an extent largely dependent on accidental circumstances. Thus at a given place the artefacts incorporated in a river deposit might be of exactly the same type, but in various states of wear" (Oakley and Leakey , 1937: 237).

Even when, as they showed, platform angles of rolled and unrolled flakes differed in the same way as shown by Paterson, they maintained *"The only question is whether these other workshop sites were of earlier date, or merely of more distant location"* (Oakley and Leakey, 1937: 237): they took the latter view.

Oakley and Leakey's study was prophetic in anticipating matters that are now considered part of modern taphonomic studies. Their suggestion also torpedoed attempts by Breuil, Paterson and others to demonstrate lithic improvement from river gravel assemblages. They were in effect saying, such trends are impossible to verify unless assemblages are of demonstrably different ages: assemblages from the same layer cannot be separated (as Paterson did in both southern England and the Soan) into chronologically-distinct entities on the basis of wear, patination or even shape.

2.5 The two traditions: artefact types and ethnicity, and the Acheulean and the Soan

Paterson investigated the problem of explaining the apparent co-existence of two lithic traditions in southern England in greater detail than he did in the Soan. His explanation was that tools are made to set patterns acquired by individuals as part of a social or ethnic group, and thus different types of stone tools or stone-working techniques indicated the presence of different groups. In a later paper on the Clactonian of southern England, he thus suggested that

> *"Acheul man has come into contact with Clacton man and absorbed some of the core and flake technical methods. This would account for the degenerate character of the core element, even though developed types of cores are present"* (Paterson and Fagg, 1940: 23).

Likewise, he even depicted *"Soan choppers of the Upper Brecklandian Acheul"* (Paterson and Fagg, 1940: Fig. 15): the implication being that some-one had gone from the Soan to Suffolk and discarded a tool. Why such long-range discard was practised was left unexplored.

It is thus hardly surprising that he identified two groups in the Soan: one using a technology based on flakes, and the other, based on (or at least, frequently using) hand-axes. That these were seen in terms of distinct groups as is apparent from his later suggestion that the Soan at one time contained two groups: one, of peaceable, vegetarian flake-users, and the other, an aggressive bunch of carnivores using hand-axes (Drummond and Paterson, 1962: manuscript prepared 1940).

2.6 Discussion

We can isolate two premises underlying Paterson's work (as well as that of many others at that time and subsequently). One was that palaeolithic *"cultures"* became progressively more refined through time; that is to say, artefacts - or at least the most distinctive ones - were always made to a set pattern, and their makers would always show the benefits of accumulated experience and expertise. The second was that distinctive methods of making tools would be acquired and passed on within the social group to which each individual belonged. The presence of tool-forms or stone-working techniques that were previously unknown in one area but used elsewhere was consequently taken as evidence for the mingling of different groups, or the intrusion of one into a neighbouring territory. To this extent, the early palaeolithic record was not seen as inherently different from that of the late prehistoric periods: time-scales were of course far longer, and the evidence was far less precise, but its *"meaning"* in terms of culture-history was much the same as for much later periods.

Modern investigators into the early palaeolithic are generally far more cautious about identifying unilinear refinement in tool-making, or ethnic groups. As Davis (1987:10) notes *"To put it baldly, a Lower Palaeolithic stone tool industry is an abysmally poor mirror of regionalised or otherwise bounded mental activity".* Interestingly, Paterson, in his final major paper on the lower palaeolithic, would have agreed with him. In his often thoughtful review of the concepts used in palaeolithic research, (Paterson 1945; publication delayed because of the war), he argued that the concept of *"culture"* was inappropriate to palaeolithic studies because there was (and always would be) insufficient data:

> *"It is exceedingly doubtful if the word can be applied to any archaeological group of complexes prior to historical times"* (Paterson 1945: 3).

So too with the concept of an *"industry"*:

> *"Even in the Lower Palaeolithic, where so often more than one technique is used, the use of 'industry' is not absolutely justifiable, since, for all we know as yet, each technique may be associated with a different complex"* (Paterson 1945: 3).

By the time he completed his studies of the Suffolk and Soan assemblages, he had in effect disowned his own conclusions.

2.7 Towards a fresh start

If, as even Paterson (1945) would appear to agree, *"It is reasonably clear that culture history in the traditional sense is neither a practicable nor a profitable pursuit in Lower Palaeolithic studies"* (Isaac 1975: 523), what should be put in its place in an area like the Soan Valley? This question has three facets:

(i) the recognition of major divisions of the entire palaeolithic record

(ii) the detection of different traditions within each major division

and

(iii) the detection of changes within each division.

Each can be taken in turn.

2.7.1 The division of the palaeolithic

As has been recognised for many years, a tripartite division of the palaeolithic into a Lower, Middle and Upper is inappropriate to areas outside western Europe, where it was first developed. Even in western Europe, it is increasingly difficult to define a boundary between the Lower and Middle Palaeolithic, and some such as Gamble (1986: 117) have proposed that only two divisions are possible, between an *earlier* palaeolithic, pre-35 ka, and a *later* one, associated with *Homo sapiens sapiens* and defined primarily by the widespread use of punch-struck blades; bone, ivory and antler tools; art; and burials with grave goods. Dennell (1983a) followed a similar though less detailed approach. Asian workers have not been idle in devising their own ways of dividing the Palaeolithic. In an Indian context, Allchin (1963) suggested that it was more appropriate to talk of an Early, Middle and Late Stone Age, defined primarily by bifaces, flakes and blades respectively. The first of these stages is problematic in the context of the Soan, as it is not clear from De Terra and Paterson's work whether bifaces constitute the earliest *"stage"* of the palaeolithic sequence in that area. Asian researchers in general seem to have shied away from any kind of tripartite division of the Palaeolithic. Ikawa-Smith (1978: 1-2), for example, noted the confusions arising over the use of terms like Lower, Middle and Upper Palaeolithic for Asian assemblages, and proposed that *"the way around it, it seems to us, is to discard the tripartite division of the Palaeolithic when we discuss Asian assemblages... It seems easier to ...propose two broad toolmaking traditions: Early Palaeolithic and Late Palaeolithic"*. The latter dates largely to 30-20,000 years ago, and is characterised by *"the proliferation of specialised tool forms (which) include parallel sided blades, some of which are retouched into standardised forms, including many kinds of engraving tools (burins), scraping tools, and projectile points"* (ibid. 2).

The same procedure will be adopted here. The Early Palaeolithic in the Soan and the Potwar is defined in this study simply as being earlier than 30,000 bp.; it cannot at present be sub-divided into chronological stages defined by the presence or absence of certain items such as (Acheulean) hand-axes, and (Levallois-Mousterian) tortoise- and disc-cores. What is clearly needed are examples of these items in secure geological contexts that are amenable to absolute dating.

2.7.2 One culture-complex or two? : flake and biface traditions

Even if much of what De Terra and Paterson (1939) wrote about the dating and development of the flake assemblages of the Soan do not bear modern scrutiny, there is still the problem of explaining the *"Abbevillian"* or *"Acheulean"* bifacial component. Questions that need asking are:

(i) do these represent a separate - if relatively short-lived - *"tradition"*, as envisaged by Paterson, and as commonly suggested for western Europe, much of Africa, and peninsula India?

or

(ii) are they simply a minor part of Paterson's Soan flake tradition, in much the same way as Ohel (1979) suggested that the English Clactonian was simply part of the local Acheulean?

At this stage, no real progress can be made in answering either question without the discovery and excavation of at least one securely-stratified Middle Pleistocene site in northern Pakistan. Isolated finds of bifaces are of some use in clarifying their time-range if found in secure, datable contexts; some examples from the Punjab will be discussed later in this volume. There is certainly no justification in assuming that the Acheulean - as defined in, for example, East Africa or India - exists independently of a flake-tool tradition merely because bifaces are absent from surface collections of flake tools.

2.7.3 The detection of change within the Early Palaeolithic

If, as suggested here, all palaeolithic material in the Soan before 30 ka is regarded as *"early"*, is it possible to recognise changes within it that transcend minor differences of, for example, raw material, local context, or collection techniques?

Such is our present ignorance that it is difficult to identify meaningful change within the Early Palaeolithic of northern Pakistan. Given the present state of palaeolithic research elsewhere, and the type of lithic assemblages likely to be found in the Soan Valley, it is likely that significant changes will be hard to detect. The point is made by reference to the lower palaeolithic record of East Africa, where there are now several assemblages from various sites that are (usually) well-dated; in association with animal remains; and often in (or at least, very near) their original place of deposition. Most have also been well-excavated. After a careful review of the evidence, the extent of change suggested by Isaac (1975) was modest by comparison with the changes inferred by previous generations of researchers. He suggested that there was through time: a greater density of stone (and bone) in sites of comparable function; a gradual thinning of core tools; and a trend towards a larger number of flake scars on core tools made from the same material.

Identification of comparable trends in the type of flake assemblages so far collected from the Soan is likely to be difficult. In the first place, most of the lithic material are flakes struck from quartzite cobbles, and lacking a standardised form. This has two consequences. The first is that of distinguishing the artefacts from the waste. This is particularly relevant in the Punjab, as will be seen later in this volume in the account of site 55, near Riwat. Simple, unretouched cortical flakes could easily have served as tools, and are not amenable to refinement or improvement in the sense used by Breuil on hand-axes. In any case, the amount of retouch on a tool depends greatly upon the type of material chosen as well as the length of time it is used: as Jones (1979) showed, quartzite tools - especially on flakes - are difficult to retouch, and often better left unsharpened. In the case of the quartzite artefacts from the Soan, retouch would not necessarily denote *"improvement"*, and may even have been accidentally induced. Conversely, an absence of retouch need not imply a *"primitive"* technology; merely that there was no point in modifying a perfectly effective cutting-edge.

If most of the assemblages from the Soan are likely to consist of unretouched quartzite flakes, it is tempting to focus attention on the few standardised types that are found. This approach, however, has its own set of drawbacks. The study of artefact types in isolation can overlook profound continuities in much of a flaked stone assemblage: for example, the type of *"pebble-tools"* found in some European lower palaeolithic localities are also found in upper palaeolithic ones, and presumably were useful when more time-consuming artefacts were not required (see Gamble, 1986:278). In areas such as Asia where stone tools were - by and large - far less formalised than in western Europe, an emphasis on specific types of tools is rash, as it will give undue attention to only a minor part of an assemblage. Even if an artefact shows, as Johnson (1972) claimed, *"foreign influences"*, the question still remains as to why these influences affected so small a part of the total assemblage.

Secondly, the shape and choice of a stone tool depends upon many factors. A major one is the circumstances of its use. Tools made in anticipation of a task may be made more carefully than those made on the spot, expediently, and then discarded as soon as they were used. Specific types of tools might not even be made or discarded under some circumstances; as Isaac (1975) noted, the percentage of hand-axes in East African Acheulean assemblages can vary from 0 to 95%, so *"type-fossils"* are not necessarily a reliable indicator of an early palaeolithic tradition, let alone its stage of technological refinement.

2.8 Surface versus excavated assemblages

De Terra and Paterson relied upon surface collections of flaked stone, mostly from the surfaces of conglomerates. Before dismissing too readily the drawbacks of surface assemblages, it is useful to note their value. Surface collections have a place in both archaeological and palaeontological research, and both would be considerably poorer without them. Surface collections can be useful if they can be compared closely with sets of excavated data; if there is little evidence of contamination; and if the deposit on and/or under which they are found can be accurately dated. Excavated material, on the other hand, can (hopefully!) be accurately dated, and is more reliable than surface collections for making comparisons between sites of the same or a different period. Although excavated assemblages are usually of higher quality, in the sense of providing reliable information on the spatial relationships of different types of material, and on the total composition of an assemblage before material is lost through exposure and erosion, they are also more expensive and time-consuming to obtain.

When surface material can be well-dated on its own merits, it can be of considerable value: for example, in building up models of land use patterns at particular periods from the distribution of specific types of pottery, or well-defined lithic assemblages. The composition of surface collections can also provide information on the likely function of sites. Likewise, the distribution of surface collections can show where archaeological material has probably been lost through erosion, as in the Koobi Fora area, where much palaeolithic material has probably been washed off low hills into stream deposits further downslope (Isaac, 1978).

For surface collections of archaeological material to be useful, however, a number of preconditions need to be satisfied. The first is that the age of either the type of material found in surface collections, or the surface on which they are found has already been established. By way of example, assemblages that are known to be mesolithic can be used in studies of land use patterns, even if the age of the deposits on which they are found is unknown; conversely, material on a late pleistocene surface clearly has to be of later date, even if it cannot be dated more precisely by its contents. Ideally, of course, both the age of the erosional surface and of the lithic material on its surface should be known. However, it is logically impossible to date an erosional surface of unknown age from surface collections that are also undated. This point is especially relevant to De Terra and Paterson's dating of the Boulder Conglomerate to the Upper Siwaliks on the grounds that they contained the earliest-looking "*Pre-Soan*" artefacts: as Rendell (this volume) indicates, these were called "*pre-Soan*" precisely because they lay on the surface of the same undated Boulder Conglomerate, and were thus thought to be early !

A second prerequiste of surface collections is that it is as essential to know when a surface was exposed as it is to know ehen it was originally formed. This point applies especially to the conglomerate exposures of the Punjab. As De Terra and Paterson (1939:286) stated:

> "*sites are so closely associated with outcrops of middle Pleistocene conglomerates that one can almost predict sites wherever this formation underlies the peneplain*".

Numerous examples are quoted in their monograph: for example, at Malakpur

> *"from the surface of these ridges atop the peneplain were collected a few flaked pebles of the pre-Soan industry. These artefacts had weathered out from the top gravel layers, and, as there was no indication of a terrace or redeposition, we take it that they occur in place in the second glacial deposit"* (De Terra and Paterson, 1939: 283-284).

Likewise, at site P6 (Khasala Khan) *"the workshop is found* **on** *limestone gravel belonging to the Boulder conglomerate"*. (ibid. 286.); at Hun, above Chauntra *"***on** *the conglomerate was found a workshop composed of early soan industries"*. (ibid 287); and south-east of Chauntra, *"a site with Abbevillian and Acheulean hand-axes"* was found *"***on** *the slope of the second range of hills"* on a gravel layer (ibid. 289-290; emphasis mine). (These localities have been discussed in greater detail by Rendell, previous chapter).

The obvious archaelogical problem with this type of surface locality is that we do not know whether stone tools are contemporaneous with the formation of these conglomerates, or with a period of subsequent exposure. If these surfaces were exposed frequently, or over long periods, surface collections of flaked stone assemblages are likely to be palimpsests, and contain material flaked at different times. This much seems to be indicated by some of De Terra and Paterson's material: for example, at Adiala, they collected some 600 artefacts that were called *"late Soan"*. However,

> *"Mixed with this are rolled early Soan tools and an Abbevillian hand-ax. This makes us suspect that there occurred repeated rewashing of certain implements from older (perhaps Potwar) gravels"* (ibid. 285).

The important point here is that a cursory inspection of an artefact-yielding conglomerate surface will not be sufficient to decide whether these stone artefacts have eroded from it; were made there from stones obtained from the conglomerate; or were made elsewhere and discarded (or transported naturally) on top of it. The first of these can be established only by excavation into the conglomerate itself, or by finding unambiguous examples of flaked stone in naturally-exposed sections of conglomerate. Detailed examination of conglomerate surfaces might show the presence of flaking debris, which in turn would indicate that stone was flaked there; if no such evidence is found, it may then be concluded that artefacts were made elsewhere, or that the *débitage* has been washed away, leaving only the larger artefacts. (In the case of older collections, it is unlikely that flaking debris would have been collected). The conclusion is, unfortunately, inescapable that in most cases, De Terra and Paterson failed to demonstrate that the artefacts they found came from *within* a conglomerate, and not from stone-working *after* the conglomerate was exposed, or even from a later deposit. One possible exception to this statement is one locality near the G.T. road, south-east of Rawalpindi, where they (1939: 283) mention that it took *"10 minutes work with a chisel to extricate some of the artefacts from their lime matrix"*; regretably, they do not provide further details as to whether these artefacts were on the surface or in a section, or whether the cementation extended below the conglomerate surface.

The possibility that some of the flaked stone found on conglomerate surfaces in the Soan Valley was derived from later deposits upslope was partially recognised by De Terra and Paterson. At Gila Kalan, for example, a conglomerate was eroding from under a loess deposit. At the base of, and in some cases, protruding from, the loess, evidence of a *"Soan workshop"* was found. However, *"A few large but water-worn flakes of pre-Soan type were found on the gravel-strewn slopes below the loess...presumably derived from the Boulder Conglomerate"* (De Terra and Paterson 1939: 292). However, as Stiles (1978: 136) points out, we should bear in mind that the *"pre-Soan"* types could be natural (see below; and Stiles 1978: 133), and that there is no evidence that they came from within the conglomerate itself. Oakley and Leakey's (1937) point that patination and wear is no guide to relative age should also be kept in mind (see above). Similar considerations apply to some of the localities visited by Stiles, such as GT (Grand Trunk) I, II and IV, where flaked stone probably originated from within silt deposits overlying exposed conglomerates (Stiles 1978: 137).

What are needed ideally in this type of work are erosional surfaces that have only recently been formed, and which are also eroding at a fast rate. Flaked stone in this type of context has a greater chance of being *"uncontaminated"* by later components than those on surfaces exposed for long periods. They are also less likely to contain stone that was flaked after exposure by, for example, heat, frost, the trampling of livestock, or the activities of bored shepherds. Better still are collections of flaked stone from **within** conglomerates, as these do not raise the question as to whether or not the flaked stone dates from the formation or the exposure of the conglomerate.

If these conditions had been satisfied, surface collections of flaked stone from the Soan could have been highly useful, if only in indicating those localities worth excavating, and in supplementing collected of excavated material. However, these conditions were not met in De Terra and Paterson's work, and thus their conclusions are largely unwarranted.

2.9 Artefacts versus ecofacts

One key question in studies of very early stone artefacts is how one should distinguish humanly from geologically-flaked stone. This question applies forcefully to De Terra and Paterson's identification of a *"pre-Soan"* flake industry, which was allegedly one of the oldest stone industries in the world. Curiously, given its importance, it was never described in detail, but mentioned in passing in only 12 references. Only four of these provide the merest hint as to their evidence of what was considered to be the oldest instances of tool-making in the subcontinent: these are references to *"coarse flakes or flaked pebbles"* (ibid. 265); *"flaked pebbles"* (ibid. 284); *"crude flakes of Cromerian type"* (ibid. 285); and *"large flakes with prominent bulbs"* (ibid. 316, when discussing the Narbada valley, India). As shown earlier in this chapter, Paterson's isolation of this group of material is largely a product of the way he grouped assemblages from the same geological context on grounds of wear and patination, and need not have any independent existence. In Stiles's (1978: 133) opinion, the *"pre-Soan"* material is not convincing as examples of humanly-struck stone.

Nevertheless, this raises an important issue, as to how early stone tool making can be convincingly demonstrated. In some cases, hominid activity can be demonstrated purely on contextual evidence, as with so-called *"manuports"* from some East African localities. In these instances, hominid agencies are invoked as the only way of explaining the presence of stone in fine-grained sediments in association with less ambiguous examples of flaked stone. Usually, however, one has to consider the flaking of each stone. This topic will be discussed at greater length in chapter 7, but for the moment it is sufficient to note that we have stressed the following criteria: the numbers of flakes removed from a stone; the numbers of directions in which a stone was struck; the percentage of the original cortex remaining; the presence or absence of bulbs of percussion and ripple scars; and comparisons with geologically-induced flaking.

Note: Stiles (1978: 139) mentions that he was told that shepherds in the Potwar flake stones out of boredom, and sharpen their axes by pounding the blade on a cobble with a hammerstone, which sometimes produces flakes. For what it is worth, we have never observed stone-flaking amongst local people, and none that we have encountered has ever recognised humanly flaked stone as such. Likewise, we have not observed shepherds sharpening their axes in the manner he describes. Usually they do so by using a cobble as a whet-stone. They do however chop firewood into bundles on a large stone, and this may induce some chipping – but not flaking – of stone.

2.10 Conclusions and summary

1) Paterson's work has to be seen in the context of similar investigations of lower palaeolithic material in the 1930's. In particular, his conclusions in the Soan have to be seen in the light of his main work on the assemblages from Clactonian sites such as Barnham St. Gregory in southern England. It was from this background that he developed his belief in a progressive and continuous development of stone-working techniques, leading to smaller and more finely-made types of tools. This background also determined the criteria he used in the Soan, and the importance he attached to patination and the amount of wear in establishing the relative age of stone tools.

2) We find no evidence that the Soan flake industry can be usefully divided into a pre-, early and late phases, as both their dating and definition are inadaquate.

3) We are unconvinced by their claims that the Soan Flake Industry existed independently of a bifacial, Acheulean tradition in the areas they studied. We find it more than a coincidence that it was Paterson who studied the two areas of the Old World where a flake- and biface-tradition were supposed to overlap in the lower palaeolithic.

4) We would instead suggest that the palaeolithic of northern Pakistan should be divided into simply an "early" palaeolithic that is pre-30 ka, and a "later" one until greater chronological resolution is obtained.

5) The major methodological problem that they failed to resolve was in determining whether stone artefacts from the surface of conglomerate exposures dated from the formation, or the exposure, of the conglomerate.

Part II

Pleistocene and Palaeolithic Investigations

Chapter 3

Potwar Plateau : Siwalik stratigraphy and Pleistocene tectonics and climate

3.1 Introduction

In the next two chapters recent work on the Siwalik series, and Pleistocene tectonics and climate in the region will be reviewed as a prelude to our own re-evaluation of the Soan area. In stark contrast to De Terra's work, the emphasis here is on the tectonic instability of the area, and on the extent to which tectonic changes during the Pleistocene have masked or obliterated any climatic record.

3.2 Himalayan Tectonics and the Siwalik Series

The remarkable sequence of fluvial deposits known as the Siwaliks provides a virtually complete record of the later stages of the Himalayan orogeny, from 14 million years (Ma) until 0.4 Ma. The Siwaliks extend from northwestern Pakistan through northern India to Burma and were deposited synchronously with the uplift and erosion of the Himalayan mountain belt (Figure 3.1). The term "Siwalik" was originally applied to the sedimentary deposits outcropping between the Ravi and the Ganges by Medlicott (1864). Pilgrim (1910) divided the "Siwalik Series" into three biostratigraphical units: Lower, Middle and Upper Siwalik. Each unit was further divided, on the basis of faunal assemblages, into biozones of the Kamlial, Chinji, Nagri, Dhok Pathan, Tatrot, Pinjor and Boulder Conglomerate (see Table 3.1). Later workers have tended to use these biostratigraphical units as lithostratigraphical ones (Lewis, 1937; Wadia, 1928), and this usage was formalised by the Stratigraphic Committee of Pakistan in 1973 (Fatmi, 1973; Shah, 1977; Shah, 1984). Many workers have commented on the difficulty of mapping Siwalik Series deposits. These fluvial molasse deposits typically show great variations in facies, both laterally and vertically. Recent attempts to establish a detailed biostratigraphy of the Chinji and Nagri formations have met with considerable success (Shah and Pilbeam, 1984), while work on the magnetic polarity stratigraphy of Middle and Upper Siwalik subgroups (Opdyke *et al*, 1979; Johnson *et al*, 1982 etc.) has developed a sound chronological framework for these deposits. The historical record of sedimentation

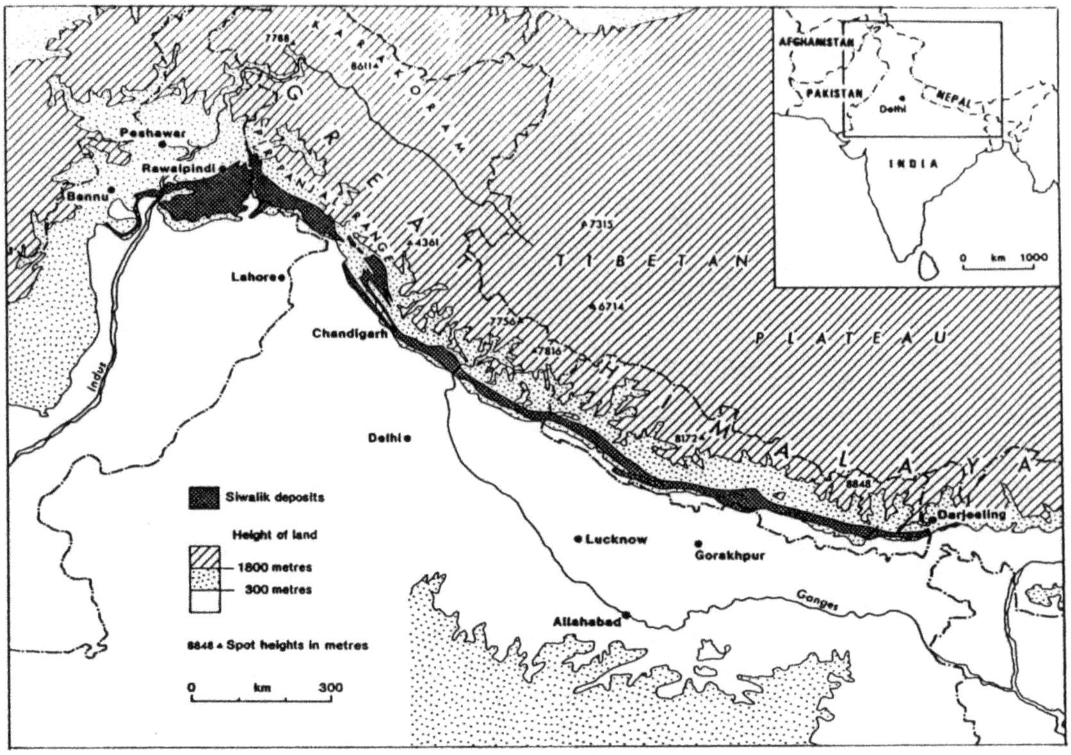

Figure 3.1: Spatial distribution of Siwalik strata

preserved in the Siwaliks is intimately connected with the uplift, folding, faulting and erosion associated with the collision of the Indian subcontinent and Asia.

The nature and chronology of this collision have recently been elucidated by Patriat and Achache (1984). Using palaeomagnetic data they have reconstructed the collision path taken by the subcontinent since it moved northwards from the latitude of Madagascar. Their analyses confirm that the collision began 50 Ma B.P. and was preceded by the accretion of the Ladakh island arc ca. 54 Ma B.P. Subduction of continental crust beneath Tibet is thought to have ended by ca. 44 Ma, when the subducting force acting on the Indian plate was balanced by the buoyancy of the subducted continental crust. At that point 300-500 km of Indian continental crust had been subducted. With the end of active subduction other mechanisms of crustal foreshortening such as folding and thrusting have been brought into play. In all a total of 2,600 ± 900 km of crustal shortening has been observed.

It is estimated that the collision between the Indian subcontinent and Asia is currently proceeding at a rate of 4-5 cm/a (Burbank, 1983; Raynolds and Johnson, 1985). Recent studies on the rate of uplift associated with this collision not only show an apparent acceleration in uplift during the Pleistocene but also exhibit substantial differences in uplift rates in different parts of the Himalayan mountain belt. Results show major differences between the Nanga Parbat area of the Karakoram with a rate of ca. 9mm/a for the period 0.5 Ma to the present (Zeitler *et al*, 1982a, 1982b) and the Great Himalaya range (uplift rate ca. 1mm/a). Other data indicate that these uplift rates have had major repercussions on the climate of the subcontinent and of Central Asia (see Chapter 3.4).

Essentially, the crustal foreshortening necessitated by the continuing collision between the subcontinent and Asia has been accommodated by uplift folding and faulting. The majority of major faults like the Main Mantle Thrust (MMT) and Main Boundary Thrust

SIWALIK SYSTEM Faunal Zones (Pilgrim, 1973)		SIWALIK GROUP Lithostratigraphic Units (Rahman, 1967; Fatmi, 1973)
UPPER SIWALIK	BOULDER CONGLOMERATE ZONE PINJOR ZONE TATROT ZONE	SOAN FORMATION
MIDDLE SIWALIK	DHOK PATHAN ZONE	DHOK PATHAN FORMATION
	NAGRI ZONE	NAGRI FORMATION
LOWER SIWALIK	CHINJI ZONE	CHINJI FORMATION
	KAMLIAL ZONE	RAWALPINDI GROUP
	MURREE SERIES	

Table 3.1: Siwalik stratigraphic divisions

(MBT) are thrust faults. The seismological evidence points to a relatively rapid decrease in the angle of the fault plane with depth. The structural geology of the Potwar Plateau is summarised in Figures 3.2 and 3.3. The activation and re-activation of faults in the Himalayan foredeep has had a profound effect not only on Siwalik sedimentation, but also on the nature of the folding and faulting of the Siwalik deposits as they finally became caught up in the Himalayan orogeny from the Late Pliocene onwards. The disruptive effect of Plio-Pleistocene tectonics is attested by the Siwalik deposits of the Potwar Plateau and by the development of the Peshawar and Kashmir intermontane basins, and is examined in more detail in the next section.

3.3 Potwar Plateau: Siwalik stratigraphy and tectonics

The Siwaliks of northern Pakistan and India comprise one of the most intensively studies areas of Neogene strata in the world. The area has been subject to more or less continuous investigation by geologists since the mid-19th century. The history of research on the Siwaliks can conveniently be divided into two phases, with the second defined by the advent of palaeomagnetic and radiometric dating techniques in the 1970's.

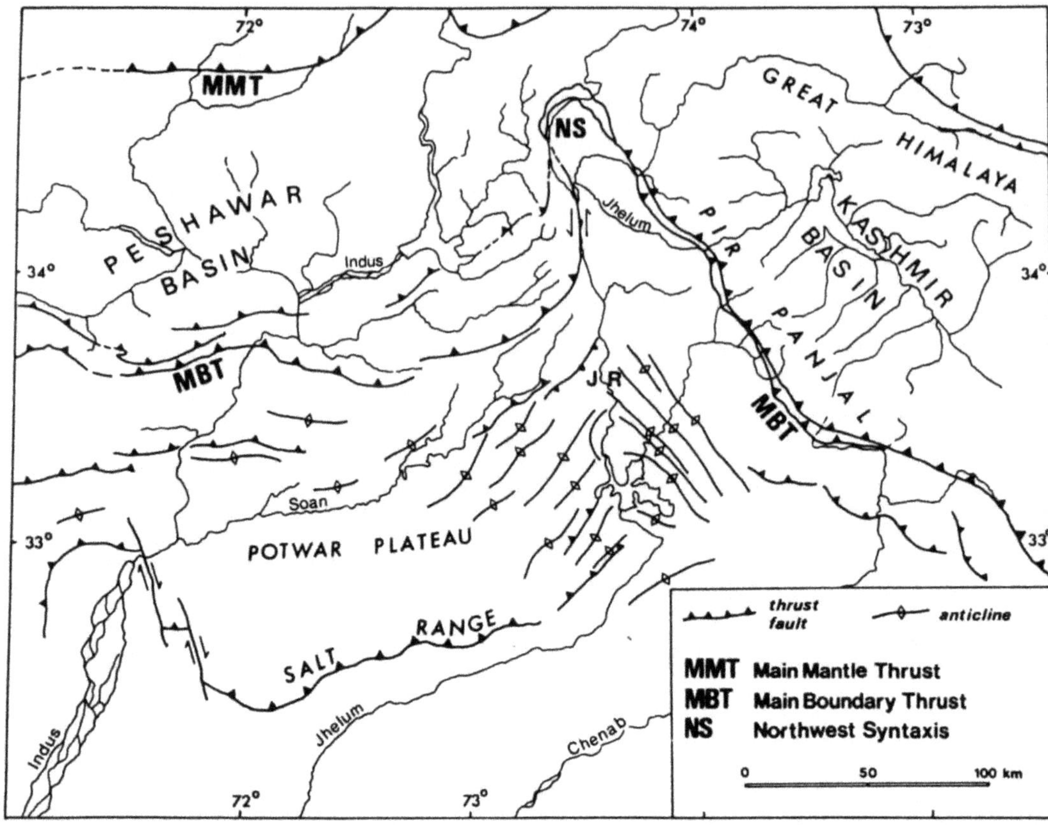

Figure 3.2: Structural geology of the Potwar Plateau and adjacent areas

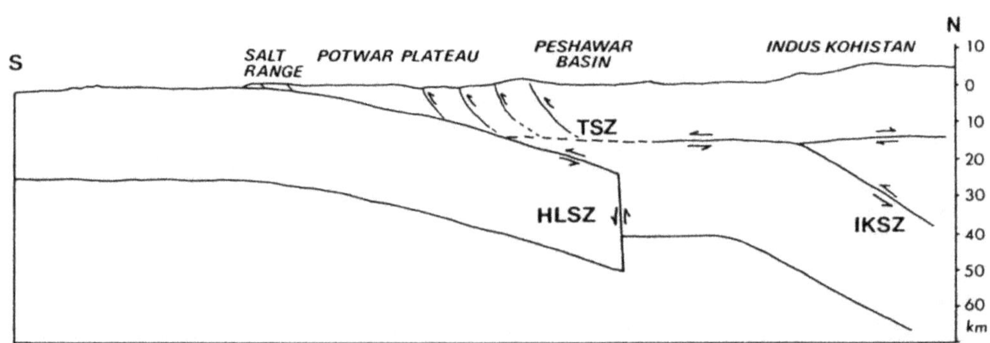

Figure 3.3: Generalised section through the Potwar Plateau and adjacent areas

3.3.1 Early work on the Potwar Siwaliks

From a very early stage, researchers working on the Siwaliks were faced with a dilemma. On the one hand, attempts to devise lithostratigraphic schemes were hampered by the lack of lateral continuity of major units. On the other hand, although biostratigraphic schemes resolved some problems, they could not be applied to areas where faunal remains were sparse or absent.

Wynne (1877) divided the Tertiary deposits of the Potwar Plateau into two series; an earlier Murree Group and a later Siwalik Group. Wynne placed the boundary between the two series between the Chinji and Kamlial formations. Other workers like Pilgrim (1910) placed the lower boundary for the Siwaliks at the base of the Kamlial formation. Pilgrim also recognised three stages within the Upper Siwalik namely Tatrot, Pinjor and Boulder Conglomerate, which he defined on the basis of vertebrate faunas. Anderson (1927) transcended the problem of defining a lower boundary for the Siwaliks by treating all the Tertiary deposits as belonging to the one "Nimandric" system. These different approaches reflect the difficulty encountered in defining a lower boundary for the Siwaliks given that the junction between the Murree and Kamlial formations appears to be gradational. Pilgrim's choice of boundary reflected the need to establish a regional stratigraphic correlation for the Tertiary deposits of the subcontinent based on vertebrate faunas. This need for regional stratigraphic correlation was not recognised by Cotter (1933:103) who commented:

> *"There appears to be no reason why a section so far distant as Sind, and differing so profoundly from those of the Potwar, should so influence our field mapping in the Potwar as to force us to abandon a natural line dividing the freshwater Tertiaries into two groups ...for the indefinite vague and gradational passage between the Upper Murrees and the Kamlials."*

Accordingly, Cotter placed his lower boundary for the Siwaliks at the base of the Chinji. He identified two units, the Nagri and Dhok Pathan, as Middle Siwalik and one unit, the Tatrot, as Upper Siwalik. This did not resolve the difficulties already mentioned of mapping Siwalik deposits in the field. As Cotter (1933) pointed out, Wadia (1928) had included deposits mapped by Cotter as Dhok Pathan (and thus, in his view, Middle Siwalik) within the Upper Siwalik group.

In the absence of faunal evidence, regional correlation of the Siwaliks in general, and the Upper Siwaliks in particular, continued to be notoriously difficult. As Anderson (1927:688) commented:

> *"The Upper Siwalik, as a whole, is the most variable division of the Nimandric system. It is not unusual to find thick zones of siltstone or sandstone which locally might be taken to have almost formational importance changing laterally into coarse conglomerate or vice versa."*

The recognition of these stratigraphic difficulties led to the formal adoption of lithostratigraphic units for the Siwaliks by the Stratigraphic Committee of Pakistan (Fatmi, 1973). The major changes instituted by the Committee were first, the inclusion of the Kamlial and the Murree formations in the newly-named "Rawalpindi Group" and second, the use of the term "Soan Formation" to replace the Tatrot, Pinjor and Boulder Conglomerate zones of Pilgrim (1910) and the "Upper Siwalik" of Medlicott (1864). The stratigraphic Committee recognised four subgroups within the Siwaliks, namely:

Chinji, Nagri, Dhok Pathan and **Soan**. Some indication of the relative thicknesses of the different units is given in Table 4.1, in which the thickness of the Upper Siwalik sediments in the Soan syncline to the south-east of Rawalpindi is particularly notable. The variability and complexity of the units merely reflect the response of fluvial systems to changes in tectonics and subsidence in the Himalayan foredeep. The time-relationships of these sedimentary sequences therefore hold the key to understanding the tectonic evolution of the Potwar Plateau during the Plio-Pleistocene. As discussed below, these relationships have recently been the subject of extensive investigation. One point that has emerged from recent research is that the boundaries between the lithostratigraphic groups recognised in 1973 are often time-transgressive, and therefore their ages vary from area to area.

3.3.2 Siwalik Chronostratigraphy

From the mid-1970's onwards, a collaborative project involving geologists from Dartmouth College, New Hampshire, and from the University of Peshawar has involved a systematic study of the magnetic polarity stratigraphy of the Siwalik sediments of the Potwar Plateau, the Plio-Pleistocene sediments of the Kashmir and Peshawar basins, and the intercalated volcanic tuffs. The presence of the volcanic material provides an invaluable absolute chronological control for the palaeomagnetic work. Some chronological control is also provided by local sequences of vertebrate faunas, although Upper Siwalik sequences, in particular, remain poorly documented.

The Siwalik sediments comprise deposits resulting from both vertical and lateral accretion. The former include overbank silts and clays and the latter comprise sands, gravel, pebbles and cobbles. The fine-grained vertical accretion deposits have relatively strong natural remanent magnetism (NRM) compared to sands, and it is these fine-grained sediments that are sampled for palaeomagnetic analysis. At least three orientated samples per sampling point are demagnetised, using alternating field or thermal techniques to remove any post-depositional overprinting, so as to obtain the stable end point from which the virtual magnetic pole at the time of deposition of the sediment may be deduced. Using this technique, a magnetic polarity stratigraphy can be built up for each sedimentary sequence. Chronological control is achieved by cross-association of field data with the timescale of geomagnetic reversals established by Mankinen and Dalrymple (1979). Where no independent chronological control is available, the interpretation of particular sections will inevitably remain equivocal. Fission-track dates for zircons from altered volcanic tuffs provide excellent time controls for many of the palaeomagnetic sections in the Potwar and adjacent areas. One group of tuffs fortuitously coincides with the Gauss-Matuyama boundary, a change from a normal to a reversed magnetic field at 2.48 Ma B.P. The source of the volcanic material is thought to be the Dasht-e-Nawar volcanic complex in Afghanistan, 75 km to the southwest of Kabul (G.D. Johnson et al, 1982). The interpretation of the stratigraphic sections is constrained by the relative abundance of lateral accretion deposits and the sampling interval adopted. In the Dartmouth/Peshawar studies, the sampling interval was of the order of 5-10m.

G.D. Johnson et al (1982) report data for 18 sections in the Potwar Plateau and adjacent areas for which fission track data are available. Other sections are reported by Keller et al (1977) for the Pabbi Hills, Visser and Johnson, (1978) for the Mangla-Samwal anticline, Opdyke et al (1979), Raynolds (1980), and N.M. Johnson et al (1982) for the Eastern Potwar, Burbank and Johnson (1982,1983) for Kashmir, and Burbank and Tahirkheli (1985) for the Peshawar Basin. The quality and quantity of the data generated by these workers have allowed the calculation of sedimentation rates, and this in turn has enabled various inferences to be made about the tectonic controls on sedimentation in the Himalayan foredeep, thereby providing a time frame for the disruption of the foredeep and the development of the intermontane basins of Kashmir and Peshawar. Overall, the quantity and quality of the palaeomagnetic and radiometric work in this area has totally transformed our understanding of its geological history and made it one of the best calibrated Neogene sequences in the world.

3.3.3 The chronology of Siwalik deposition and associated tectonic evolution of the Himalayan foredeep

Raynolds and Johnson (1985) provide a detailed summary of sedimentation rates in the eastern Potwar associated with the development of the Jhelum re-entrant and the migration of the foredeep. They present data for 15 sections for which good chronological control exists. Deposition rates have been calculated for a number of different time-slices from 3.4 Ma to the present. These data show a broadly southward migration of the zone of maximum accumulation (the depocentre), at a rate of 20m/ka, with maximum rates of sediment accumulation in excess of 50cm/ka. The rate of accumulation is a function of the interaction of the rates of subsidence and sediment supply. Raynolds and Johnson (1985) infer that a southward migration of the Himalayan foredeep has been in progress since 3.4 Ma, and they present an idealised model of cumulative sediment accumulation through time, indicating that, as an area becomes involved in the migration of a foredeep, the rate of sediment accumulation is expected to pass through three phases. The first features relatively slow accumulation near the distal margins of the foredeep; the second, more rapid accumulation at the depocentre and finally, the rate of accumulation decreases as uplift and deformation are initiated in the proximal foredeep. In the case of the eastern Potwar, the migration path of the depocentre appears to have been constrained by the developing Jhelum Re-entrant (JR) and the Northwest Syntaxis (NS).

Most importantly, this chronostratigraphic study has also resolved the time-framework of the uppermost Siwalik facies - Pilgrim's (1910) "Boulder Conglomerate Stage". Raynolds and Johnson demonstrate that this unit is diachronous, with a southward advance of coarse facies in the eastern Potwar at a rate of 30m/ka since 3.4 Ma B.P., rather than synchronous, as earlier workers suggested. This discovery has immense repercussions on Pleistocene research in this area. Put bluntly, there is no "Boulder Conglomerate Stage" as such. **The "Boulder Conglomerate" cannot be correlated with any time period or event (such as glaciation) outside the region for the simple reason that its age differs from one part of the Potwar to another.** In the Soan the Upper Siwalik conglomerates are ca. 2 Ma whereas in the Pabbi Hills, 100 km to the south east, they date to 1 Ma. One should, however, be cautious in interpreting Raynolds and Johnson's generalisation of a southward migration of coarse facies in the eastern Potwar since the so-called "Boulder Conglomerates" exhibit distinct spatial variation in both nature and composition.

The tectonic implications of the sedimentary history of the Potwar plateau and adjacent areas have been explored by Burbank (1983) and Burbank and Raynolds (1984). It is apparent from the Dartmouth and Peshawar studies that the area of accumulation of Siwalik sediments been disrupted during the past 3-4 Ma. Burbank and Raynolds (1984) argue that this disruption of the foredeep has occurred as a series of pulses of rapid deformation associated with the development of the Main Boundary Thrust (MBT), and with intervening periods of relative quiescence rather than, for example, as a slow but continuous deformation during the Pleistocene, as Wadia (1928:345) seemed to suggest.

The oldest Lower Karewa sediments in the Kashmir Basin date from 4 Ma B.P. (Burbank and Johnson, 1982,1983), while the oldest ones in the Peshawar Basin date from about 2.8 Ma (Burbank and Tahirkheli, 1985). Both basins are bounded to the south by northward dipping thrust complexes. Crustal convergence is, however, accommodated in different ways in the two areas and these differences are reflected in the development of the Northwest Syntaxis (NS) (Burbank, 1983).In their synthesis of information from the dated sequences in the Potwar and the intermontane basins, Burbank and Raynolds (1984) identify a southward migration of thrust faulting, and uplift during the period 3.0-1.8 Ma ago with the locus of thrusting moving 50 km to the south. On the basis of geophysical data presented by Seeber et al (1981), Burbank (1983) hypothesises the development of a detachment surface below the Peshawar Basin extending southwards to the Salt Range during this period. The sequential southward migration of disruption of the foredeep is documented by dated sequences in the Soan syncline, where the folding event occurred 2.1-1.9 Ma B.P.; in the Bhangala and Ganda Paik anticlines, folded 1.6-1.5 Ma B.P.; the Mahesian anticline, near Dina, folded 0.5 Ma B.P.; and in the southwestern portion of the Rhotas anticline (Basawa Kas) and the Pabbi Hills anticline where folding commenced 0.4 Ma B.P. (Burbank and Raynolds, 1984, Raynolds and Johnson, 1985). These data are summarised in Figure 3.4 which shows an interpretation of seismic data along a section through the Eastern Potwar and Salt Range by Pennock *et al* (1989). Pennock *et al* demonstrate that ca. 24 km of crustal foreshortening appears to have taken place between the Soan syncline and the Pabbi Hills anticline since 5.5 Ma. They calculate an average rate of shortening of 7mm/a over the last 2.5 Ma, corresponding to 15 % of the convergence rate between India and Eurasia.

The recent dramatic uplift of both the Pir Panjal and the Attock ranges dates from ca. 0.4 Ma (Burbank, 1983). The pulse of uplift of the Pir Panjal on the southern flank of the Kashmir Basin is particularly noteworthy with 1400-3000 m of uplift at a minimum rate of 10 mm/a - a rate comparable to the uplift of the Nanga Parbat region during the same period (Zeitler *et al*, 1982a, 1982b).

On the basis of this synthesis, two major phases of disruption of the foredeep have been identified, the first from 3.0-1.9 Ma and the second from 0.4 Ma to the present. This chronological framework has important implications for studies of Siwalik and post-Siwalik sediments in the Potwar, most notably that the unconformity or disconformity between Siwalik and post-Siwalik sediments is time-transgressive. For this reason alone, earlier attempts to correlate local sequences lithologically were doomed and have to be replaced by magnetic polarity and faunal sequences.

3.4 Pleistocene climatic change and tectonics

The establishment of the dynamic causal mechanism for the southwest monsoon is the key to understanding recent climatic change in the northern subcontinent.The rates of uplift associated with the Pleistocene phase of the Himalayan orogeny might be expected to have had a significant effect on the climate of the foredeep area, the intermontane basins and the Tibetan plateau. Evidence that the Great Himalayan range was at the same level as the Tibetan plateau 2 million years ago (Wang *et al*, 1982) implies that the southwesterly monsoon could have penetrated into Central Asia at that time. Subsequent

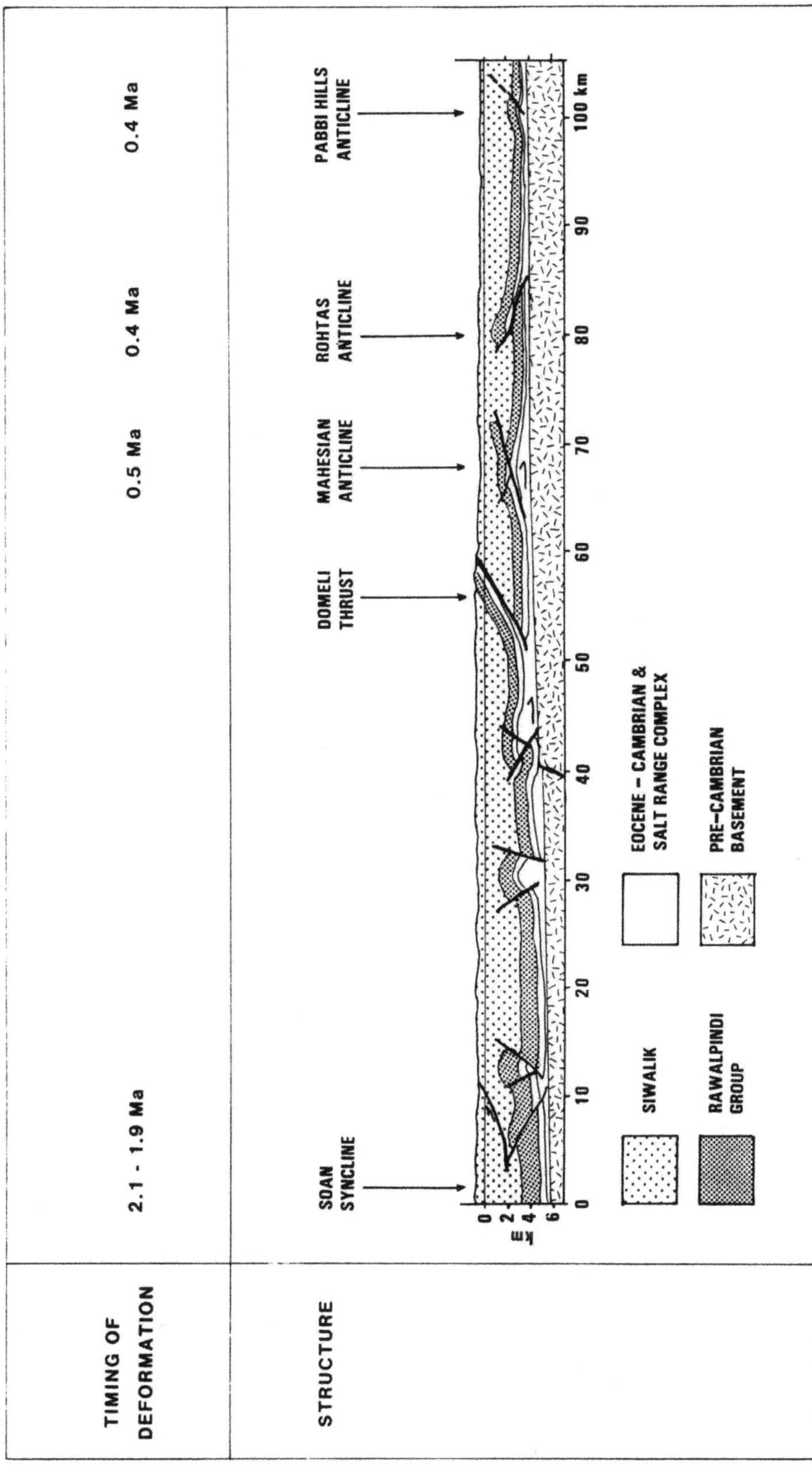

Figure 3.4: The timing of foredeep deformation (section after Pennock *et al*, 1989).

uplift of the Great Himalayan range has blocked this penetration and this change can be expected to have had a marked effect on the albedo of the Tibetan Plateau. Similarly, the Late Pleistocene uplift of the Pir Panjal would have cut off the Kashmir Basin from the influence of the southwesterly monsoon. The role of the Tibetan plateau, a land area of 2,400,000 km^2 at a height of 5 km, in forcing climatic patterns has been widely discussed. Duplessy (1982) points to current conditions in which the presence of a snow-free Tibetan Plateau actually accelerates the onset and increases the intensity of the Asian monsoon owing to the plateau's rapid heating during the spring and the resultant pressure difference between the high Asian continental mass and the Indian Ocean. Manabe and Hahn (1977) modelled a significant weakening in the southwesterly summer monsoon in response to an increase in surface albedo over an ice-covered Tibetan Plateau. This view is corroborated by data for the last 22 ka derived from sediment cores from off the southwestern coast of India which show an arid phase 22 ka - 18 ka, with a very weak southwesterly airflow, followed by a humid phase with progressive northward penetration of monsoon rainfall culminating in a maximum at 11 ka (Van Campo, 1986). The gradual penetration of the monsoonal rainfall over the northern part of the subcontinent is associated by Van Campo with the northward migration of the Intertropical Convergence Zone. Data from Rajasthan confirm the existence of an arid phase ca. 18 ka B.P. (Agrawal and Gupta, 1988). In contrast, data from Kashmir show both a significant increase in broad-leaved elements in the Butapathri bog profile (Dodia *et al*, 1985) and the development of the Burzahom palaeosol dated at ca. 18 ka B.P. and intepreted by Agrawal (1988) as indicating the onset of a moister, wetter climatic regime in Kashmir. Agrawal (1988) also argues that the proximity of Kashmir to the Tibetan Plateau forced an "early" deglaciation at about 18 ka i.e. during the last glacial maximum in northwestern Europe.

Prell (1984) establishes a strong correlation between monsoonal upwelling and changes in solar radation at the frequencies of orbital precession, and demonstrates that changes in solar radiation appear to force changes in monsoonal circulation at timescales of 10^3 to 10^5 years. He concludes his paper by noting that:

> "*... no unique latitude of critical response can be identified in the radiation data. This observation rests on the fact that the response lies not in the radiation data as a function of latitude but rather in the interaction of the earth's surface with the incoming radiation. Hence, the great area and great elevation of the Tibetan Plateau and the surrounding regions are the key dynamic link in determining the strength of the monsoon*" (Prell 1984:364).

Palaeoclimatic research in Asia is clearly at a turning point, with the emphasis moving away from the search for correlation with global (ocean core) trends towards the establishment of major regional variations about those trends.

3.4.1 The Quaternary Glacial Record in the North-West Himalayan area

As noted in Chapter 1, the establishment of a record of Quaternary glaciations has focussed on the recognition of points of maximum glacial advance, the number of glacial advances, and glacial or glacially-related deposits. Most early workers estimated that glaciation had extended down to a level in the range of ca. 2440-1800m. Coulson (1938) concluded that glaciation extended to 4,500 ft (1370 m) in Hazara. More recently, Porter

(1970) has identified a lower limit for Pleistocene glacial advance in Swat-Kohistan at 5,000 ft (1525 m) and found traces of only three main periods of glacial advance in Swat-Kohistan. Deposits identified by earlier workers as moraines have been reevaluated. Some of the Kashmir Basin moraines identified by De Terra and Paterson (1939) have been reinterpreted as tectonically-controlled fanglomerates (Burbank (1983), Burbank and Johnson (1983), Singh (1982)). Holmes (1988) recently undertook a complete re-evaluation of the glacial record in Kashmir and concludes that the extent of Quaternary glaciation in Kashmir was smaller than was previously supposed, with no tills identified below 2200 m. The unexpected north-east to south-west trend of increasing altitude for reconstructed equilibrium lines for the Pleistocene glaciers is explained in terms of the recent uplift of the southern (Pir Panjal) flank of the Kashmir Basin (Holmes and Street-Perrott, 1989)

Since moraines were restricted to the higher parts of the Pir Panjal and Swat-Kohistan, there is no direct evidence of glacial advance in the Peshawar Basin and Potwar Plateau areas. Interest has consequently focussed on a number of deposits associated with glacial activity: erratic blocks, other glacially-derived material, and loess.

Erratic blocks

Early work on the erratic blocks of the western Potwar was reviewed in Chapter 1. None of the workers quoted there claimed that the erratics had been transported by a glacier and only De Terra and Teilhard de Chardin (1936) and Coulson (1938) attempted to link them to a fluvioglacial deposit. Catastrophic flooding of the Indus appears to provide a reasonable explanation of the origin of the erratics. Such floods are frequent and well represented in the historic record. The distribution of the blocks does, however, suggest that the Indus may have followed a slightly different course at certain times during the Pleistocene.

Glacially-derived material

Evidence of glacial or glacially-derived sediments from within the Siwalik series is limited to De Terra's "Boulder Conglomerate" and to the Bain Boulder Bed. As pointed out in chapter 1, De Terra related the unit which he mapped as "Boulder Conglomerate" to the second (alpine) glaciation, but his mapping and dating of the Boulder Conglomerate are both wrong. Within the Soan valley, the unit mapped by De Terra is now interpreted as a Post-Siwalik valley fill (the **Lei Conglomerate**; see chap. 4.5).

The Bain Boulder Bed outcrops in the Marwat-Kundi range which now encloses the southern portion of the Bannu Basin in the North-West Frontier Province. Morris (1938) considered that the Bain Boulder bed was a tillite on the basis of :

> *"(1) the complete absence of bedding and grading within the deposit; (2) the heterogeneous composition of the deposit; (3) the characteristic angularity and freshness of the mineral fragments of the matrix; (4) the abundance of non-resistant fissile rock types as compared with normal fluviatile Siwalik conglomerates; (5) the abundant patches of finely comminuted rock-debris included in the matrix and altered but little from the composition of the outcrops whence they were derived; and (6) the occurrence of occasional scratched boulders within the bed."* (Morris, 1938:402).

At the type section in Bain Pass, the unit certainly stands out as distinctly different from the surrounding fluvial sediments. Morris interpreted the Boulder Bed:

> *"as being composed of the englacial material of a floating ice-tongue which*

in early Pleistocene times descended from the heavily glaciated highlands of Waziristan and Afghanistan, pushed out into the waters of a Siwalik Lake, and in due course melted." Morris (1938:403).

A new interpretation of the Bain diamictite has recently been suggested by Khan *et al* (1985). They interpret the deposit as a volcanic lahar derived from the Dasht-i-Nawar caldera 350 km to the north-west in Afghanistan. They point out that Morris's criteria for establishing the Boulder Bed as a tillite are by no means diagnostic and could also apply to debris-flow deposits. Lahars are debris-flows which are associated with volcanic activity (Crandell, 1971) and are characterised by

"lack of internal stratification, crude size gradation from coarse to fine upward in places, irregular-shaped small air spaces in matrix, and propensity to flow as massive thick lobes down valleys without leaving much behind except as a thin verneer" (Khan *et al*, 1985:61).

Modern analogues include lahar flows from Mount Rainier, Washington. One problem is that if the Bain Boulder Bed is accepted as being a lahar, its estimated volume of 30 x 10^9 m^3, makes it the largest known deposit of this type in the world (Khan et al, 1985). A second problem is that the path of the lahar is now unclear because tectonic disruption of the Bannu Basin Siwalik sequence and related deposits during the Late Pleistocene has greatly transformed the topography of the area.

The Bain Boulder Bed differs radically in appearance from the fluvial silts and sands with which it is interbedded. Although there are some problems with the lahar hypothesis, it remains the most likely explanation for this unusual deposit. Morris's hypothetical ice tongue floating out across a vast early Pleistocene lake seems merely fanciful by comparison.

Thus evidence of a glacial origin for erratic blocks, the "Boulder Conglomerate" and the Bain Boulder Bed appears to be lacking or, at best, equivocal, leaving loess as the only potentially glacially-derived sediment in the Potwar area.

Loess

Although most authorities now accept that loess in the strict sense is aeolian in origin, there is still considerable debate about the origin of the predominantly silt-size particles of quartz which form the bulk of loess deposits and the climatic conditions under which it was deposited. As seen in Chapter 1, De Terra and Paterson (1939) interpreted the Potwar loess as glacial by default.

3.4.2 The glacial origin of loess

While Smalley has championed the cause of a glacial origin for silt-sized quartz grains (Smalley, 1966; Smalley and Vita-Finzi, 1968; Smalley and Cabrera, 1970, 1971), other workers notably Warnke (1971), Whalley *et al* (1982), and Pye (1984, 1987) have tended to favour a desert origin. Recent work on the release of silt-sized quartz as rocks weather at the earth's surface has caused Smalley and Krinsley (1978) to recognise the importance of this source of silt, and Nahon and Trompette (1982) argue that much of the Quaternary loess is derived from reworking of weathered profiles. Perhaps unsurprisingly, no single, or indeed simple, explanation of the origin of quartz silt has a monopoly of the truth. One particular problem is that while the shape and surface textures of sand-size grains of quartz appear to be environmentally diagnostic (Bull, 1981), the same may not be true for silt-sized grains (Goudie *et al*, 1984). Smalley and Cabrera's (1970) conclusions that angularity and the presence of adhering particles are somehow exclusive to glacial grinding are rightly challenged by Warnke (1971).

In trying to decide whether a particular loess deposit has a glacial origin a number of related questions need to be answered. These questions involve not only the availability of a source of silt-sized quartz but also the appropriate conditions for entrainment, transport and deposition of the silt.

(1) Is loess actively forming today?

This question is more difficult to answer than might first appear, owing to the timescales and rates of deposition involved. Rates of loess deposition during the Quaternary are in the range 10^1 - 10^3 mm/ka (Pye 1987). When expressed on an annual basis, some of these rates result in accumulations which would be very difficult to measure. However dust storms are concentrated in both space and time and it is possible to measure rates of both loss (from source areas) and gain (in areas of deposition known as dust sinks). Goudie (1978) gives current rates of dust accumulation in the range 10^{-1} - 10^4 mm/ka, based on measurements compiled from ocean cores as well as land-based sampling. There is some evidence that dust storm activity was higher at certain times during the Quaternary. The concentration of fine aeolian particles in a Greenland ice core ca. 18ka B.P. was, for example, as much as 100 times greater than at present (Goudie 1983:519). The general lack of well-developed Holocene loesses also points to higher dust storm activity during periods of the Pleistocene.

The use of dust storms and dust deposition rates as modern analogues of active loess formation presupposes a comparability of particle size distributions. One characteristic of loess deposits is that the bulk of the grains have equivalent spherical diameters in the range 20 - 50 microns. Grain size data for dust deposits show a comparable size range (Goudie 1978, Table 10), although some far-travelled dusts are much finer. Since agricultural activity in loess areas appears to be one of the mechanisms currently responsible for dust storm activity, it is hardly surprising that the size distributions of loess and resultant dust should be broadly comparable. One key factor is that grains smaller than 20 microns have a much higher mobility than larger silt grains and may be transported over distances of up to 300 km, while particles finer than 10 microns can have a worldwide distribution (Pye, 1984, 1987).

(2) What size ranges of material are produced by (a) glacial grinding (b) abrasion in deserts (c) salt weathering of sand-size quartz (d) weathering processes releasing silt-sized quartz?

A process has to be envisaged that will produce large amounts of quartz particles of the appropriate size. Glacial grinding was originally favoured by Smalley (1966) and Smalley and Vita-Finzi (1968) on the grounds that it is the only process whereby sand with a mode size of 500 microns could be reduced to silt with a mode size of 50 microns. Smalley and Vita-Finzi (1968) produce a figure comparing the hypothetical size distributions of particles produced by glacial grinding and by spalling, as a result of grain impacts under desert conditions. Their contention that the size distribution of glacial sediments can include a large portion of silt-sized particles is supported by some field evidence (Dreimanis and Vagners, 1971). The size distribution resulting from grain impacts, based on Kuenen's (1969) work is distinctly bimodal, with rounded sand grains and fine debris with the intermediate (loess) size range absent. More recently, work by Whalley *et al* (1982) on aeolian attrition has contradicted Kuenen's finding. They were able to demonstrate experimentally that a range of silt-sized particles could be produced by abrasion. Pye and Sperling (1983) showed, again experimentally, that salt weathering of regolith sands produced substantial quantities of silt-sized grains. They also comment that

> *"The shapes and surface textures of grains affected by salt weathering are dominated by angular breakage surfaces which are indistinguishable from those produced by frost action and crushing"* (Pye and Sperling, 1983:60).

Some of the basic premises of Smalley's (1966) assumptions have been undermined by the discovery of structural weaknesses in quartz grains which can be exploited by physical and chemical weathering processes (Moss and Green, 1975; Riezebos and Van der Waals,1974). On the basis of a study of siltstones in the geological record, Nahon and Trompette (1982) argue that silt-size particles released by weathering processes form the major component of Quaternary loess. In a major review paper, Smalley and Smalley (1983) seek to refine and redefine the issue of particle production. They concede that *"cold weathering processes"*, which often involve both frost and salt, *"must be considered as a major method for producing loess material"* (1983: 54). Although a glacial origin for loessic silt now looks rather equivocal, it is nevertheless apparent that most major loess deposits in Europe, Central Asia and China accumulated during a time of widespread glaciation. We must therefore pose one final question:

(3) Is the type of atmospheric circulation prevailing under glacial conditions more important for the mobilisation and deposition of loess than the glacial activities that produce silt-sized quartz?

It could be argued that during a period of glaciation, weathering and reworking of sediments by fluvial action are dominant processes in unglaciated areas. More silt is made available for aeolian transport by exposure on braid bars or alluvial fan systems, and atmospheric conditions favour entrainment and transport. Smalley and Smalley (1983) appear to be moving towards this kind of explanation, while Pye (1987) concludes that the key requirements for loess formation are: (1) the existence of bare unstable geomorphic surfaces composed of poorly sorted sediments with a high silt-clay ratio, and (2) a fairly high frequency of strong turbulent winds.

The consensus amongst geologists appears to have moved away from the glaciogenic origin of loess towards a much more equivocal position. Recent studies of the loess deposits of Asia, particularly those of China, have been partly responsible for this change in stance. It thus seems wise to be cautious about assuming that the Potwar loess is, by definition, a glacially-derived deposit.

3.5 Summary

1) Work on Siwalik stratigraphy before the 1970's attempted to define boundaries either lithostratigraphically or biostratigraphically. Neither method proved satisfactory, given the lateral variability of this fluvial molasse sequence. Difficulties were compounded by the lack of good marker horizons.

2) Work from the 1970's onwards has concentrated on establishing a reliable magnetic polarity stratigraphy for the Potwar. As a result, this area now has one of the best calibrated Neogene sequences in the world.

3) The results of this recent work present a very different account of the Siwaliks from that deduced from earlier studies. Today the Siwalik sequence is interpreted almost entirely in

terms of how river systems responded to the Himalayan orogeny rather than to changes in global climate. It is currently impossible to infer any climatic change, or to identify any particular climatic event from the Siwalik sequence. Some details of palaeoclimate are likely to emerge from studies of faunal remains and from palaeosol development within overbank sequences in the Pabbi Hills, but potentially the most important indicator of palaeoclimate in the Potwar is the post-Siwalik loess.

4) As far as the present study is concerned, the most important point to emerge from recent work is that the so-called "Boulder Conglomerate Stage" is strongly time-transgessive; its age varies significantly across the Potwar, and it cannot be use to denote any particular phase of the Late Pliocene or Pleistocene.

Chapter 4

Soan Syncline : Siwalik and Post-Siwalik stratigraphy and tectonic evolution

4.1 Introduction

In the Soan syncline, to the southeast of Rawalpindi, some 4880 m of Siwalik strata are exposed, of which 1700 m have been mapped as Upper Siwalik (Gill, 1952). The synclinal trough is strongly asymmetrical with its axis running from the north-east to the south-west. The northwestern limb dips steeply, and is almost vertical, in contrast to the gently dipping southeastern one. Table 4.1 includes data from measured sections on the two limbs of the fold. The syncline is bounded to the northwest by a complex folded series of Murree beds and to the southeast by the Riwat Fault. The area was mapped by Wadia (1928) and later by Gill (1952), whose map forms the basis of Figure 4.2. Gill overcame the problems of establishing Siwalik stratigraphy in the absence of good vertebrate faunal localities, by using the technique of strike traverses in the field, and by making use of air photographs. He was therefore able to monitor facies change along the strike. Such detailed mapping is an essential prerequisite if an understanding of the Siwalik sequence in this area is to be achieved.

4.2 Problems with Siwalik Stratigraphy

Notwithstanding the fact that the Soan syncline has been recognised by many workers as a key area for studies of the Siwaliks, there has been a substantial amount of disagreement and confusion (see Chapter 1) about the sequence and, in particular, the presence or absence of a major unconformity between Middle (Dhok Pathan) and Upper Siwalik sediments. This disagreement appears to stem from the asymmetrical nature of the synclinal trough. The marked angular unconformity between steeply-dipping Siwalik and horizontally-bedded post-Siwalik strata on the northwestern limb of the fold means that it is easy to distinguish between the two sets of deposits This contrast in dips is quite clearly seen at Adiala, south of Rawalpindi on the right bank of the Soan (Fig 4.1. In contrast, the uppermost Siwalik beds on the southeastern limb have a relatively gentle dip of 5° – 8° and are much harder to distinguish from the post-Siwalik deposits, particularly

		thickness in metres	
		north flank JARAI KAS	**south flank** KAS DOVAC
POST-SIWALIK	POTWAR LOESSIC SILT	0	60
	LEI CONGLOMERATE	0	105
UPPER SIWALIK	SIWALIK CONGLOMERATE	790	790
	BROWN SANDSTONES AND RED CLAYS	914	1070
DHOK PATHAN	VARIEGATED CLAYS THIN WHITE AND BROWN SANDSTONES	506	400
NAGRI	NAGRI SANDSTONES RED AND DRAB CLAYS	826	665
CHINJI	CHINJI RED SHALES	1440	1030
	RIWAT BEDS	-	260
	TOTAL THICKNESS SIWALIK STRATA	4476	4215

Table 4.1: Stratigraphy of the Soan syncline (after Gill, 1952)

Figure 4.1: Siwalik and post-Siwalik strata on the northern limb of the Soan syncline at Adiala

when the Siwalik material is the major source of supply for the post-Siwalik deposits.

The Upper Siwalik Sub-Group is the most variable unit within the Siwaliks in terms of lateral facies change. Owing to the disruption of foredeep sedimentation, regional correlations of the Upper Siwalik are inevitably problematic. In 1973 the term "Soan Formation" was adopted for the Upper Siwalik Sub-Group. The type section, described by Kravtchenko (1964), lies to the southwest of the central portion of the Soan Syncline near Mujahad Village (Lat: 33° 22′ N; Long: 72° 47′ E). The choice of this type section for the Upper Siwalik Sub-Group was unfortunate. Lindsay *et al* (1980:311) comment:

> *"The Soan Formation is poorly fossiliferous in the type area, and our palaeomagnetic study indicates the type Soan Formation is not equivalent in age to strata that yield Upper Siwalik fossils, but is equivalent to strata of the Middle Siwalik Dhok Pathan Formation."*

Since the boundaries between the sub-groups are known to be diachronous this choice of type section may be less of a problem than it seems. Nevertheless, the fact that there is a problem at all reflects the complexity of Upper Siwalik stratigraphy.

4.3 Stratigraphy of the Upper Siwalik in the Soan Syncline

Although the appropriateness of the choice of the Mujahad Section as the type section for the Soan Formation has already been questioned, it is probably worth recording the characteristics of the section as a basis for the comparison with the Soan river sections. According to Fatmi (1973), the type section comprises:

> *"alternations of pale pinkish, orange, brown soft claystone, brown, grey sandstone and compact conglomerate. The conglomerate is dark grey in colour containing pebbles and boulders of 'Margala Hill Type' limestone."*

In contrast, the section along the Kas Dovac, a small left-bank tributary of the Soan, on the southeastern flank of the Soan Syncline is described as comprising:

> *"an alternation of soft greenish-grey, brown, coarse-grained sandstone, pinkish-red siltstone and claystone with cream coloured massive boulder conglomerate. The pebbles and boulders of quartzite, porphyrite and sandstone ranging in size from 2 to 10 inches (5–25cm)."* (Fatmi, 1973).

The difference in the composition of the conglomerates from the two sections is worth noting.

In the course of mapping the Siwalik deposits within the Soan Syncline, Gill (1952) noted that the sequence of deposits failed to conform to the "standard" Nagri, Dhok Pathan and Tatrot, Pinjor, Boulder conglomerate model. He mapped the Upper Siwalik as comprising the upper portion of a "Brown Sandstones" unit, and a "Siwalik Conglomerate" unit, and placed his Middle-Upper Siwalik boundary within the Sandstones unit on the basis of faunal evidence cited by Wadia (1928). Gill described the brown sandstones unit as comprising

> *"monotonous alternations of deep brownish-red silty clays and brown sandstones ... the grade of the sand rock varies from fine sand, grit and gravelly sand to lenses of conglomerate."*

While the Siwalik Conglomerate in the Kas Dovac section is:

> *"not composed entirely of conglomerate ... the two conglomerate horizons are separated by a zone of deep red and pinkish clay and silt with thin brown sandstone beds."*

He considered that the upper portion of the Brown Sandstones was equivalent to the Tatrot stage, while the Conglomerate belonged to the Pinjor stage. He placed a major unconformity between the Siwalik conglomerate and the post-Siwalik Lei conglomerate. Recent studies of the magnetic polarity stratigraphy of a number of sections within the Soan have provided a new chronological framework for these deposits.

4.4 Magnetic Polarity Stratigraphy

Studies of the magnetic polarity stratigraphy of three sections of Siwalik sediments within the Soan Syncline provide a chronological fix on the deformation of the syncline (Raynolds, 1980; Moragne, 1979; McMurty, 1980). The sections at **Dhok Saiyidan, Shahpur** and **Soan** (Moragne,1979; Raynolds and Johnson, 1985 Section 1-3) constrain the folding event to between 2.1 Ma and 1.9 Ma. The location of these sections is shown in Figure 4.2 and the polarity stratigraphy is summarised in Figure 4.3.

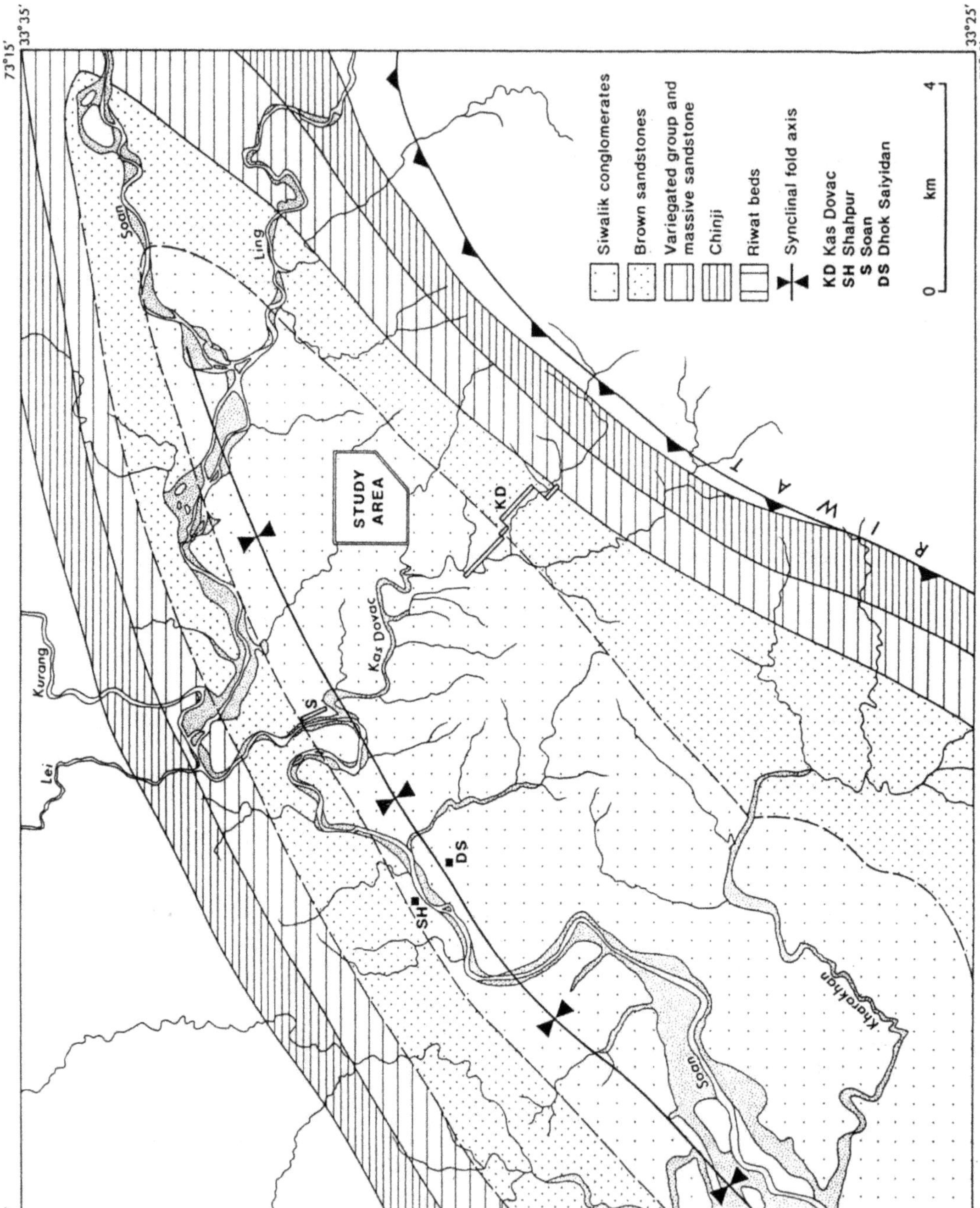

Figure 4.2: Geological map of the Soan Syncline, after Gill (1952)

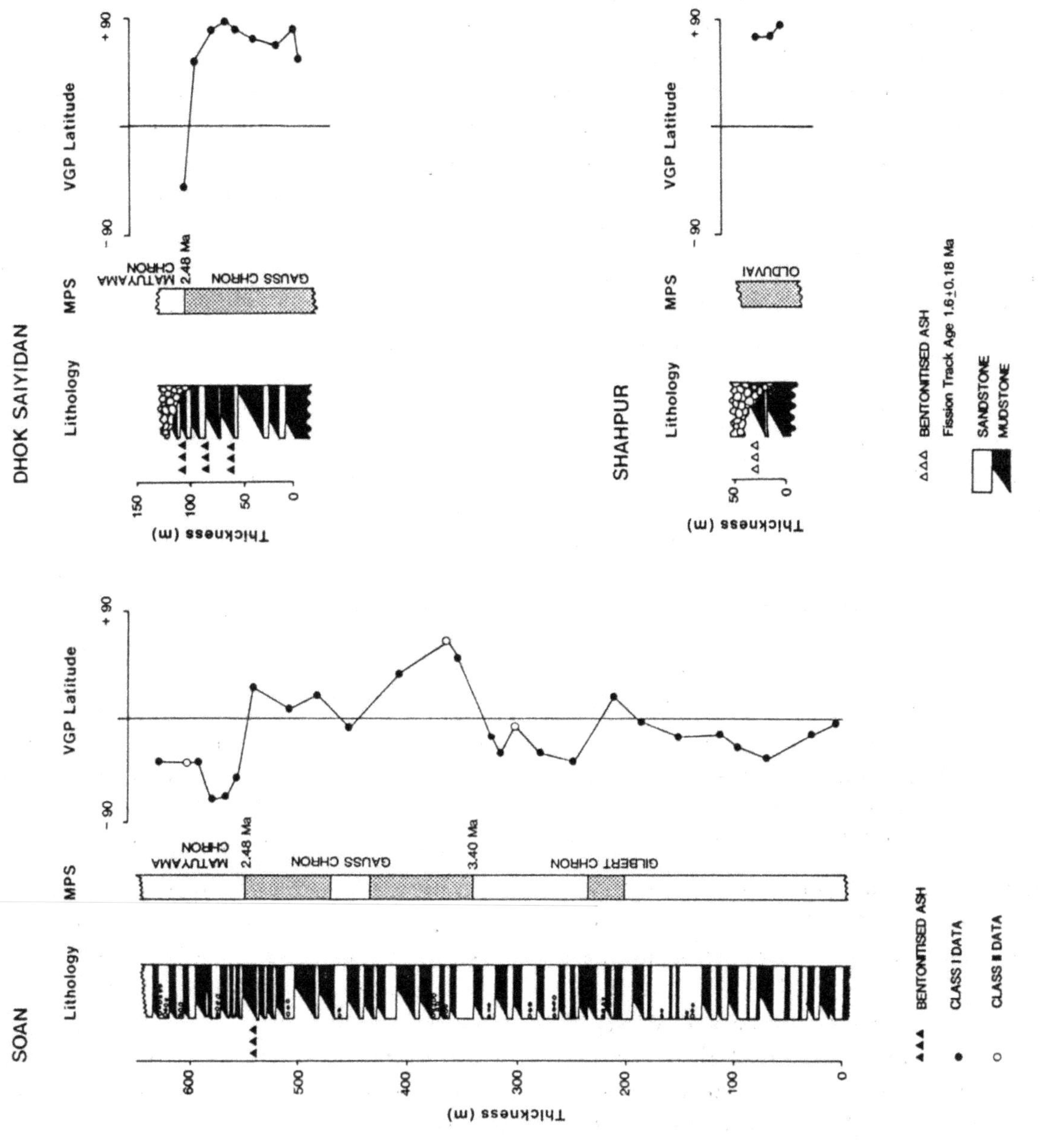

Figure 4.3: Magnetic polarity stratigraphy of sections at Soan, Dhok Saiyidan and Shahpur, after Raynolds & Johnson (1985)

The lower portion of the Siwalik series exposed in the Soan syncline was sampled by Raynolds (1980). He measured a 1650 m thickness of Siwaliks along the **Kas Dovac** (see figure 4.2 for location of section). The basal portion of his section includes the Top White Sandstone (TWS) unit, a distinctive, laterally continuous horizon, which was also mapped by Gill. Raynolds mapped limestone conglomerates in the middle portion of his section, and an unconformity just below the top of the section with quartzite-rich conglomerates above. The magnetic polarity stratigraphy of the Kas Dovac section remains somewhat equivocal (see figure 4.4). Raynolds argues that the uppermost part of his section probably belongs to the Gauss chron and that the lower portion of that chron is missing, this gap in the sequence being marked by the unconformity. The upper quartzite-rich conglomerate outcrops to form a distinctive line of low hills. The change in the source material of these Upper Siwalik conglomerates is presumed to reflect a change in the drainage pattern, while the unconformity reflects an erosional hiatus on the southeast flank of the syncline associated with renewed activity along the Riwat fault.

Moragne (1979) measured one 650 m section on the northern limb of the syncline along the Soan River just to the north of the confluence with the Kas Dovac, and where the dip of the beds approaches the vertical. A volcanic ash yielding a poorly constrained, unpublished fission track age of 1.9 ± 0.4 Ma (Raynolds, 1980) outcrops at about 110 m below the top of this Soan section and coincides with a change in polarity in the section from normal to reversed which is interpreted as the Gauss/Matuyama boundary. The comparatively large error bars associated with this date make such an interpretation plausible. The same ash horizon was traced to the southern limb of the syncline at Dhok Saiyidan, where the beds dip at 9° towards the synclinal axis. At Shahpur, across the Soan river from Dhok Saiyidan, on the northern limb of the syncline, Moragne discovered a horizontal sequence of sediments overlying the folded Siwalik deposits with a strong angular unconformity. The unconformable sequence contains a different bentonitised volcanic ash that has been dated by fission track methods to 1.60 ± 0.18 Ma (G.D. Johnson *et al*, 1982). The ash occurs within a normal polarity zone within the horizontally bedded post-Siwalik sequence. An erosional disconformity separates this horizontal ash-bearing sequence from the overlying limestone-rich channel fill (the Lei Conglomerate). These dates tightly constrain the formation of the syncline to between ca. 2.1 and 1.9 Ma. We have accepted Raynolds and Johnson's (1985: Fig.17) intepretation of the age of the fold event.

Raynolds (1980) argues that the tectonic evolution of the Soan area comprises two phases. The first took place 3.5–2.5 Ma when the Riwat fault was active, and uplifted the southern part of the area, thus resulting in the gap in sequence during the lower Gauss chron. This movement ceased by ca. 2.4 Ma. The second was between 2.1 and 1.9 Ma. In this phase, the northwestern part of the area was strongly compressed, beds were tilted almost vertically, and then subjected to a massive amount of erosion before the deposition of the horizontally-bedded ash-bearing sequence. Although deformation appears to have ceased by 1.6 Ma ago, the palaeomagnetic data suggest that the syncline has undergone a 20° anticlockwise rotation between 1.8 Ma and the present (Burbank and Raynolds, 1984). Their data on this rotation are, however, poorly constrained with an uncertainty of ± 15°. Our data for the Riwat sections provide a much more tightly constrained fix on this rotation (see Chapter 5).

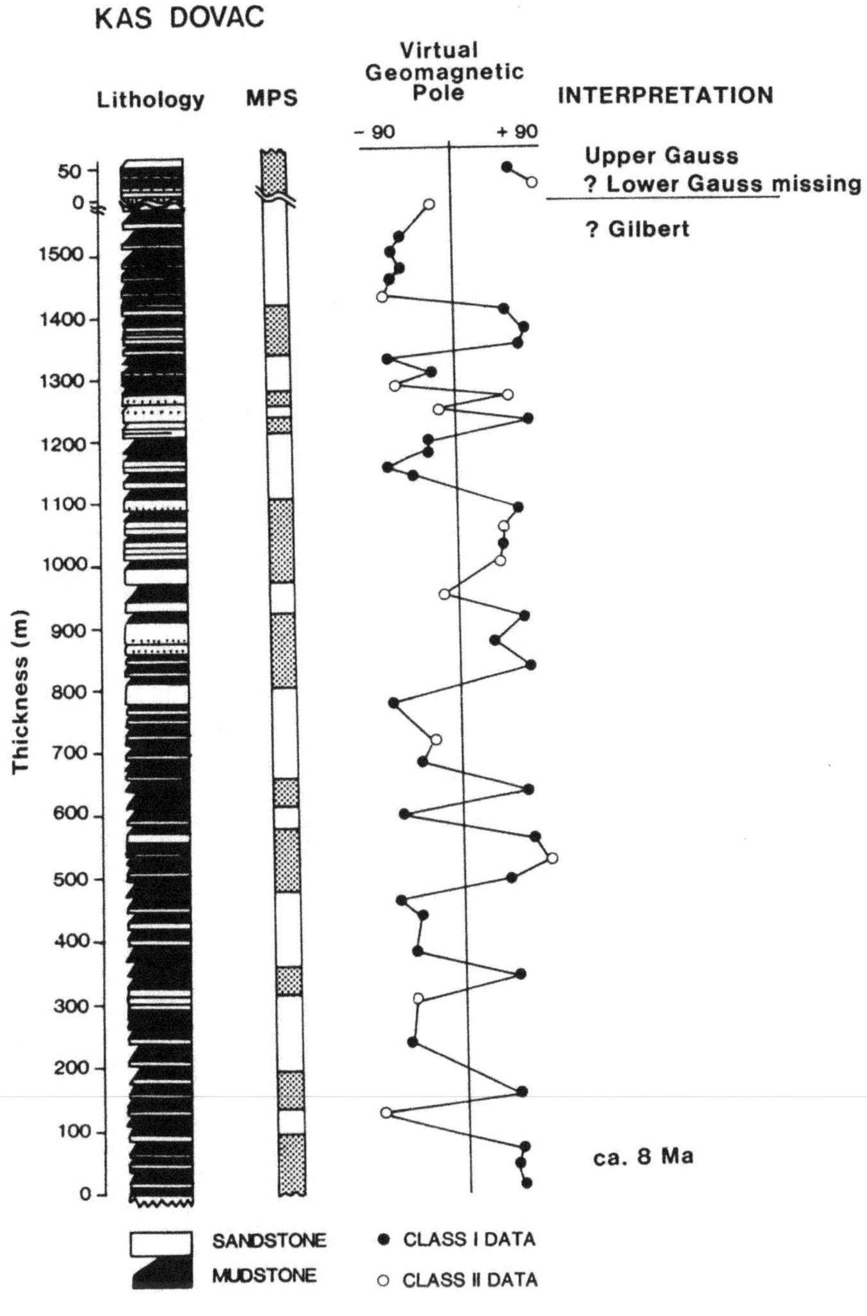

Figure 4.4: Magnetic polarity stratigraphy of the Kas Dovac section, after Raynolds (1980)

4.5 Stratigraphy of post-Siwalik Deposits in the Soan Area

The important point to note is that the boundary between the Siwalik and post-Siwalik deposits of the Potwar is time-transgressive. Siwalik sedimentation continued in the Eastern Potwar until well into the Brunhes chron (<0.73 Ma) whereas it had ceased in the Soan area by 1.9 Ma. Siwalik deposits are separated from post-Siwalik deposits by an unconformity. In the Soan area, two main post-Siwalik deposits have been recognised by Gill (1952) namely : the **Lei Conglomerate** and the **Potwar Loessic Silt**.

The **Lei conglomerate** is a valley fill resting unconformably on tilted and eroded outcrops of Siwalik sediments. The deposit extends southwestwards from Rawalpindi down the axial trough of the Soan syncline and overlaps the vertical to subvertical Siwalik strata on the northern limb of the fold. Gill (1952:386) notes that

> *"Its facies is very variable. In the Soan valley the conglomerate is mostly a very poorly graded gravelly conglomerate of Eocene limestone ... The beds of conglomerate are markedly lenticular and intercalated with beds of sand and silt of pale brown to ochre colour".*

The formation has a maximum thickness of 100-110 m. At some localities, such as north of Adiala village, in the Soan valley south-west of Rawalpindi, and at Rakh Dhungi, between the Soan and Sil rivers, the nature of the post-Siwalik sediments becomes more complex. At both localities the clast-supported Lei Conglomerate is underlain by thick sequences of horizontally bedded silts and sands. The palaeomagnetic analysis of the Shahpur section of the Lei conglomerate suggests that the deposit dates to the later part of the Matuyama reverse polarity chron (Moragne, 1979; Raynolds and Johnson, 1985); in other words, later than the end of the Olduvai subchron (1.67 Ma) but before the beginning of the Brunhes normal polarity chron (ca. 0.73 Ma).

"Potwar loessic silt" is a term that was used (De Terra and Teilhard de Chardin, 1936; De Terra and Paterson, 1939) to describe the deposits of silt, sand and, in some areas, fine gravel which cover the eroded surfaces of the Siwalik strata and the Lei conglomerate. Thicknesses of silt in excess of 10 m are exposed in the Haro valley to the north and west of Hasan Abdal and in the Soan valley to the southeast and southwest of Rawalpindi. Although the origin of the silt has been the subject of much controversy, and Theobald (1877), for example, favoured a lacustrine origin, most workers appear to have agreed on an essentially aeolian origin, albeit with some evidence of fluvial reworking in places (Wadia, 1926; Cotter, 1933; De Terra and Paterson, 1939; Gill, 1952). De Terra and Paterson (1939) considered that the main phase of loess deposition in the Potwar occurred during the Riss glaciation.

4.6 The loess of the Potwar Plateau

Loess is widely distributed in the northern part of the Potwar Plateau, from the Haro valley in the west to the Gujar Khan area in the east (Figure 4.5). In the middle Soan valley and in the Haripur and Taxila areas of the Haro valley, the loess is loess in the strict sense of an aeolian deposit composed of fine quartz-rich silt. In other parts of the western Potwar the silts appear to have been extensively reworked by fluvial agencies.

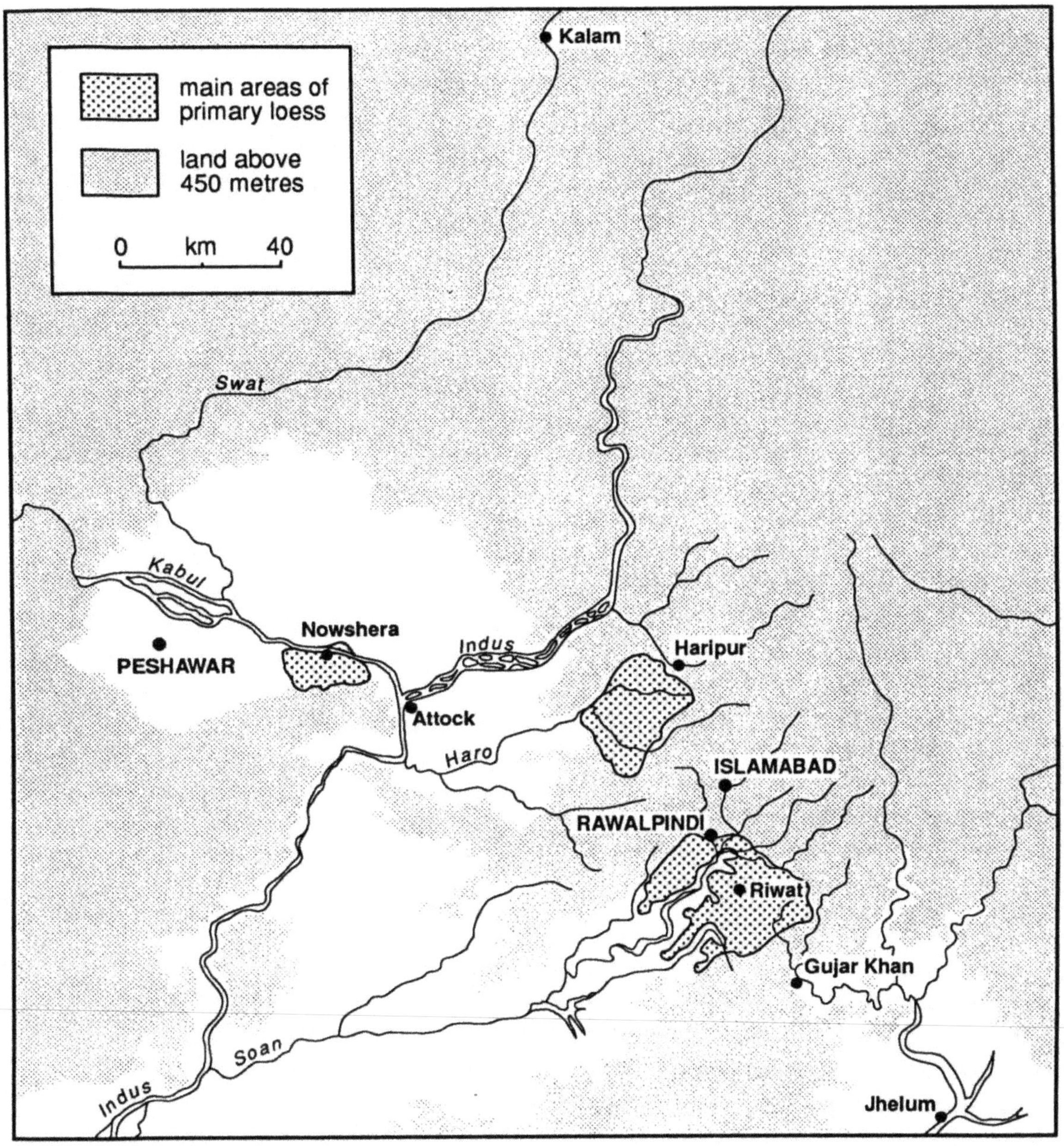

Figure 4.5: Distribution of loess in the Potwar Plateau and adjacent areas

Samples of loess were collected both extensively within the Soan area, in order to try to identify any larger scale spatial trends in grain size or composition, as well as intensively within the Riwat study area. The results of the study of the Riwat loess will be discussed more fully in chapter 6, and at present we can confine our attention to the general features of the loess cover of the Potwar. A pilot study of Peshawar Basin loess was also undertaken, for the purposes of comparison (Rendell, 1988).

The Potwar loess appears to be remarkably uniform in terms of grain size distribution, mineralogy and major oxide concentration. The grain size of the material is fine with median diameters in the range 8 - 22 microns. The sand fraction (>63 microns) accounts for only 10-13 percent of the samples while the clay-size fraction lies in the range 7-20 percent (Figure 4.6). The deposit is calcareous with carbonate contents in the range 14-18 percent. Some of the material towards the base of the thicker sections is partially decalcified. Quartz is the dominant mineral, with subsidiary feldspars and clay minerals (notably kaolinite and illite). A comparison of X-ray diffraction plots for the <20 micron and <2 micron fractions of the same samples, show that the calcite component is predominantly silt-size, that quartz is present in both fractions, and that clay minerals dominate the <2 micron fraction (Figure 4.7). The geochemistry of the loess shows remarkable uniformity. Data from inductively-coupled argon plasma spectrometry analysis, determined on a carbonate-free basis, show that composition is dominated by SiO_2 (65-66 percent), Al_2O_3 (14-15 percent), total Fe oxides (5-7 percent) with no other oxide exceeding 5 percent. The calcium carbonate component is primary and this view is corroborated by SEM studies of the loess microstructure, and by preliminary measurements of $\delta^{13}C$ and $\delta^{18}O$ on two of the Soan samples which place the carbonates in a "marine" rather than a pedogenic carbonate grouping (Wintle, pers.comm.). The most likely local source for primary carbonate is the Margala Hills Eocene limestone.

4.6.1 Spatial trends in Potwar loess

One way of determining the origin or at least the direction of transport of a loess deposit is to try to detect spatial trends in grain size or composition. In the case of the Potwar loess, extensive sampling was undertaken in the Haro and Soan valleys.Data for percent clay size (<2 micron) and median diameter (D_m50) of the grain size distributions were plotted against distance east-west and north-south. As can be seen from Figure 4.8, there are no obvious trends. Nor is there a detectable trend in the thickness of deposits, although this is complicated by substantial fluvial reworking of loess in the lower Haro valley. In the Soan area thickness is dictated by the irregularities in the underlying topography. The lack of any obvious distance-decay effect in loess thickness may in part reflect the nature of the vegetation cover on the surface on which loess accumulated. A recent study by Tsoar and Pye (1987) has demonstrated that a strong distance decay effect is seen for forested areas, while the effect for grass cover is far weaker. The little pollen evidence that we have confirms that the loess surface was grass-covered.

The striking aspect of the Potwar loess is its uniformity, in terms of grain size, mineralogy and geochemistry and, in particular, its comparative fineness. A pilot study of the magnetic fabric of the loess in the Riwat area is currently being undertaken to see whether any palaeowind directions could be detected using fabric anisotropy. Possible sources of the loess are discussed in chapter 6.5.

A chronological framework for the loess deposits has been established using thermoluminescence dating. The nature, development and application of this relatively

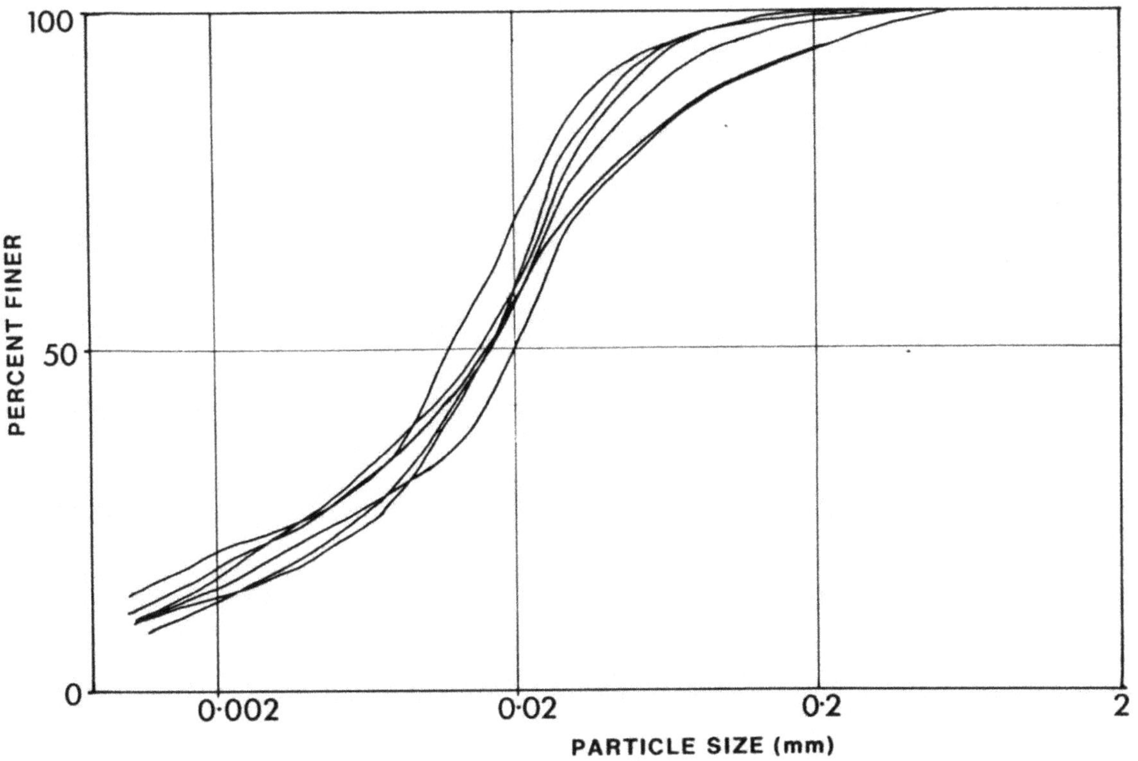

Figure 4.6: Grain size distribution of Potwar loess

new dating technique are discussed in chapter 6.3.

4.7 Summary

The main points to emerge from this chapter are:

1) The dominant geological structure in the Middle Soan valley is a synclinal fold in the Siwaliks. The Siwalik sediments forming the northwestern limb of the fold dip almost vertically and are easily distinguished from the overlying, horizontally-bedded post-Siwalik deposits. In contrast, the Siwalik strata on the southeastern limb have a relatively low angle of dip, as the syncline axis is approached, and are consequently more difficult to distinguish from the post-Siwalik sediments.

2) There have been several studies of palaeomagnetic sections within the Soan syncline during the last decade. These studies date the Soan syncline was folding event to between 2.1 and 1.9 Ma; since then the syncline has also undergone an anticlockwise rotation of the order of 16°. The dating of the folding event is constrained by radiometric dating of volcanic ash horizons within the Siwalik and post-Siwalik sequences.

3) The main post-Siwalik deposits in the Soan valley are the Lei Conglomerate and the so-called Potwar loessic silt. The base of the Lei conglomerate probably dates to the latter part of the Matuyama reverse polarity chron (ca. 1.6 - 0.73 Ma) and differs radically from the underlying Upper Siwalik quartzite Conglomerates, in being composed of pebbles and cobbles of Margala Hills limestone. The Potwar loessic silt is largely aeolian in origin. The loess is uniform over the Potwar in terms of grain size and mineralogy. A detailed study of the loess in the Riwat area is presented in chapter 6.

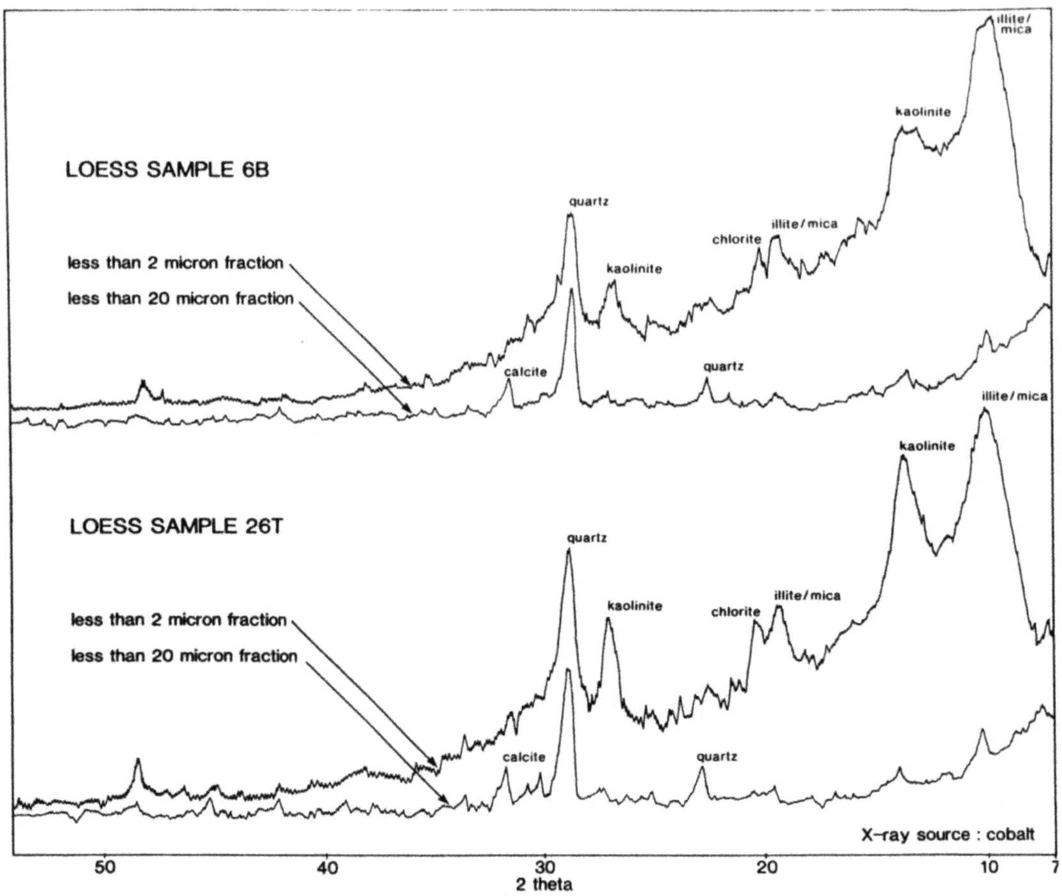

Figure 4.7: X-ray diffraction of Potwar loess

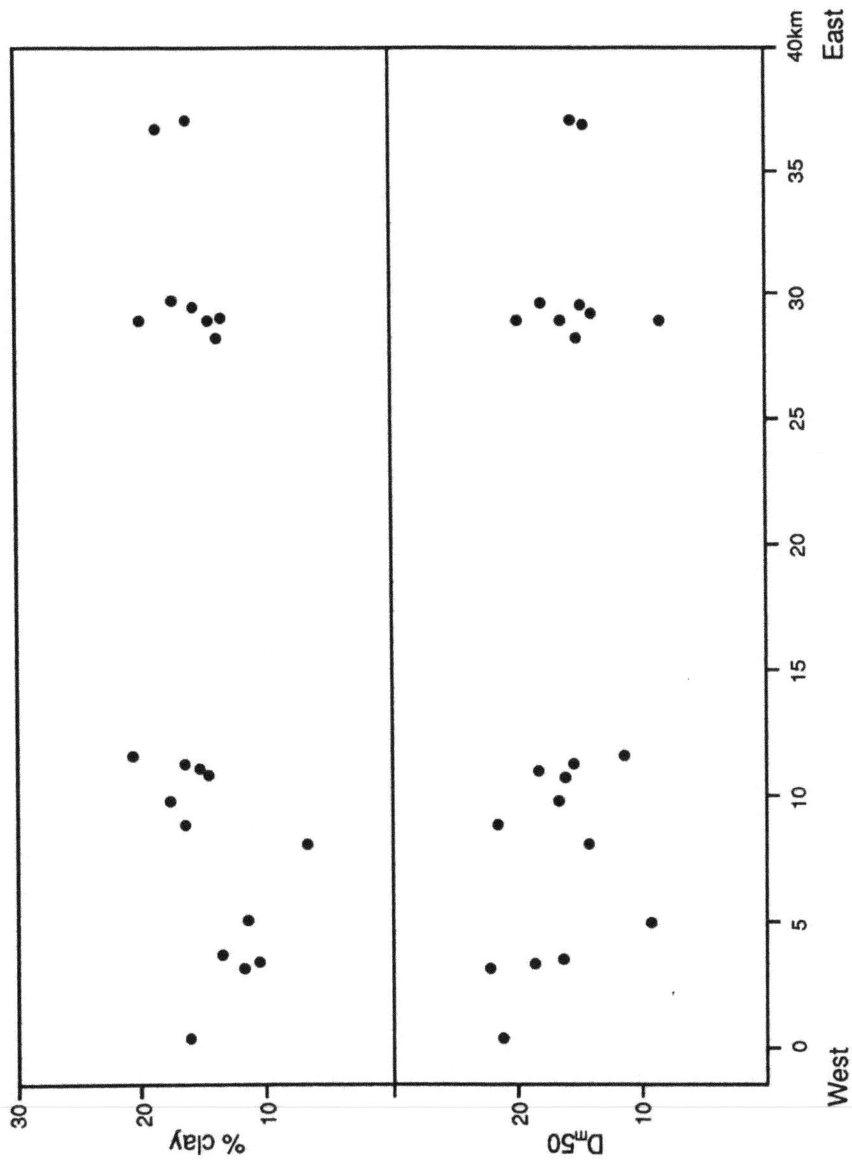

Figure 4.8: Spatial characteristics of Potwar loess

Chapter 5

Riwat: The Siwalik Sequence

5.1 Introduction

In Chapter 1 we criticised De Terra for not being explicit about his approach to unravelling the geological sequence in the Soan valley. Therefore it is necessary for us to be honest about our approach. We set out to try to find artefacts or faunal remains in datable, stratified contexts. The geological component of this work therefore involved the establishment of a chronological and palaeoenvironmental framework for the archaeological materials. The absence of radiometric dating techniques in the 1930's forced De Terra to use regional correlations of generalised, and often composite, sections as a means of establishing a chronological framework. He thus emphasised the general patterns of deposits and landforms within a region and attempted a synthesis of data collected at isolated localities. Given the short time that he spent in the field area, he would have been unable to walk out his sections and establish the kind of spatial continuity that is vital for the type of synthesis that he attempted. We approached the problem of studying the Soan sequence at a completely different scale. We tried to establish local sequences by using detailed mapping to resolve the field relationships of spatially variable sedimentary units. We have not attempted to devise a regional sequence based on comparative geological stratigraphies; as we have already seen, regional correlations in the Potwar are plagued by several problems including the laterally-discontinuous as well as time-transgressive nature of the deposits. In contrast to De Terra, we have been able to circumvent some of these problems, by recourse to palaeomagnetic and radiometric dating. The decision to focus on a detailed local sequence, and not to be particularly concerned with regional correlation, was also influenced by the paucity of other well-dated Pleistocene sequences in Asia. It is worth noting that this work has been undertaken at a time when Pleistocene chronologies and palaeoenvironmental reconstructions are being re-evaluated at both global and regional levels.

The choice of Riwat area as the focus for detailed study arose out of the results of the first long season of fieldwork in the Soan area. During the period December 1979 - February 1980 one of us (HR) visited all the main De Terra and Paterson sites within the Soan Valley, with the exception of Balawal and Gila Kalan. The aim was to become familiar with the geological and geomorphological settings of these palaeolithic sites and to look for archaeological material stratified within datable deposits. By the end of that first season it was apparent that: De Terra's relative chronology was incorrect; that field relationships between Siwalik and Post-Siwalik deposits were complex, particularly on southern limb of the Soan syncline; that TL dating represented the only hope for dating the later part of the sequence; and that the area around De Terra's Milestone 163 site was most promising in terms of the quality of exposures of Siwalik and post-Siwalik sediments, and of the palaeolithic material found there.

5.2 The stratigraphy of the Upper Siwalik deposits at Riwat

A detailed study of Siwalik and Post-Siwalik sediments was made in an area 1.4 km x 2.0 km to the north of Riwat on the southern limb of the Soan syncline (Figure 5.1). The Riwat study area lies 6-8 km to the south-east of the synclinal fold axis, and is peripheral to the main zone of deposition of the Lei Conglomerate (see Figure 4.2 for location). The study area is bisected by the Grand Trunk (GT) road. The contact between the Siwalik and post-Siwalik deposits is best exposed in an area of deeply dissected terrain to the south of the road. In order to reconstruct the surface of the eroded Siwalik sediments it was first necessary to make a topographic map of the study area. This mapping was done over a period of three field seasons using a mixture of plane-tabling, triangulation and levelled traverses. A line of low hills, marking the ridge-forming outcrop of an Upper Siwalik gritstone/conglomerate horizon lies close to the eastern edge of the study area. In general terms, both the thickness and the complexity of the post-Siwalik sequence increase with distance westwards across the study area.

Within the study area, the Siwalik strata dip at angles of from 30° degrees to as little as 4° to the northnorthwest, towards the synclinal axis. Successively younger strata have gentler dips. Gill (1952:385) commented that:

> *"the dip of the beds begins to decrease rapidly within the outcrop... This decrease in dip is a purely structural feature and is not related to conditions during deposition. This is confirmed by the quasi-vertical attitude of the same range of beds on the north flank (of the syncline)".*

A thickness of ca. 150 m of Upper Siwalik strata is exposed in the study area. These beds consist of alternating silt, silt/clay and sand horizons, with one major gritstone/conglomerate horizon. The silts and silt/clays are characteristically reddish brown (5YR 5/3 - 5/4) to light reddish brown (2.5YR 6/4 - 5YR 6/4) in colour and vary from thinly bedded to blocky. Some horizons also show evidence of bioturbation. The sand horizons have sharp but planar basal contacts and are poorly cemented and incoherent with no obvious structures. Most of the sand horizons are of the order of 3-9 m thick, whereas the individual silt and silt/clay units are ca. 0.5 m thick.

The dominant horizon within this sequence is a 4-5 m thick gritstone/conglomerate horizon. Recent incision has exposed this horizon in a deep gully at the base of 70 m section of Upper Siwalik strata, and the horizon can be traced southeastwards, until it becomes ridge-forming (Figure 5.2). The gritstone/conglomerate horizon also becomes thicker and more complex when traced laterally in a southwesterly direction to the Kas Dovac. The horizon is unusual in containing pebbles and cobbles that are predominantly composed of quartzite. A detailed section through this horizon is discussed in chapter 7. The post-Siwalik Lei conglomerate has clasts that are predominantly composed of Margala Hills (Eocene) limestone. The gritstone/conglomerate horizon is discontinuously cemented with calcite. The lateral extension of this horizon forms the uppermost bed in

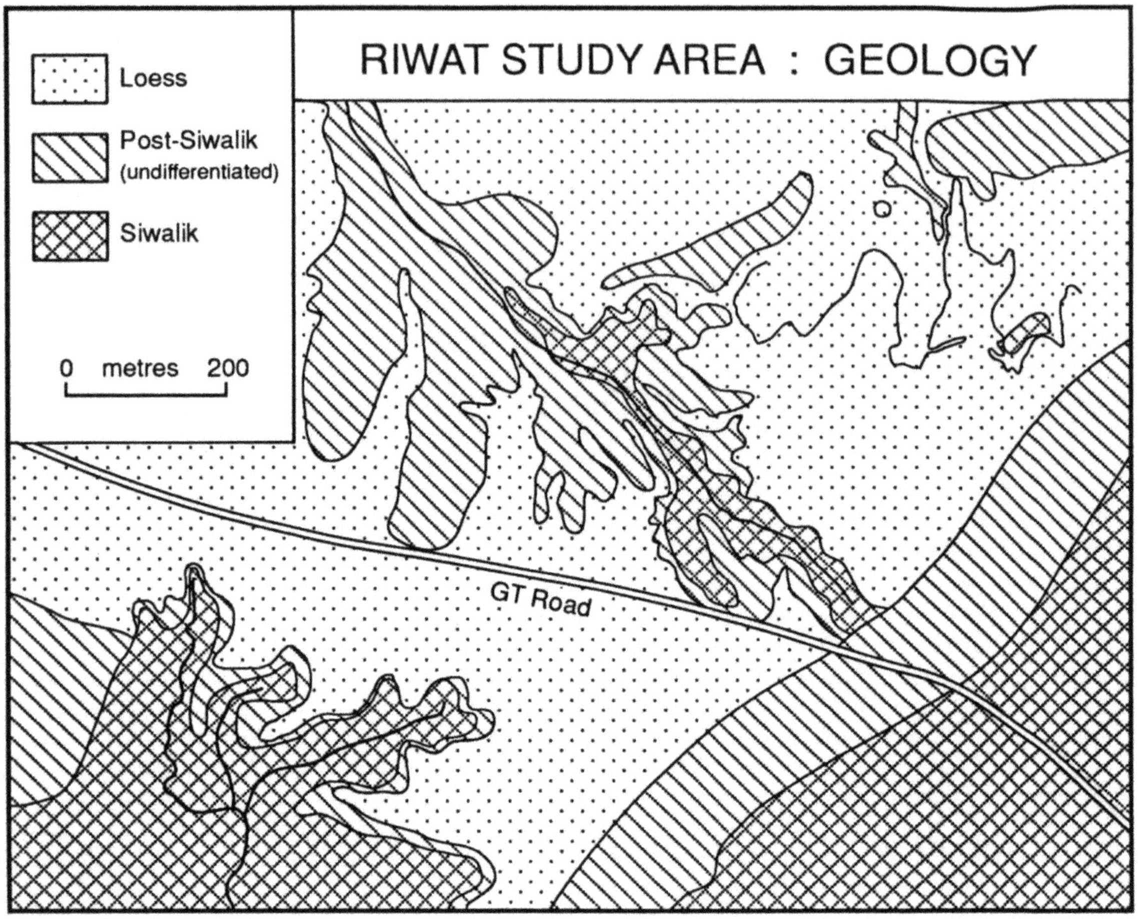

Figure 5.1: Riwat Geology

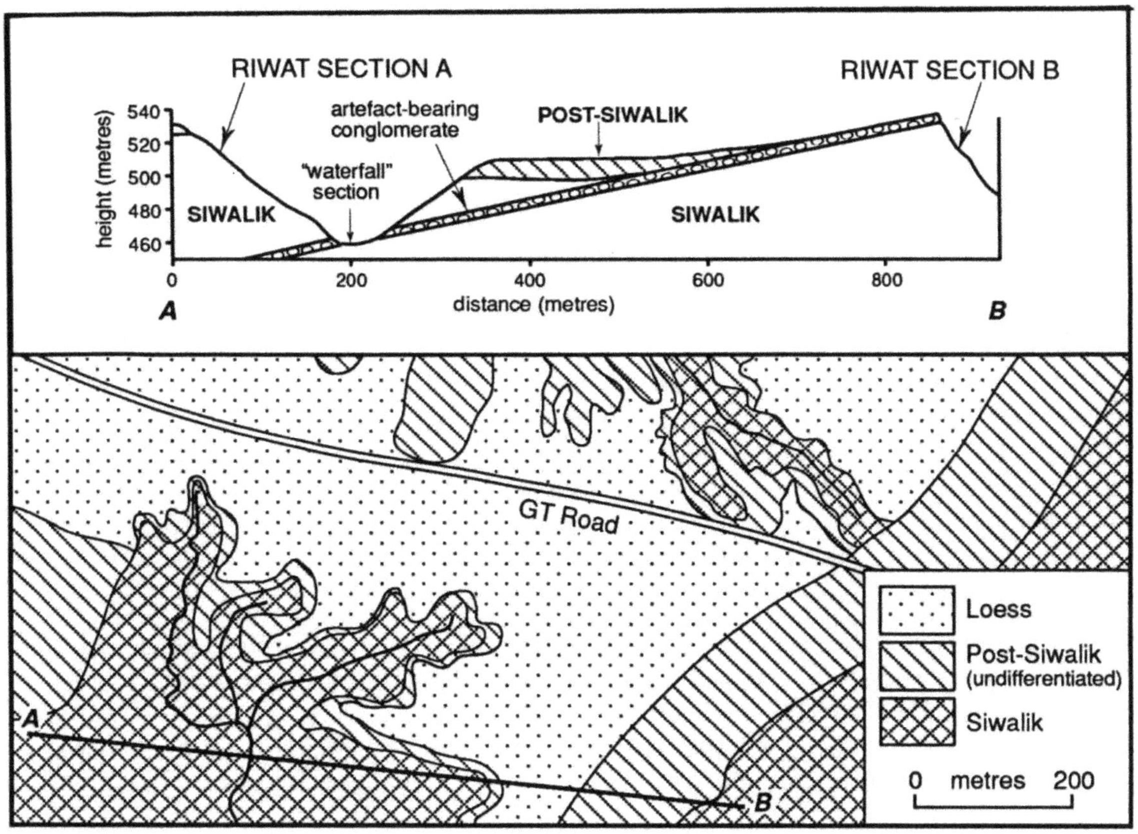

Figure 5.2: Riwat : location of palaeomagnetic sections

Raynolds' (1980) Kas Dovac section, and for reasons that will be discussed in more detail later, we suspect that the horizon is strongly time-transgressive.

In the Riwat sections the base of the gritstone/conglomerate marks an erosional hiatus. The contact with underlying beds is sharp and undulating and the basal unit of the gritstone contains large rolled mudballs, with long axes up to 14 cm, and quartzite cobbles, with long axes measuring up to 16 cm. With the discovery of artefactual material within part of the gritstone/conglomerate horizon it became imperative to try to estimate its age. Although the Upper Siwalik in the Soan is notoriously unfossiliferous, a fossilised rhinocerous jaw was discovered in 1988 *in situ* in the sand unit immediately above the gritstone/conglomerate horizon.

We stress that this gritstone /conglomerate is totally integral to the Upper Siwalik sequence in the Soan valley. The horizon can be traced laterally for several kilometres; is ridge-forming; is tilted conformably with other Siwalik strata; and is overlain by a 70m thickness of Upper Siwalik sands and silts.

5.3 Magnetic polarity stratigraphy and dating of the Upper Siwalik section at Riwat

The sampled sections and their location are shown in Fig 5.2. The thickness of the composite section is 119m. The upper portion of the section underlies the high ridge of De Terra and Paterson's Milestone 163 locality. Orientated samples were collected from 44 stratigraphic levels above the gritstone conglomerate horizon (Section : Riwat A), and from 27 levels beneath it (Section : Riwat B). In contrast to the Dartmouth work, where samples were taken at intervals of 5-10 m, a sampling interval of 0.5 m was adopted where material composition permitted. The mean sampling interval for the Riwat sections was ca. 1.7 m, compared to ca. 27 m for Raynolds' (1980) Kas Dovac section. The sand horizons and some of the silty sands have such a low concentration of magnetic minerals, and a consequently low magnetization, that sampling had to be limited to the silt and silt/clay horizons. Paired samples were collected at each level and were then cut into 8 cm^3 cubes. At least two cubes were cut from each sample to give a minimum of four palaeomagnetic specimens per sampling level. Overall at least 284 sample cubes were prepared for the Riwat sections.

Work began with samples from the Riwat A section. The first stage in the palaeomagnetic analysis was to measure the natural remanent magnetization (NRM) of all samples. The variation of NRM values down the stratigraphic column is shown in figures 5.3 & 5.4. Most of the values lie in the range 0-10 mA/m but significantly higher values may be noted at certain points. We interpret these fluctuations as reflecting variations in the type and concentration of magnetic minerals. In order to investigate the stability of the remanent magnetization and to resolve the separate components in samples showing multi-component remanence, specimens from each stratigraphic level were subjected to incremental demagnetization.

The first step was to attempt to optimise the demagnetization procedure. Paired specimens were selected from six stratigraphic levels and subjected to either thermal or alternating field (AF) demagnetization. The AF demagnetization was carried out at steps of 2.5 or 5 mT and the thermal demagnetization at steps of 50°C until the magnetization became too weak for reliable determinations to be made using the Molspin magnetometer. During this pilot study it became apparent that the thermal demagnetization was more effective in defining a demagnetization stable end point, a stable characteristic magnetization (SCM), for these sediments (Rendell *et al*, 1987). As a result, samples from the remaining stratigraphic levels were demagnetized thermally at increments of 50°C up to maximum temperatures in the range 600 - 650°C. Data were placed in four categories of reliability on the basis of the degree to which an SCM could be defined.

It should be emphasised that the pilot study was undertaken in order to select the optimum method for demagnetization rather than to establish parameters for the "blanket" demagnetization of the remaining specimens, as has tended to happen in the Dartmouth studies. Our confidence in the quality of the data set is established for individual samples on the basis of the paths followed on the Vector End Point (VEP) plots during demagnetization. Further discussion and examples of VEP plots are given in Rendell *et al* (1987).

The categories of reliability are as follows:

Category A. After removal of a weak viscous component of magnetization, a well-established demagnetization stable end-point is defined by a linear trend through the

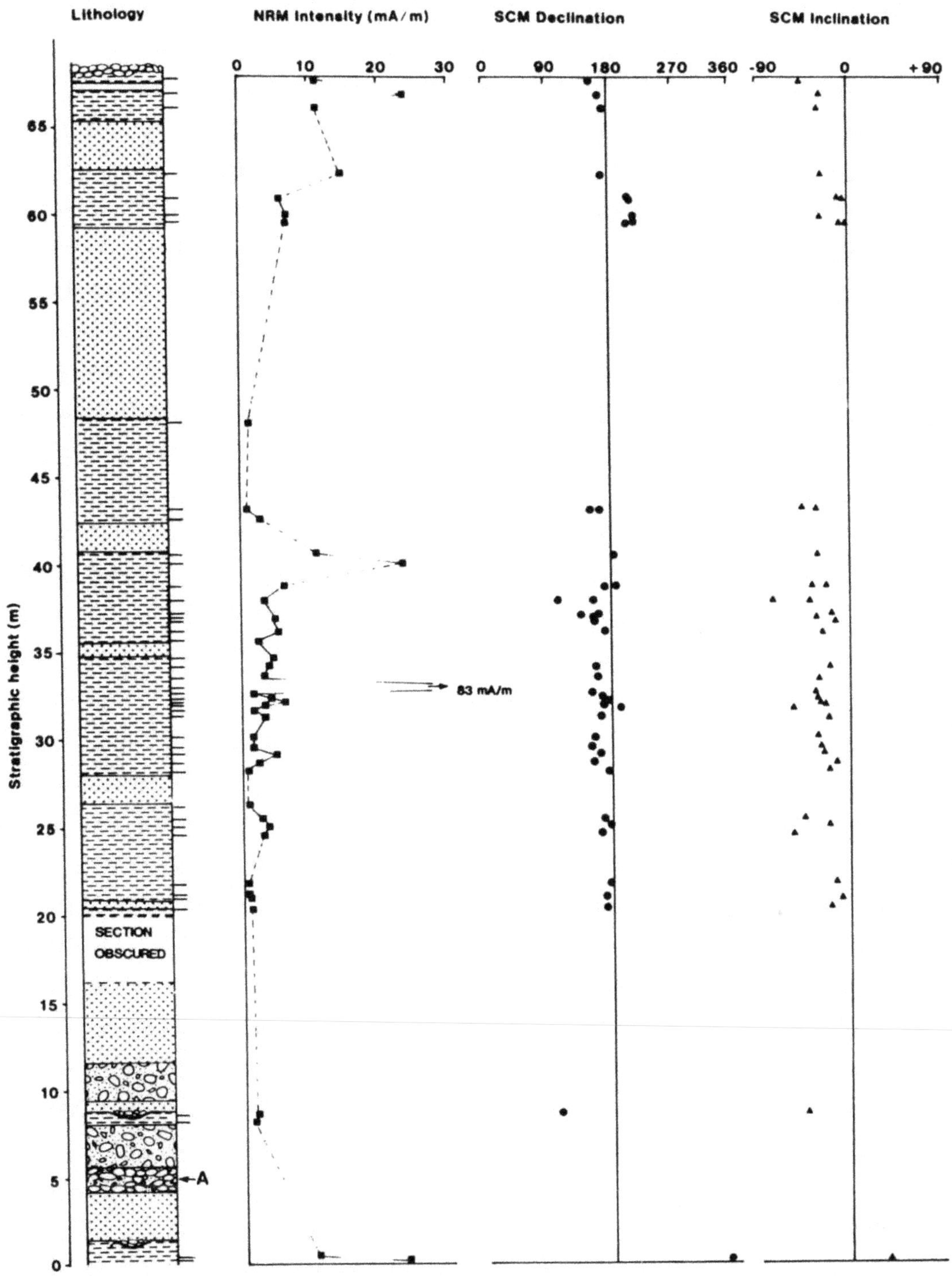

Figure 5.3: Riwat A : magnetic polarity stratigraphy

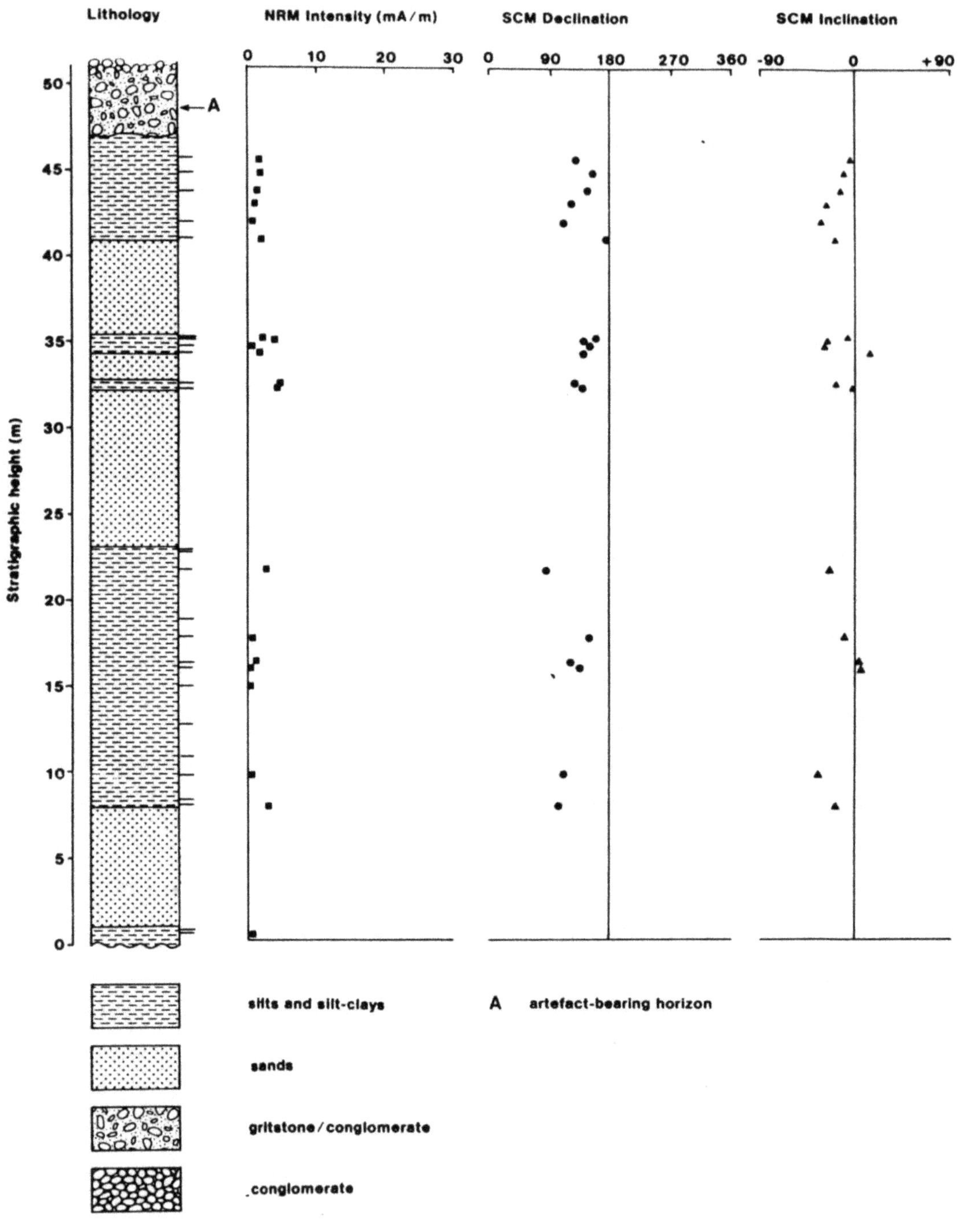

Figure 5.4: Riwat B : magnetic polarity stratigraphy

origin of the vector end-point (VEP) plot. The polarity is established beyond question, and the specific direction of the SCM is regarded as being reliable to better than 5°.

Category B. A linear trend through the origin of the VEP is evident after thermal demagnetization above about 200°C, but is less well-defined than for category A vectors. Although the polarity of the SCM is again established beyond question, the specific magnetization is only reliable to 10°.

Category C. Although a clear directional trend is apparent towards a particular stable end-point, this point is either not reached or is obscured by instrumental noise. Although the polarity of the SCM is established, the direction of specific magnetization is poorly defined with an uncertainty of greater than 10°.

Category D. Either the initial NRM intensity is too weak or the demagnetization data are too noisy to permit a reliable definition of the polarity.

The data for category A stable characteristic magnetisations of the sections above and below the conglomerate/gritstone horizon are given in figures 5.3 & 5.4. In almost every case samples showed the southerly declinations and negative inclinations characteristic of reverse polarity magnetisation. The only normal polarity magnetization was exhibited by one sample from the lowest stratigraphic level in Section A, immediately beneath the gritstone/conglomerate horizon. We have assigned the strata from the composite Riwat section (Sections A and B) to the most recent period of reverse polarity, the Matuyama Chron, 0.7 - 2.48 Ma. The position of our section, and its associated artefactual material, within that chron is considered below.

5.4 The age of the artefact-bearing gritstone/ conglomerate horizon

The gritstone horizon lies in the uppermost portion of the Upper Siwalik sediments exposed in the Soan syncline. It participated both in the folding event and in the rotation of the syncline about a vertical axis. It lies within a series of reverse magnetised sediments which can be assigned to the Matuyama reverse polarity chron. The age of the folding event, dated to 2.1 - 1.9 Ma by Johnson *et al* (1982) and Raynolds and Johnson (1985), and discussed in Chapter 4, tightly constrains the age of the conglomerate horizon exposed in the Riwat sections to the early part of the Matuyama chron, with a youngest possible age of 1.9 Ma. However, the conglomerate horizon can be traced laterally to Rajwal village and to the upper portion of Raynolds' Kas Dovac section (Figure 5.5). Here the conglomerate forms part of the 50 m sequence of "**unconformable**" strata at the top of the section. Raynolds has three palaeomagnetic samples from within the sequence of which two clearly show normal polarity magnetization. Beneath the unconformity the clasts in the coarser facies are predominantly of limestone, and the polarity of the magnetization is reversed. Raynolds (1980) interprets the beds above the unconformity, but below the quartzite conglomerate, as belonging to the upper portion of the Gauss chron, while the beds immediately beneath the unconformity are assigned to the Gilbert Chron, with the unconformity corresponding to a ca. 1 Ma gap in the sequence. Raynolds supports his interpretation with McMurty's (1980) results for the Rajwal section, in which the lateral continuation of the conglomerate sequence includes an ash-bearing complex, associated with a magnetic polarity transition from reverse below to normal above. On the basis of stratigraphic correlation of the Rajwal sequence with the Dhok Saiyidan and Soan sections of Moragne (1979), Raynolds concludes that the polarity transition in the

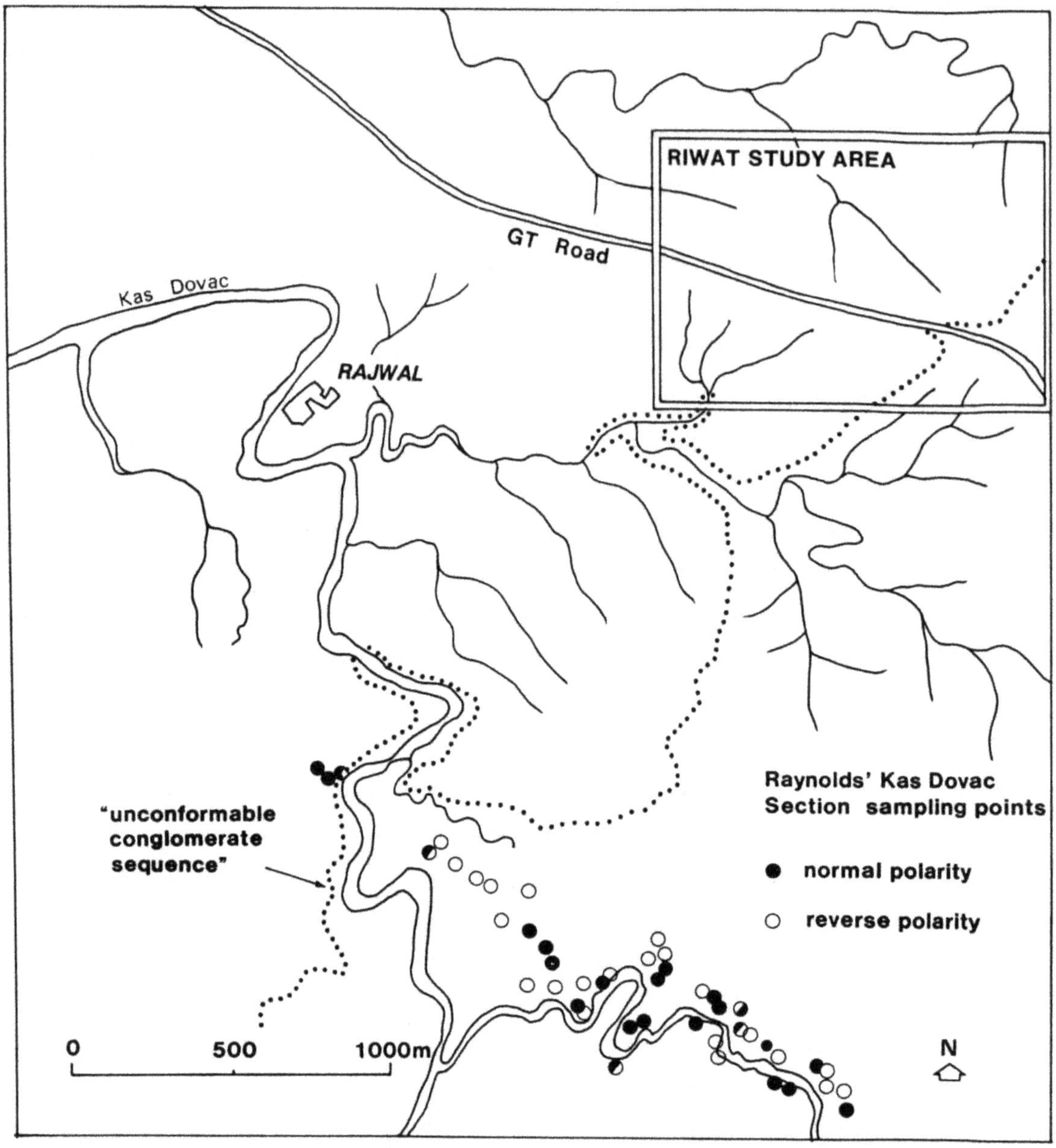

Figure 5.5: Spatial relationship of the Riwat and Kas Dovac sections

Rajwal sequence represents the Gauss/Matuyama boundary. If Raynolds is correct, this indicates that the base of the conglomerate unit is diachronous and that it gets younger northeastwards along the strike from the Kas Dovac section. A detailed programme of sampling was undertaken in March 1988 in order to test this hypothesis and to define the diachronous change more closely, and work on the samples is currently in progress.

5.5 The evidence for the rotation of the Soan Syncline tectonic block

Burbank and Raynolds (1984) estimate that an anticlockwise rotation of the Soan Syncline of 20° (with an uncertainty of 15°) has occurred since 1.8 Ma. Our data allow us to

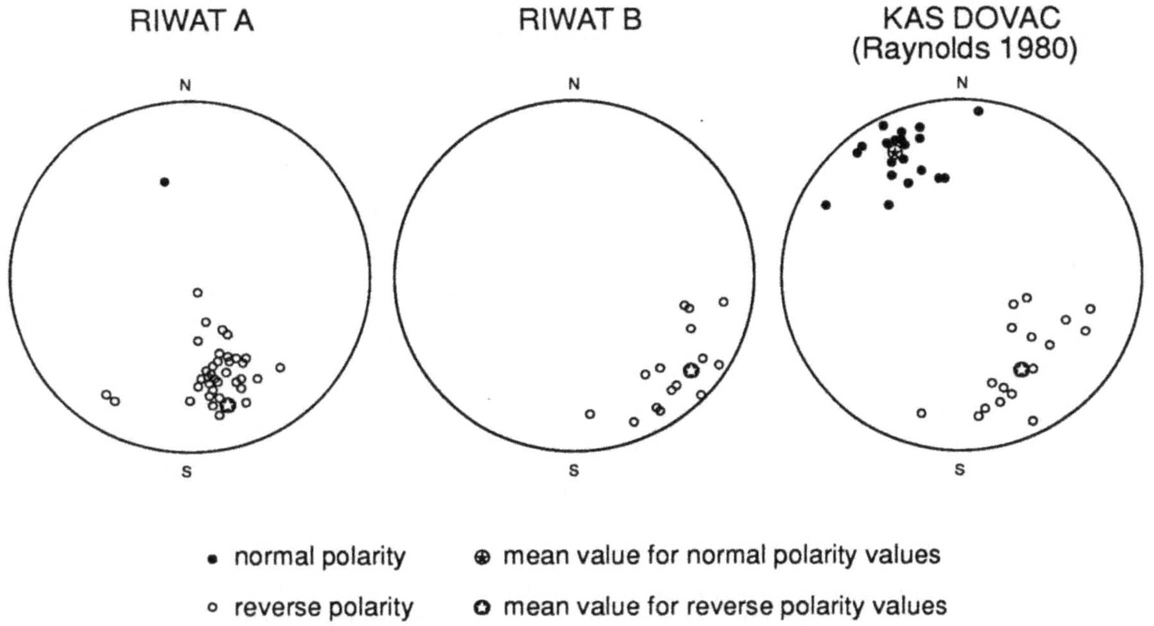

Figure 5.6: Stereographic plots of SCM data

confirm this rotation and, given a greater number of samples, to constrain it much more tightly. Data from the Riwat A section, above the gritstone/conglomerate horizon, show a rotation of 16 °, with an uncertainty of 4°. Those from the Riwat B section show a greater degree of rotation, as do Raynolds' data for the Kas Dovac section, although in both cases the degree of scatter is much higher (see Figure 5.6). If the degree of rotation does increase systematically down the profile, or if data from above and below the conglomerate horizon exhibit consistently different degrees of rotation, then this tells us something significant about the history of the deformation and rotation of the Soan Syncline. It appears that the rotation had started by the Upper part of the Gauss Chron, and that a total anticlockwise rotation of at least 30° has taken place since then.

There is some evidence that the rotation may not be confined to Soan syncline, and may apply to the whole area bounded by the Salt Range Thrust. Data from Bhangala and Ganda Paik sections in the Eastern Potwar, with a folding date of ca. 0.6 Ma, show a slight anticlockwise displacement from north to south (Raynolds, 1980: Fig.26), whereas our data for two unpublished stratigraphic sections in the Pabbi Hills, which lie immediately outside the Salt Range thrust, show no rotation.

5.6 Palaeoenvironmental interpretation of the Upper Siwalik

Gill (1952) mapped the Upper Siwalik in the Soan valley so as to include "Brown Sandstones" which he related to the Tatrot faunal zone and the "Siwalik Conglomerate" related to the Pinjor. Although the gritstone/conglomerate unit within the Riwat sequence is an important, relatively resistant, ridge-forming horizon, it is certainly a minor component of the measured sections in terms of thickness. The Upper Siwalik sections at Riwat are dominated by fining-upwards cycles of sands, silts and clays. This sequence does change laterally, and 1.8 km southwest of Riwat Section A, in the Kas

Dovac, the conglomerate unit is far more dominant, reaching thicknesses in excess of 60m, and comprising up to four conglomerate horizons interbedded with sands and silts.

The conventional view of the Upper Siwalik Sub-Group is that it is dominated by conglomerate facies. As discussed in chapter 4, Raynolds and Johnson (1985) chart the southward migration of the Boulder Conglomerate during the Plio-Pleistocene in the eastern Potwar, with the coarsening of facies reflecting the increasing proximity of the Himalayan mountain front. The sediments in the Riwat area do not conform to a picture of coarse facies dominance. The fact that the Soan folding event took place in the late Pliocene, and that the post-Siwalik sequence in the Soan is contemporary with the clast-dominated Upper Siwalik sequences in the eastern Potwar at Dina and Sakrana (Raynolds, 1980), may go some way towards explaining the absence of coarse facies in the sections.

The Upper Siwalik sections at Riwat are dominated by floodplain deposits of silt/clay, silt and sand. The sections exhibit the characteristics of a floodplain area, peripheral to major channels. The sand horizons may be interpreted as crevasse splays or minor channel sand bodies. The gritstone/conglomerate in which the artefacts were found is a channel deposit with a sharp transition at its base, and which grades upwards into another floodplain sequence. The artefacts have all come from a matrix-supported conglomerate horizon at the base of this sequence, interpreted as a channel bar. Channel bars would have provided a source of quartzite pebbles in an otherwise relatively stone-free environment.

5.7 Summary

This chapter presents our own observations of one small area of the Soan Valley, near Riwat and south-east of Rawalpindi. The main points to emerge from this investigation of the Upper Siwalik deposits are:

1) The Upper Siwaliks are predominantly composed of alternating units of sands, silts and clays, but with one major conglomerate horizon. Artefacts found in this horizon in 1983, and subsequently, are discussed in chapter 7.

2) The 70m section above the artefact-bearing horizon was sampled in 1986. A further 47 m section outcropping beneath the horizon was sampled in 1987. In all, more than 280 samples were taken for palaeomagnetic analysis from a total of 71 stratigraphic levels, at a mean spacing of 1.7m. The analysis of these samples has revealed that, with the exception of the stratigraphic level immediately below the artefact-bearing horizon in section RIWAT A, all stratigraphic levels exhibit reverse polarity. We interpret this as indicating that the section belongs to the most recent major period of reverse polarity, namely the Matuyama Chron (0.73 - 2.48 Ma).

3) The dating of the section is made more precise by considering its tectonic context. We have demonstrated that the section forms an integral part of the Upper Siwalik sequence which is known to have been folded during the period 1.9 - 2.1 Ma. The normal polarity event beneath the artefact bearing horizon is an excursion which may be related to the Réunion event, rather than to later (Olduvai or Jaramillo) or earlier (Upper Gauss) events.

4) We also conclude that there is every indication that the artefact-bearing horizon is strongly time-trangressive and that it increases in age southwestwards from the Riwat study area.

5) Our study also refines previous estimates of the amount of rotation that the Soan

syncline has undergone both before and since the folding event. Our data indicate a counter-clockwise rotation of $16° \pm 4°$ since 1.9 Ma and a rotation of up to $30°$ during the last 3 Ma.

Chapter 6

The post-Siwalik sequence at Riwat

6.1 Introduction

In the central part of the Soan valley, the post-Siwalik sequence is dominated by the limestone-rich Lei conglomerate. On the northern flank of the syncline at Shahpur, the Lei conglomerate overlies a horizontally-bedded, ash-bearing sequence of sands, silts and clays (Moragne 1979). As discussed in Chapter 4, this ash is dated to 1.6 ± 0.18 Ma (Johnson *et al* 1982). The base of the Lei Conglomerate is marked by a sharp undulating erosional contact. Raynolds (1980) regards the Lei conglomerate as dating to the later (post-Olduvai) part of the Matuyama chron, between 1.6 and 0.7 Ma. Near Soan Bridge on the GT road, the limestone conglomerate is interbedded with sands and silts and reaches a thickness of 60-70 m. The thickness of the conglomerate and the proportion of coarse facies are both highly variable laterally, as might be expected in this area. The Lei Conglomerate overlies and overlaps the eroded surface of the folded Siwalik strata. It is clear from the field relationships of the Siwalik and post-Siwalik strata that a tremendous amount of erosion occurred in the axial trough after folding and prior to the deposition of the Lei conglomerate and its associated finer sediments.

6.2 Stratigraphy of the Riwat post-Siwalik Sequence

In the study area, part of the loess sequence is underlain by fluvial and colluvial post-Siwalik sediments. The junction between the Siwalik and post-Siwalik sediments is often difficult to establish in the field. Mapping is however aided by a subtle but consistent difference in colour between the two sequences. The Siwalik silts and silt clays are characteristically reddish brown to pale reddish brown in colour (5YR 5/3-5/4, 5YR 6/4) whereas the post-Siwalik sediments are consistently reddish yellow (7.5YR 6/6 - 6/8). The loess surface provides an excellent datum for assessing the nature of the erosion surface cut in the Siwaliks, and the thickness and degree of dissection of the post-Siwalik sediments. This analysis of the stratigraphy is a considerable aid in palaeoenvironmental reconstruction. A map of the study area, and detailed geological sections are gives in

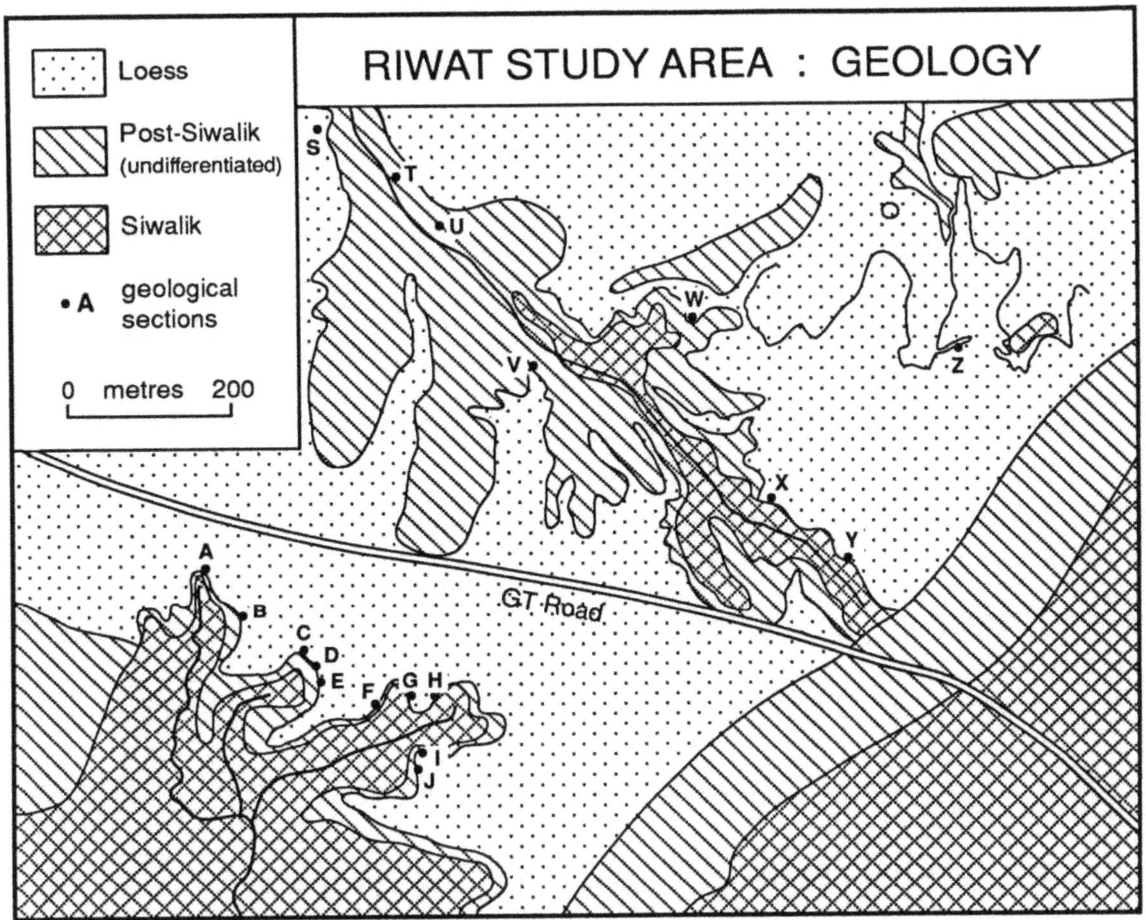

Figure 6.1: Location of sections

Figures 6.1, 6.2 and 6.3.

Two distinctly different sequences of post-Siwalik strata are revealed in gully sections to the north and south of the GT road. North of the GT road, the post-Siwalik sequence is dominated by loess (see Figure 6.2). The eroded surface of the Siwaliks is covered with a spread of quartzite pebbles and cobbles either on, or derived from, the dipslope surface of the prominent Upper Siwalik gritstone/conglomerate horizon, considered in the previous chapter. The eroded Siwalik surface dips northwards. Exposures of Siwalik strata are limited as current erosion processes have not incised through the total thickness of overlying post-Siwalik sediments. The loess cover thickens rapidly northwards increasing from less than 1 m to 10 m over a distance of 540 m. Beneath the loess a fluvial and, in some places, an aeolian sequence is revealed. Occasional inliers of spreads of post-Siwalik pebbles and cobbles protrude through the loess cover to form local topographic highs. These spreads of quartzite pebbles are undoubtedly derived from the Upper Siwalik conglomerates. They bear no relation to the Lei conglomerate in terms of composition, the latter being composed almost entirely of limestone clasts. Two distinct pebble/cobble horizons are seen in the post-Siwalik sequences to the north of the GT road. A lower quartzite conglomerate outcrops at or near the base of sections S, T and V, and overlies the older loess at section T. An upper conglomerate horizon outcrops immediately beneath the thinning loess cover to the north of section S. In both cases these clast-supported conglomerates are horizontally bedded.

To the south of the GT road, a much more extensive and varied sequence of Siwalik and

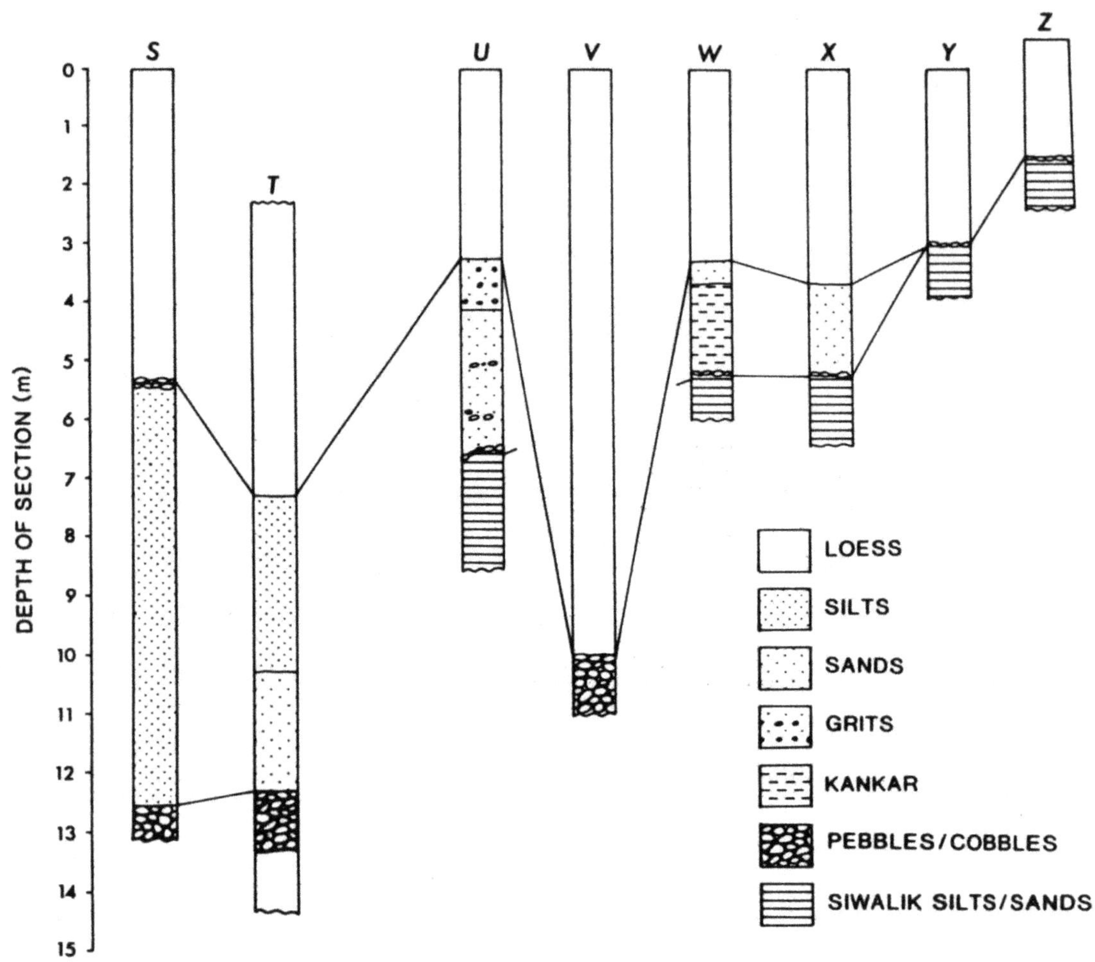

Figure 6.2: Sections north of GT road

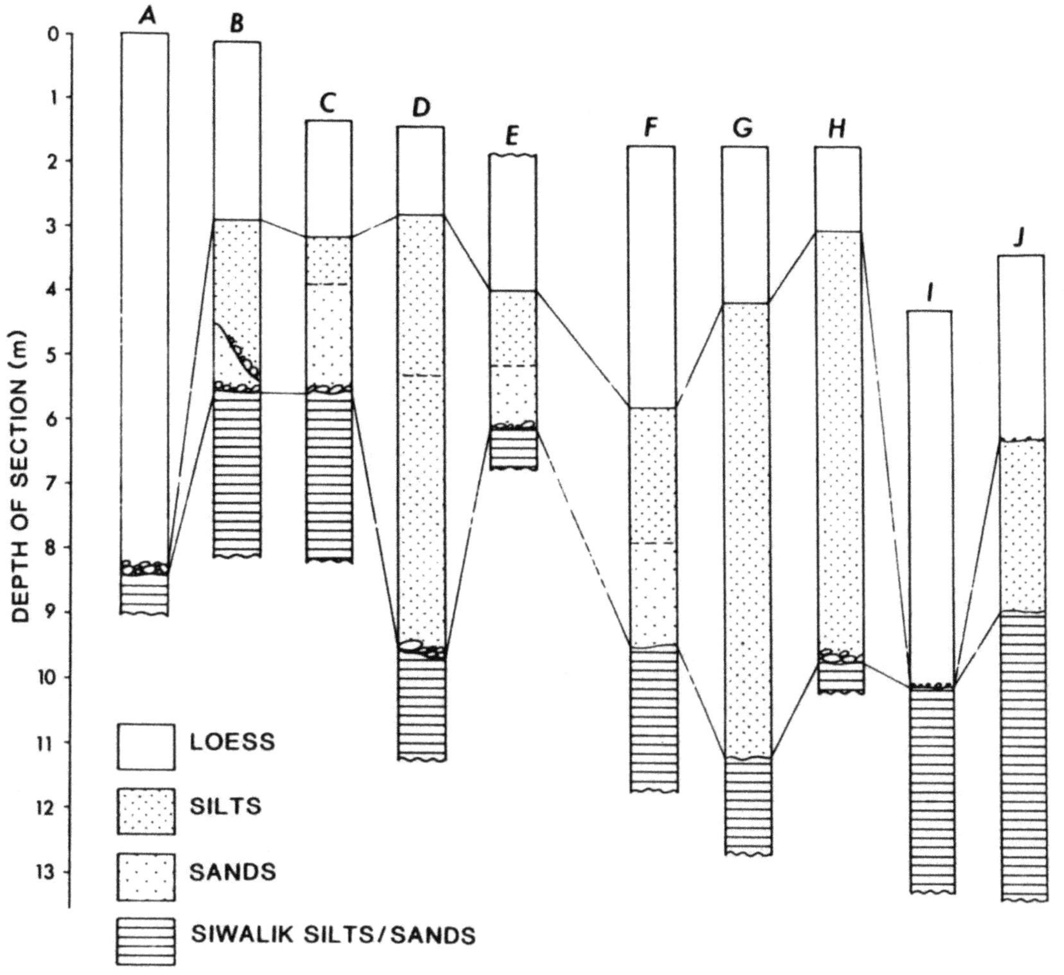

Figure 6.3: Sections south of GT road

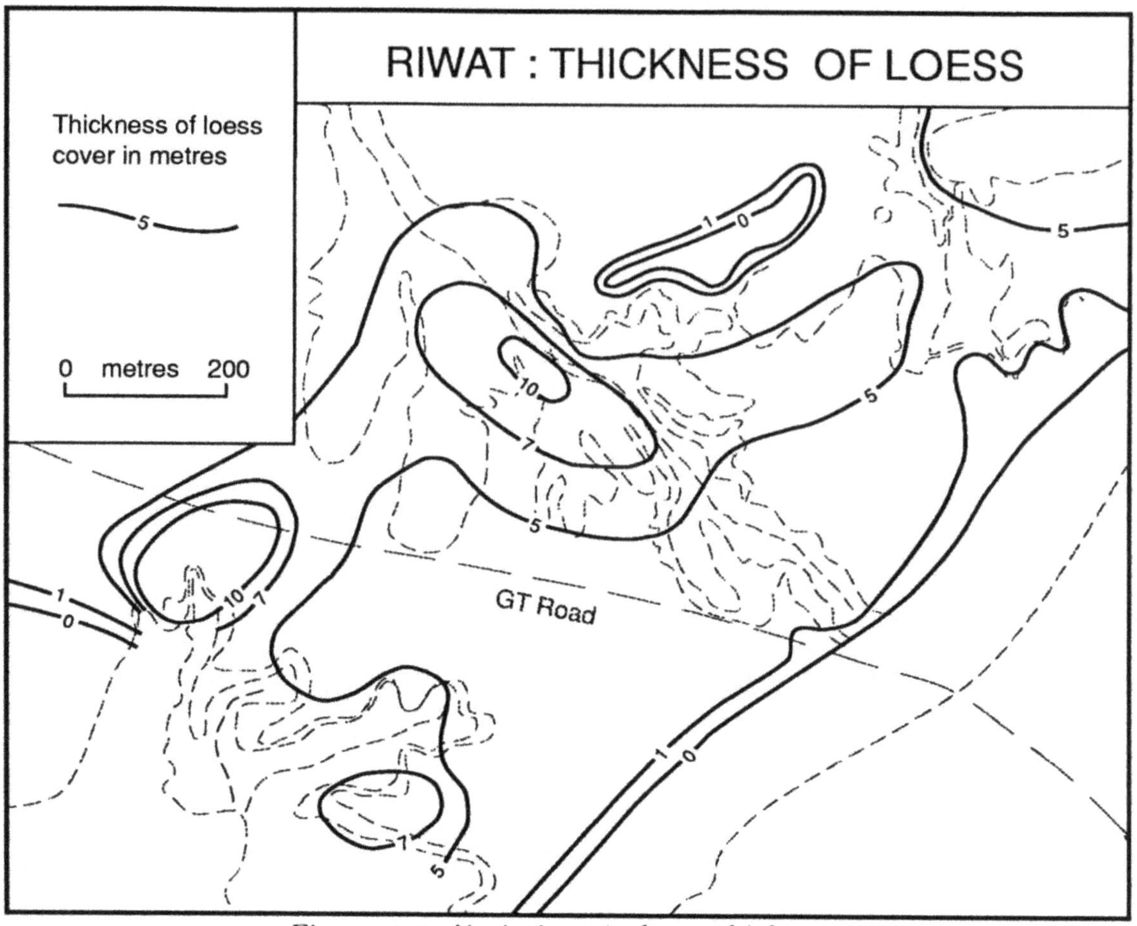

Figure 6.4: Variations in loess thickness

post-Siwalik sediments is exposed (see figure 6.3). The thickness of post-Siwalik sediments varies from two to 20 m. The surface of the eroded Siwaliks is irregular, while variations in the thickness of loess and the underlying alluvial and colluvial sediments indicate a phase of erosion prior to the deposition of the loess cover (see figure 6.3).

The high ridge formed by the Siwalik outcrop on the western flank of the study area is covered by a spread of quartzite pebbles and cobbles. Where revealed in section, it is apparent that this uncemented clast-supported conglomerate layer unconformably overlies the Siwalik strata and must represent a lag deposit that developed during an early phase of incision following the Soan syncline folding event.

A distinctive and spatially continuous gritstone horizon is exposed along the gully side between sections B and C within the post-Siwalik sequence. This gritstone horizon yielded a large number of palaeolithic artefacts, notably fresh cores and flakes. This horizon, and its relationship with other artefact-bearing horizons, is discussed in detail in Chapter 7.

6.2.1 The loess sequence at Riwat

Loess deposits overly the eroded surfaces of the Siwaliks and the Lei Conglomerate. In the Riwat area the effect has been to fill topographic depressions so as to produce a smooth surface of virtually constant elevation. The loess surface slopes at a low angle (1° 18′) towards the Soan river. The variations in topography beneath the loess cover

Figure 6.5: Photograph of the relatively thin loess cover in part of the Riwat study area

	number of samples	% silt-clay < 63 microns	% clay < 2 microns	D_m50 microns	% carbonate
SOAN AREA					
T LOESS	21	90.3 ± 5.9	15.5 ± 3.7	16.1 ± 3.8	17.4 ± 4.8
B LOESS	15	78.9 ± 17.1	14.8 ± 2.8	27.4 ± 24.2	17.8 ± 6.3
RIWAT AREA					
T LOESS	23	86.6 ± 11.1	15.9 ± 3.9	18.8 ± 6.3	13.9 ± 8.2
B LOESS	12	78.2 ± 12.5	23.9 ± 4.3	20.8 ± 17.8	6.6 ± 5.3

Notes:
T LOESS - sample taken from within upper 2 metres of section
B LOESS - sample taken from within 1 metre of base section

Table 6.1: Characteristics of Potwar loess

are summarised in Figure 6.4. The loess deposits include both primary loess, and loess showing local reworking by rainsplash and surface flow, and by slumping. The thickness of deposit is highly variable ranging from <0.2 m to 14 m (see Figs. 6.5 and 6.6). The thicker deposits show a poorly developed stratification. The loess is a very pale brown colour (10YR 8/3-4, 7/3-4). A darker colour is observed towards the base of some of the deeper profiles (>4m), where the loess is a light yellowish brown (10YR 6/4) to brown or yellowish brown (10YR 5/3-4). At some sections this darker horizon is associated with an increase in the sand-size fraction of the loess. Grain size analysis of 35 samples of loess from the study area reveals that samples from near the base of the sections have both a higher sand-size and a higher clay-size component, and a lower carbonate content than loess samples from the top of the profiles. These data are summarised in Table 6.1, in which data for Soan valley loess generally are provided for comparison. If the differences in grain size and carbonate content are indeed systematic, then they point to some decalcification and translocation of clays. In order to examine evidence of weathering and soil formation processes within the loess, one deep section (section V, Figure 6.2) was sampled in detail. The lower portion of the loess exposure Section V appears weathered (Figure 6.7). It is important to note that the majority of the thinner and stratigraphically-younger loesses are unweathered.

During the detailed mapping of the study area, a second, stratigraphically older, loess deposit was identified in sections both north and south of the GT road. Outcrops of the deposit can be traced over a distance of ca. 1 km in the study area. This "older loess" is separated from the overlying loess by either an erosional haitus or a fluvial gravel. The loess is compact (bulk density:1.7 gm/cm^3), decalcified and very pale brown in colour (10YR 7/3). At one of the sections (Section T Figure 6.2) it exhibits distinctive vertical partings. It is identical to the overlying loess in terms of grain size distribution and geochemistry.

Figure 6.6: Photograph of the R89 10m loess section

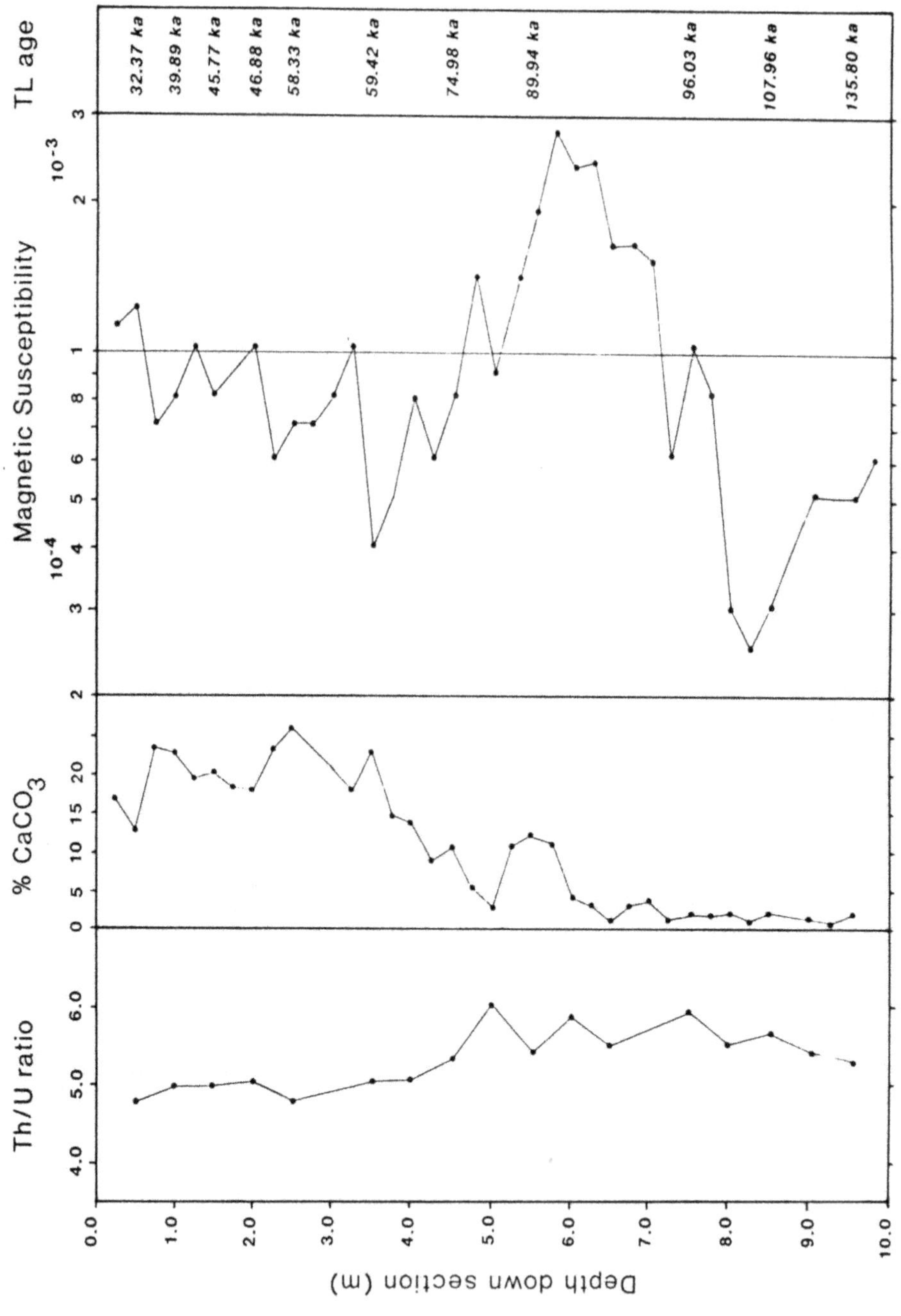

Figure 6.7: Detail of 10m loess section R89

6.3 Chronology of the post-Siwalik and thermoluminescence dating of the loess sequence

The problem of establishing a chronology for the loess deposits in the Soan area and for associated sediments and palaeolithic artefacts involved the search for a suitable dating technique. The loess contains nothing that could be dated by radiocarbon techniques, as it lacks both charcoal and humic horizons. Nor were any hearths or other materials amenable to conventional thermoluminescence dating found. The only option appeared to be the development of TL dating techniques for the sediment grains themselves, to date the deposition of the sediment, and to provide time constraints for the associated palaeolithic materials.

6.3.1 Thermoluminescence (TL) dating

The application of TL techniques to the dating of sediments has undergone a dramatic increase during the last decade, pioneered by Wintle and Huntley (1979, 1980). Although scientists in the USSR had been using the technique since the early 1960's, their work never achieved widespread recognition, and the validity of some of their assumptions has subsequently been questioned (Dreimanis *et al*, 1978). An increasing number of papers at the International TL meetings have been devoted to TL dating of sediments. During the most recent conference, held at Cambridge in July 1987 some 30 percent of the papers on TL were concerned with sediment dating.

The basic assumptions underlying TL dating of sediments are straightforward. Naturally occurring minerals such as quartz or feldspar have defects in their crystal structures. When these minerals are exposed to a flux of ionizing radiation, from the decay of naturally-occurring isotopes of uranium, thorium and potassium and from cosmic rays, electrons are released and a proportion of them become "trapped" at defect sites. On thermal excitation (i.e. heating) the trapped electrons escape and a certain proportion of them recombine at centres within the crystal with the emission of a photon (i.e. light). The temperature at which light is emitted is a measure of the trap depth, and the amount of light emitted is a measure of the size of the trapped electron population.

In order to obtain a TL date, one condition must be satisfied, and two other parameters must be measured:

1. Initialisation: The trapped electron population must have been reduced to zero in antiquity.
2. The laboratory dose equivalent to the dose received in antiquity (the Equivalent Dose (ED)) must be measured, and:
3. The environmental dose rate must be calculated.

The TL age is given by:

$$\text{TL age} = \frac{\text{ED}}{\text{annual radiation dose}}$$

or:

$$\text{TL age} = \frac{\left(\frac{\text{Natural TL}}{\text{TL per unit of radiation dose}}\right)}{\text{annual radiation dose}}$$

In the relatively simple case of TL dating of pottery, the heating involved in firing the pot reduces the trapped electron population to zero. The rate at which the potential TL signal increases is then a function of the environmental dose rate and the sensitivity of the natural dosimeters (the mineral grains within the pot) to radiation dose. Unless sediments have been baked by contact with volcanic material or by proximity to hearths, the establishment of a zero point for age determination can be a problem. It has been shown that most, but not all, trapped electrons are evicted from the defects if the mineral grain is exposed to light. The nature and degree of light exposure ("optical bleaching") prior to burial is one of the keys to the success of TL dating of sediments (or indeed to its lack of success for some samples!). Other problems that are exclusive to TL dating of sediments arise out of some of the obvious differences between sediments and pottery. One problem is that even the oldest pottery (dating to ca. 8,000 years ago) is not very old at all on a geological timescale. Non-linear response to dose and saturation problems seen for sediment samples in excess of 50 - 100 ka are not a problem for pottery dating. Similarly the estimation of mean moisture content of a sediment over 100,000 years is a non-trivial problem. The establishment of an effective "zero point" for a sediment may also be difficult. This last problem explains why most TL workers have preferred to try to date aeolian samples, where exposure to light is guaranteed, rather than samples from other sedimentary environments.

Provided one can be confident about the transport history of a sample, and its likely degree of exposure to light, the main problem for TL dating remains the establishment of the Equivalent Dose (ED). There are three main methods for doing this:

(a) the Additive Dose or N + beta - I_0 method. (b) the Partial Bleach method. (c) the Regeneration method.

In each case the aim is to discover the laboratory radiation dose that is required to give a TL signal which matches the natural TL (NTL) accumulated since deposition and burial of the sediment grains. The three methods are illustrated in Figure 6.8. Each of the methods requires the preparation of a large number of subsamples. For the fine-grain dating discussed in this chapter, 30 - 40 subsamples, in the form of fine powder covering 10 mm diameter aluminium discs, were prepared for each sample.

The Additive Dose Method involves the construction of a TL growth curve for the particular sample and the establishment of the level of residual signal, the zero point.

The Partial Bleach Method was developed on the assumption that limited exposure to light may have resulted in insufficient bleaching. The method is based on the measurement of the reduction in TL, both the NTL and NTL + laboratory beta doses, caused by a standard short exposure to light. The intercept of the lines fitted to the (NTL + beta) and the (NTL + beta + bleach) defines the equivalent dose.

The Regeneration Method is based on the laboratory simulation of a sample's TL history. After the measurement of the NTL a number of the sample discs are exposed to light and then given additive beta doses. The dose required for the "regenerated" TL to exactly match the NTL is the ED.

6.3.2 Laboratory Procedures

All the TL measurements carried out on the Pakistan loess employed the fine grain technique (Zimmerman, 1971). After pretreatment with dilute HCl to remove carbonates, and size separation by sedimentation, aliquots of the 2-10 micron fraction of the loess

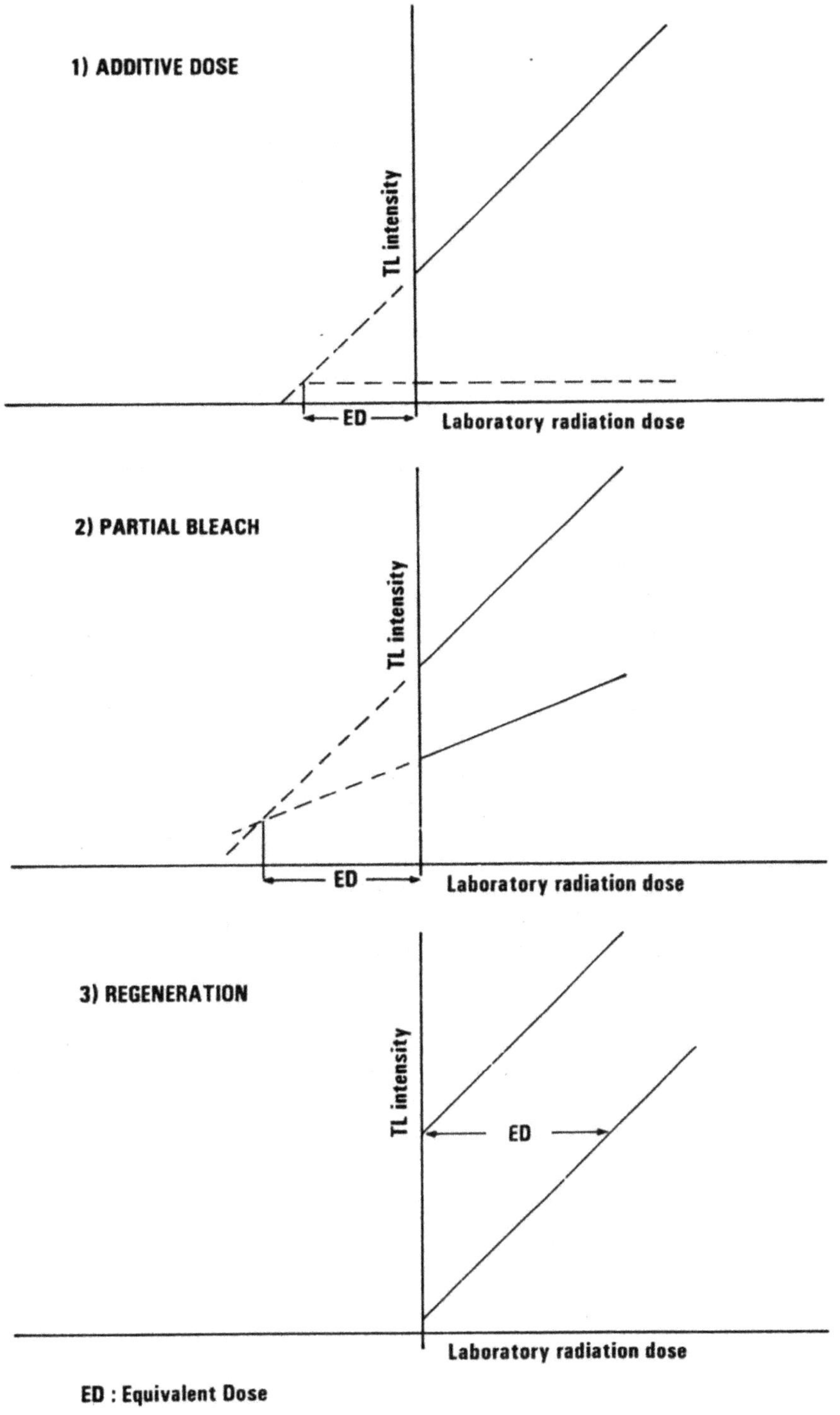

Figure 6.8: Methods of Equivalent Dose determination

samples in suspension in acetone were pipetted on to clean 10 mm diameter aluminium discs and allowed to evaporate to dryness at room temperature. Sample discs were heated at a constant rate to a maximum temperature of 420°C in an argon atmosphere. TL emissions were measured using an EMI 9635Q photomultiplier tube and a combination of broad band optical filters. Initially emissions in the blue part of the spectrum were monitored using a Corning 5-58 (blue) and a Chance Pilkington HA3 (heat-absorbing) filter. All subsequent measurements were made through a Schott UG11 (ultra-violet) filter, plus an HA3. TL intensity and temperature data were recorded using an Intertec Superbrain microcomputer. Laboratory radiation doses were given using a ^{90}Sr beta source or an ^{241}Am alpha source. Optical bleaching was performed using either a 300 watt Wotan lamp, with sample discs unshielded or under glass, or exposure to daylight.

Equivalent doses were calculated for data from the higher temperature part of the TL glow curves. Normally the traps responsible for TL signal at temperatures above 270°C are regarded as stable, in the sense of not losing trapped electrons, over the time period of interest for the dating. A more or less constant value for ED over the temperature range 270 - 350°C is a prerequisite for obtaining a date (Aitken, 1985).

6.3.3 Initial results

At the start of the dating programme on the Pakistan loess material, a number of experiments were undertaken to investigate the response of the material to additive laboratory doses, and its bleaching behaviour. Initial results pointed to problems with bleaching (Rendell *et al*, 1983; Bowall *et al*, 1987). In particular, it was noted that samples exposed to light showed an increased TL sensitivity to dose on subsequent laboratory iradiation. Changes in TL sensitivity were also shown to be transient. Such changes in sensitivity or, more correctly, TL efficiency invalidate the Regeneration Method of ED measurement for the Pakistan material. Only in the case of the stratigraphically older loess was the Regeneration Method employed so as to obtain "minimum age estimates". The responses of these samples to additive laboratory doses on to the NTL, showed that the samples were approaching saturation.

The Additive Dose Method uses bleaching simply to determine the level of the unbleachable (residual) TL. This method may overestimate the ED, and therefore the TL age, in cases where:

(a) the sample has not been zeroed by light exposure prior to burial so that the laboratory bleaching "overbleaches" the sample, giving a value of I_0 that is too small.

(b) the sample is in a non-linear region of response to dose (so that simple linear extrapolation will yield an ED that is too large).

(c) the responses to the alpha and beta/gamma/cosmic portions of the environmental dose are different.

Given the problems anticipated with the Regeneration Method it was decided to concentrate on the Additive Dose approach and to use the Partial Bleach Method as a cross-check (Rendell and Dennell, 1987). The confidence with which the Additive Dose technique could be used was tested by constructing TL growth curves for stratigraphically "young" samples of the loess and examining their linearity. Growth was linear for additive doses of up to 600 Gy (equivalent to an additional 120 ka of dose, assuming an annual dose of 0.005 Gy).

The annual dose was calculated for each sample based on thick-source alpha counting or neutron activation analysis for the uranium and thorium isotopes, and either inductively-coupled argon plasma spectrometry or atomic absorption spectrophotometry for the determination of potassium content. The use of on-site calcium fluoride dosimeters was used as a cross-check and as an aid to estimating the water content of the sediment on an annual basis. As a result of an investigation of water content histories of the loess (Rendell 1983, 1985), a mean water content of 5 percent $\pm$ 5 percent was assumed for all the samples dated.

6.3.4 Results of TL dating programme

The TL dating programme had two main aims: first, to establish a general chronology for the Potwar loess and second, to provide a dating framework for the palaeolithic excavation at Riwat. In order to test the reproducibility of the TL technique, a few samples were processed in triplicate. The results were in agreement to within 5 percent. The resultant TL dates have uncertainties of $\pm$ 10 percent after sources of both systematic and random error have been taken into account (Aitken and Alldred, 1972; Aitken, 1976). A complete catalogue of all TL dates and dosimetry data is given in Appendix G.

Some of the dates for the loess in the Riwat area are summarised in Figure 6.9. These results emphasise the variability of deposition rates across the Riwat study area as the uneven topography was gradually smoothed out to present a "summit surface cover" of very low surface relief. The rates of deposition of loess calculated on the basis of the TL dates range from 8cm/ka for section R89 to 27cm/ka for section RK. These differences must reflect localised redistribution of the loess in topographic lows. It is also clear for the results that the section R89 actively accumulated loess for far longer than any of the other sections sampled within the study area. The R89 section also enabled the EDs, and, therefore, TL age estimates to be compared down profile. The results from this analysis have been discussed in detail elsewhere (Rendell and Townsend, 1988). They reveal that the use of the Regeneration Method is entirely inappropriate for the Pakistan material, since the technique yields constant ages of ca. 74 ka for the whole of the lower portion of the section. In contrast, the results from the Additive Dose Method show age monotonically increasing down the profile.

6.3.5 A note on the sampling of the 10 m loess section

One of the key sections for the dating programme was the 10m section (R89). This section also provided data on changes in the nature of the loess with depth. Given the desirability of sampling long, clean vertical sections, a few words are appropriate as to how this was achieved at Riwat. Samples were taken at intervals of 0.25m for general analysis, and 0.5m for dating, by RD on abseil, using a double 11mm diameter nylon rope, two jumar clamps, and a sit harness. A prussik loop was used as a back up. The most difficult part of the procedure lies in attaching the clamps to the rope early in the abseil. If this causes undue difficulty, it is better to complete the abseil and then jumar up the rope from the ground. The belay consisted of two ice-screws and a large loess block, behind which the rope was looped. This procedure is not recommended, and a large metal stake and/or a small tree should be used in work of this kind. Needless to say, a safety helmet is essential to protect the sampler from material dislodged by the ropes. Fig. 6.10 shows an early stage of the sampling.

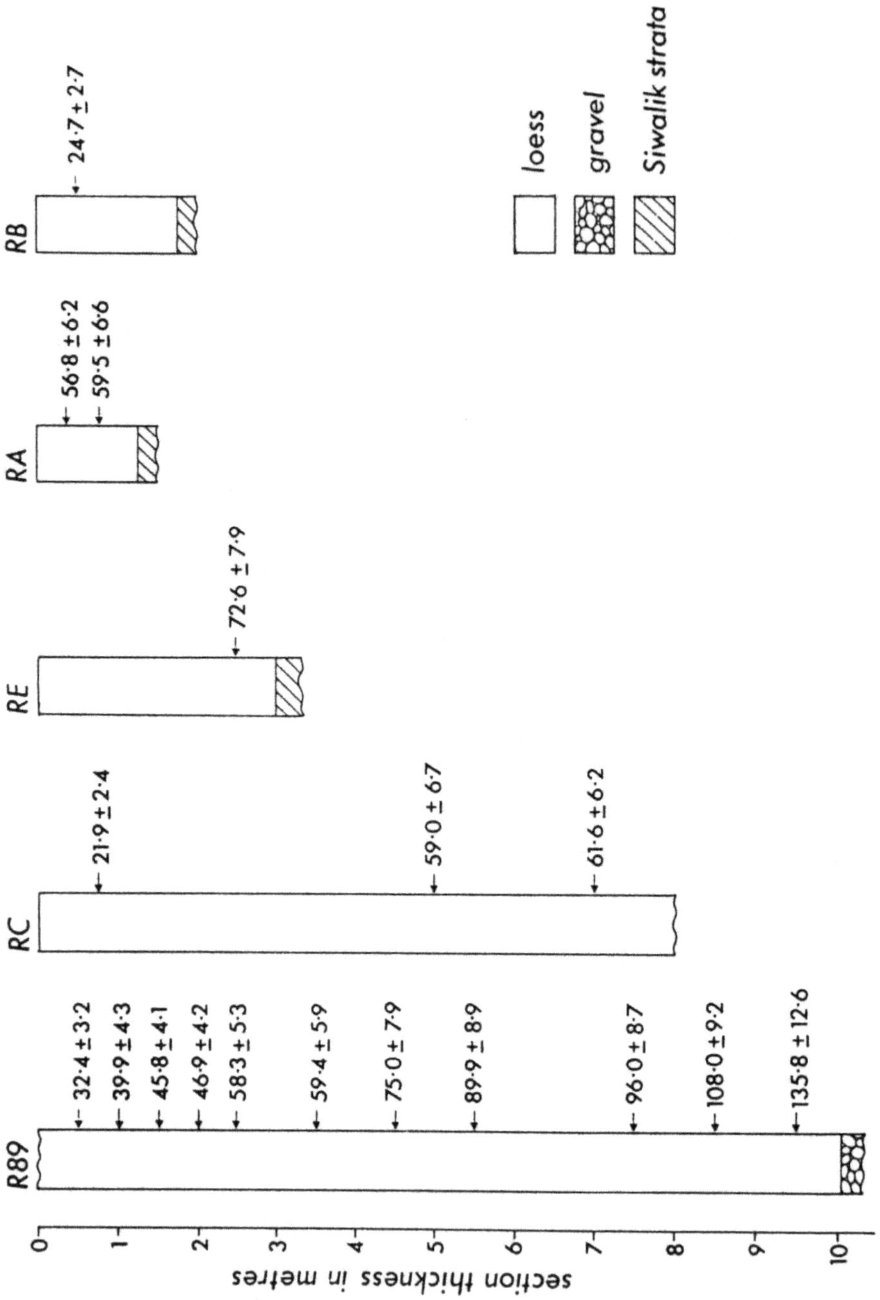

Figure 6.9: TL ages for Riwat area

Figure 6.10: Photograph of sampling of the R89 section

6.4 Chronology of loess deposition in the Middle Soan Valley

Although sampling was concentrated in the Riwat study area, samples were collected for analysis at other selected sites within the middle Soan valley. On the basis of the results obtained it is apparent that the stratigraphically "youngest" loess dates to 18 ka BP, that loess deposition and preservation were widespread between 18ka and 75 ka but that, between 75 ka and 130 ka, loess is only preserved in isolated pockets. The implication here is that loess was being deposited throughout the Potwar area, but that much of it was being actively reworked and eroded and therefore that preservation was the exception rather than the rule.

The stratigraphically older loesses yielded minimum ages of 170 ka BP, as determined by the Regeneration Method. It must be stressed that these dates may severely underestimate the true age of this older loess. The loess cover is severely attenuated in places across the Riwat study area as is shown in Figure 6.9, and this problem is particularly acute where the dating of Riwat Site 55 is concerned. This problem is discussed more fully in chapter 9.

6.5 The palaeoclimatic significance of the loess sequence in the Soan Valley

Loess puts in a very late appearance amongst the complex of Siwalik and post-Siwalik deposits in the Soan area. Yet loess remains one of the few indicators of palaeoclimate within the Soan sequence. The Pakistan loess sequence does not really stand comparison with the thick and complex sequences of loess exposed in China and in Soviet Central Asia. Even the sequence in Kashmir typically includes thicknesses of the order of 30m. One can speculate as to the reasons for the absence of a thick loess cover in Pakistan.

In many areas of the world, the availability of source material for loess is seen as a limiting factor, but in Pakistan it appears likely that the critical conditions are those favouring entrainment of silt particles by wind, and those favouring the subsequent preservation of loess deposits. Several aspects of the loess problem are discussed below. Loess is an exceptional, aeolian, deposit in a sequence dominated by fluvial activity, and perhaps the reason for this is that the preservation of loess deposits has been poor.

6.5.1 Source of loess

Vast quantities of silt have been generated throughout the Miocene, Pliocene and Pleistocene by processes of weathering and erosion in the rising Himalayan mountain chain. The processes of uplift and erosion have continued until the present and provide the ultimate source of silts for the Potwar loesses. In common with other Pleistocene loesses, the source area for the bulk of the silt material is presumed to be local (i.e. within 100 - 200 km). Likely source areas include: the braided course of the Indus river as it leaves the mountain front north of Attock, and the exposed surfaces of alluvial fans along the mountain front. By contrast, the nearest source of "desert" loess lies 600 km to the south, and is not a realistic source for the Potwar loess. The critical conditions for the formation of Potwar loess would appear not to be a direct function of silt availability, but to have more to do with conditions of entrainment, transport and deposition.

6.5.2 Loess and climate

If the uplift of the Tibetan Plateau has provided an essential driving mechanism for the Asian monsoonal circulation, then the weakening of the monsoon during the Pleistocene requires some explanation. As discussed in chapter 3.4, the albedo of the Tibetan Plateau is thought to play a key role in determining the strength of the southwest monsoon, with a snow-covered plateau resulting in a weakened monsoonal circulation. During the present interglacial, for example, fluvial activity dominates the Potwar area and loess deposits are preserved in isolated areas above the level of the main river valleys.

In northern Pakistan, the dominance of the southwest monsoon may hold the key to phases of loess deposition. It is possible to envisage loess entrainment, transport, deposition and preservation associated with locally drier periods and weaker monsoonal circulation. The most recent phase of active erosion appears to be associated with the re-establishment of the northward penetration of the southwest monsoon in the Late Pleistocene. This picture is corroborated by the work on Indian ocean sediments (Van Campo, 1986) which shows an onset of wetter conditions after 18ka. The most recent phase of erosion and incision of the loess sequence dates from 18 ka to the present. The low latitude of the Potwar (32-34° N) may help to explain this relatively early deglaciation.

The most recent phase of loess deposition (18ka-75ka) corresponds to oxygen isotope stages 2 to 4 and therefore to glacial conditions in the northern hemisphere. It is also apparent, and perhaps rather surprising, that loess deposition, albeit with little or no preservation, took place throughout oxygen isotope stage 5. Syndepositional weathering of the loess at the R89 section is envisaged to explain the complete decalcification of the lower part of the profile which appears to have been deposited during Stage 5.

6.6 The palaeoenvironmental interpretation of the post-Siwalik sequence at Riwat

Several distinct phases can be identified in the post-Siwalik sequence at Riwat. These are summarised below and in Table 6.2

1. A phase of incision and dissection of the tilted Siwalik strata, leaving an irregular surface with high local relative relief in places. The depth of incision shows a general increase in the direction of the Soan river.

2. Sections on the south side of the GT road reveal evidence of colluvial activity. The erosional junction with the Siwaliks is overlain by chaotic colluvial deposits including huge partly cemented, case-hardened sandstone blocks. These are overlain in turn by alluvial sands and silts. A gritstone horizon forms a laterally continuous outcrop near the top of this sequence. This gritstone horizon yielded an artefact assemblage of cores and flakes that dates to before 75 ka. The pre-loess surface is complex, with the post-Siwalik cover removed almost completely in some places prior to loess deposition.

3. To the north of the GT road, in an area largely marginal to fluvial activity, the major post-Siwalik, but pre-loess, deposit consists of a colluvial spread of quartzite pebbles over the eroded Siwalik surface. The sequence increases in complexity northwards and

	SOUTH OF GT ROAD	NORTH OF GT ROAD
	incision	incision
18 ka		
	loess deposition	loess deposition
75 ka		
	deposition of gritstone fluvial erosion/deposition (erosion Post-Siwalik sequence)	deposition/weathering loess (R89 section) or erosion/deposition
135 ka		
	fluvial erosion/deposition	fluvial erosion/deposition
>170 ka	loess deposition	loess deposition
	colluvial deposition	colluvial/alluvial deposition
	reworking of quartzite cobbles derived from erosion of Siwalik Conglomerate	
	erosion of Siwalik strata	

Table 6.2: Riwat study area : Late Pleistocene Sequence

includes the stratigraphically-older loess cover and several horizontally-bedded lenses of clast-supported quartzite conglomerates.

4. The period from 130 ka to 75 ka included erosion and loess deposition. Loess from this phase is preserved in isolated pockets and is generally decalcified.

5. A phase of loess deposition dominated the period from 75 ka until just after 18 ka. During the later part of this period the Riwat Palaeolithic site (Site 55) was gradually covered by loess.

6. Since the end of loess deposition, erosion has been dominant, producing the deeply dissected landscape that is visible today.

6.7 Summary

On the basis of this study of the post-Siwalik sequence at Riwat, the following conclusions may be drawn:

1) The post-Siwalik sequence is a complex mixture of colluvial, alluvial and aeolian deposits and covers an unevenly eroded surface cut in Siwalik strata.

2) The dominant post-Siwalik deposit is loess. The loess sequence has been dated by thermoluminescence. Much of the loess cover dates to the period 18 ka to 75 ka, and does not belong to the penultimate (Riss) glaciation, as originally proposed by De Terra and Paterson (1939).

3) The loess sequence is slightly out of phase with global trends, with deposition ceasing at the height of the last glacial maximum in north-west Europe, but occurring more or less continuously from isotope stage 5e until ca. 18ka.

4) The loess record is interpreted as reflecting changes in the strength and penetration of the southwest monsoon, during the last 150,000 years.

Chapter 7

The artefact-bearing gritstone/conglomerate horizons at Riwat

7.1 Introduction

As stated earlier, a major aim of this investigation was to find instances of humanly- or hominid-struck stone artefacts in secure geological contexts that could be well-dated. Our consideration of Paterson's work in the 1930's, and our impressions gained early in this study, forced the conclusion that a small amount of well-dated evidence was considerably more valuable than a large amount of undated, or poorly-dated material. We noted in chapter 2 that all the palaeolithic material collected by Paterson (and most subsequent workers in the area) came from surface collections that were usually on conglomerate exposures. As a result, it has always remained unclear whether this material related to when a particular conglomerate was deposited, or to when it was subsequently re-exposed. It has been equally unclear whether these assemblages belong to one particular period, or are simply palimpsests resulting from long usage of a particular outcrop as a source of raw material.

The problem in the Soan Valley is not one of finding palaeolithic material: this is straightforward, and often an hour or so of walking can result in surface finds of hand-axes, prepared cores, blades and other types of material. Early in our investigations, we stopped collecting surface material, as there seemed little point in filling the already-limited museum storage facilities with large amounts of undatable material. Rather, the main problem in the Soan Valley is one of finding palaeolithic material that can be dated: for this one needs to find artefacts within a datable geological sequence. We concentrated on two kinds of artefact-bearing horizons. The first was within or overlain by loess. In some cases, the loess cover was sufficiently thin to be removable by excavation. The main results of this kind of investigation will be presented in chapters 8 - 11. The other was artefact-bearing horizons of cemented or partly cemented sediments, notably gritstones and conglomerates. The advantages of this type of sedimentary context are that i) they are more resistant to erosion, and thus artefacts contained in them are likely to be exposed slowly over several years before they are eroded out of context; ii) in deep sequences, artefact-bearing horizons can potentially be dated by TL if they are comparatively recent, and by palaeomagnetism and/or radiometric techniques if older.

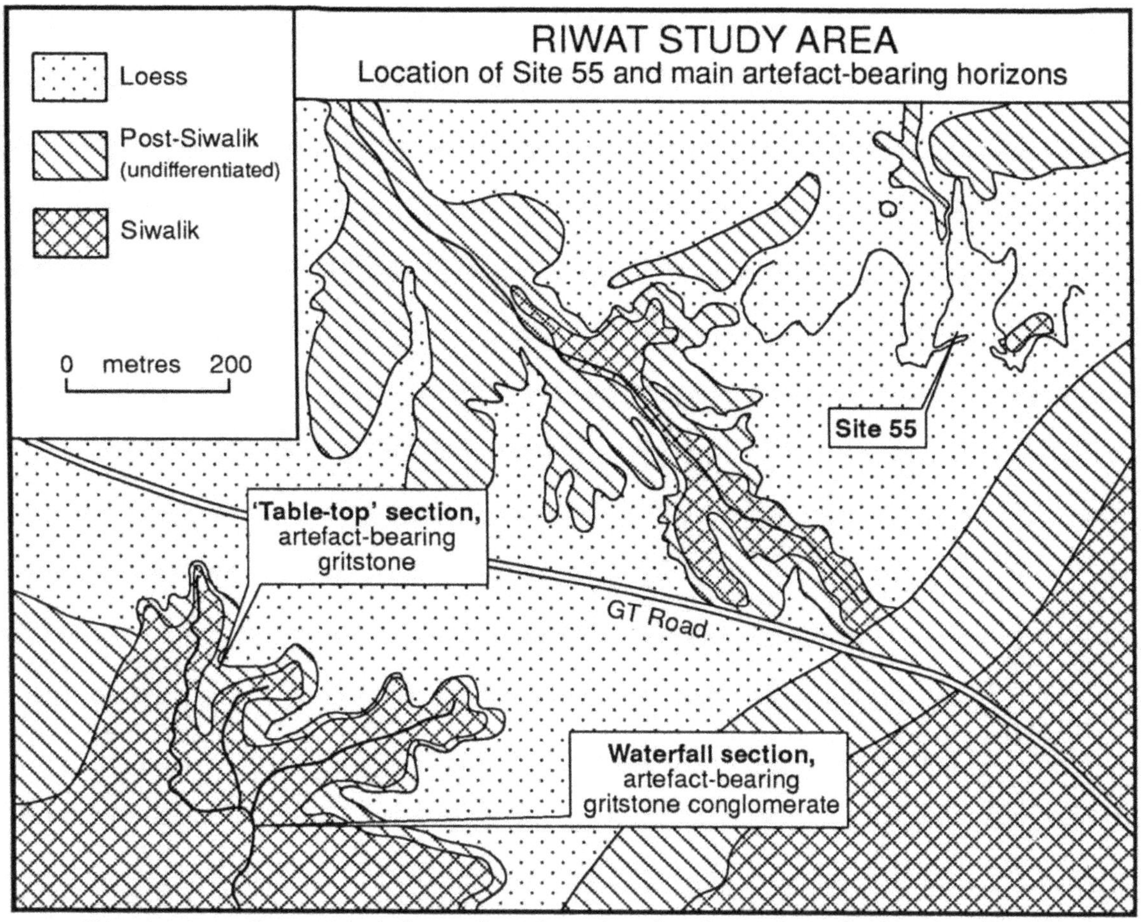

Figure 7.1: Location of artefact-bearing horizons in relation to site 55

On the other hand, they have some important drawbacks as sources of evidence: i) they inevitably contain geologically-flaked material, which can often be difficult to distinguish from humanly-flaked stone; ii) excavation of conglomerates is difficult, if not impossible, because of the hardness of the matrix, and often the depth of overlying sediment; iii) it is most unlikely that archaeological material will be in its original context; and iv) fluvial deposits provide only a minimum age estimate for their contents, as artefacts and animal remains can be reworked and redeposited over long periods of time.

Material from two artefact-bearing horizons will be considered in this chapter. Both lie on the south side of the G.T. road, in the deeply- dissected area considered in chapters 5 and 6. (Fig. 7.1) They lie at different depths, and are clearly distinguished from each other as is indicated in Fig. reflocsect. The first is the post-Siwalik gritstone mentioned previously (section 6.2); it is horizontally-bedded, has small clasts of sandstone and quartzite clasts, and lies at a higher stratigraphic level than the second.

7.2 The upper gritstone horizon

As Fig. 7.3 shows, this conglomerate is exposed as a "table-top" towards the top of a deep section. At this particular place, it forms a platform ca. 4 metres square, perched on a thinner column of pink and buff silt-clays and buff silts. Besides serving as a useful land-mark, it also provided shade on hot days. We visited it first in 1981, and noticed

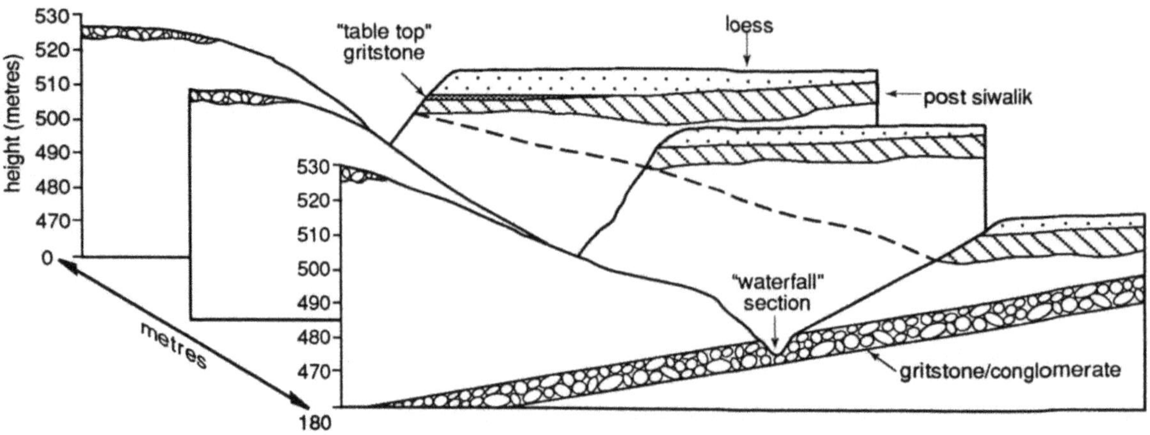

Figure 7.2: Relative positions of the artefact-bearing horizons

a block ca. 1m by 1m by 0.5m had fallen off the side nearest the G.T. road, and was lying at the base of the supporting column of silts. A flaked stone was seen embedded in its surface (Fig. 7.4). When chiselled out, it was found to be an elongate cobble, from which one end had been removed; the face had been flaked in several directions, and flake scars were fresh and clear. It can be classed as a hemispherical disc core. The surface of the "table-top" was also examined in 1981, and a number of small, sharp honey-coloured quartzite flakes were found on its surface. Unfortunately, they were indeterminate in shape, and consequently of little value in terms of dating specific types of palaeolithic artefacts.

This upper gritstone horizon was surveyed in greater detail in 1985. The gritstone outcrop can be followed along the same side of the gully slope from the table-top for a distance of ca. 80m before it lenses out. Often, this upper gritstone is exposed as a series of large blocks up to 2m thick that are in the process of becoming detached from the rest of the horizon, in much the same way as the "table-top". Each of these blocks was inspected for artefacts. None was found, except for a rolled hand-axe.

The upper gritstone dates from the later part of the Middle Pleistocene, or the early part of the Upper Pleistocene: in approximate terms, it is thus probably around 100 - 200,000 years old. The loess horizon overlying the gritstone yielded a TL date of 74.3 ± 8.3 ka, thus providing a minimum age for the underlying gritstone. Although post-Siwalik, the age of the gritstone may, of course, be substantially older than the loess. Of the two artefacts found in it, the disc core is more useful, as its condition is fresher, and its age may therefore be nearer to that of the gritstone itself. However, the hand-axe may also be of the same age; as Oakley and Leakey (1937) pointed out long ago in connection with Paterson's work, the degree of freshness or rolling of stone artefacts in river deposits need not be an indication of relative age, but simply an indication of how far it has been transported. It is thus an open question as to whether this example indicates that hand-axes were made at the time when this gritstone was deposited. We found older examples of hand-axes in 1983 at Dina and Jalalpur in the Jhelum basin ca. 60 km to the east, and in early Middle Pleistocene contexts that were ca. 500-700,000 years old (Rendell and Dennell, 1985; Dennell *et al*, 1985; Dennell, 1985). If the example from this upper gritstone horizon at Riwat is regarded as close in age to the gritstone itself, this evidence implies that similar type of artefacts may still have been manufactured a substantially later date.

Figure 7.3: Photograph of "table-top" section

Figure 7.4: Photograph of artefact *in situ* in gritstone

7.3 The lower gritstone/conglomerate horizon

The story of investigations of this horizon is a long and often complicated one that requires detailed treatment in view of the importance of the discoveries, and the interest (and controversies) they have since engendered (e.g. Dennell, 1989; Hemingway and Staepert, 1989). In view of these factors, it is worth setting out the main chronology of our work:

7.4 Chronology of the investigations of the lower gritstone/conglomerate horizon at Riwat

(i) 1983: hominid-struck pieces were discovered in the lower gritstone/conglomerate horizon, the age of which was then unknown.
: the central portion of the field season was spent on a reconnaissance of the Dina, Jhelum and Jalalpur areas.

(ii) 1984: no fieldwork took place.

(iii) 1985: the horizon near where the 1983 discoveries were made was re-examined, without further results; Rendell undertook the preliminary palaeomagnetic sampling of the overlying sediments
: most of that field season was spent on survey in the Jhelum basin, looking for fossils and artefacts at Dina, Rohtas and Jalalpur .

(iv) spring 1986: the first three month field season in the Pabbi Hills, collecting fossil and stone off erosional surfaces between 1.2 - 1.4 Ma old.
: Rendell undertook the detailed palaeomagnetic sampling of the section above and immediately below the lower gritstone/conglomerate horizon at Riwat (reported as RIWAT A section; see chapter 5.3)

(v) autumn 1986: the palaeomagnetic analyses were completed, and Rendell and Hailwood were satisfied with a date of 1.9 - 2.1 Ma for the lower gritstone/conglomerate horizon.

(vi) 1987: a second three-month field season took place in the Pabbi Hills, this time collecting fossil and stone on surfaces dating back to 2.0 Ma in the light of the Riwat date.
: Rendell sampled a further 51m section beneath the artefact- bearing horizon (reported as Riwat section B: see chapter 5:3), and walked out the horizon in a south-westerly direction for several kilometres, and to the point where the top of Raynolds' (1980) Kas Dovac section was reached.

(vii) 1988: the entire horizon at Riwat was re-surveyed, resulting in the discovery of a flake (R88/1), 40m from where the first piece (R001) was found in 1983, and in the same horizon. This section was drawn at a scale of 1:20 by Dennell and Dr. L. Hurcombe, and the latter also made a cast of the socket left by the removal of piece R001. Rendell took gritstone samples from around the area of the socket to check that the gritstone around piece R001 had not been re-worked. She also sampled a further 30m section beneath the horizon, and re-sampled Raynolds' section and intervening sections to try to tie the Riwat sections to those in the Kas Dovac.
: the results to date were exhibited at the Royal Society annual science soirée.

We feel, therefore, that we have not been overhasty in presenting the results obtained over six years. With this background in mind, the results can be presented in more detail.

7.4.1 The discovery of hominid-struck pieces in the lower gritstone/conglomerate horizon at Riwat

The lower gritstone/conglomerate horizon at Riwat is exposed at the base of the deep gully already described in the previous chapter. As one descends the gully along a foot-path and then along the bed of a small (and usually dry) stream channel at the base of the main fork of this gully, one passes below the upper gritstone horizon and a complex sequence of silts and sands up to 70m thick, capped unconformably by a conglomerate that was identified by De Terra and Paterson as a degraded fragment of their T1 terrace (Milestone 163 site; see chap 1). Eventually, one comes to a confluence with a large left-bank tributary. Immediately downstream, there is a small water-fall that has cut through a harder horizon of gritstone. This water-fall has been formed by the stream slowly eroding headwards as it cuts back upstream. It is in fact composite, and has two tiers: when the stream channel is in flow, water first falls over a rim ca. 2 m high, then along a flat sill of gritstone some 10 m long, before falling another 1.5 - 2.0 m over the second rim. Thereafter, the stream bed is more level, and proceeds along a boulder-strewn and slightly incised bed to its eventual junction with the main channel of the Kas Dovac. This stream is normally dry, and flow is ephemeral. After heavy rains in spring and autumn, it flows swiftly for a few hours at most, given the limited catchment size of only 0.15 km2. Although we have never observed this stream during the summer monsoon, inspection of marooned brushwood and grass suggests that downstream of the water-chute, it is unlikely to be more than a metre deep.

The gritstone through which it cuts is part of the lower gritstone/conglomerate horizon. The first piece we noticed was found embedded in the lower of the two gritstone tiers forming the water- chute, slightly to the left (facing downstream) of the lip of the water-fall (Figs. 7.5 and 7.6). Before we removed this piece, we inspected it carefully. Our attention had been taken by two flake scars on the exposed side. Judging from the angle with which these must have been detached, they were evidently detached before the piece was incorporated into the gritstone itself. Further flaked surfaces could be seen partly exposed on one side of this piece, but in part still embedded in the gritstone. In addition, further flake scars were only discovered on removal of the piece, as they were wholly embedded within the matrix. The piece was photographed, and then carefully removed with a hammer and cold chisel. As seen in Figs. 7.7 and 7.8, it has been flaked on three sides, resulting in the removal of at least eight flakes, as well as most of the cortex. Size details are given in Table 7.1. A search was made for further pieces in the horizon further downstream. The right bank is steeper, and after 20 m or so, gives way to an almost vertical section that is only poorly consolidated, and with several layers of sands, as well as pebble stringers. No artefacts were found in this section in 1983 (but see section 7. 4 for the discovery of 1988). The left bank is very different. Here, the ground slopes back more gently. In places, the gritstone/conglomerate horizon has been undercut, resulting in the detachment of large blocks that have often slipped a little downslope (see Fig. 7.9). These are up to 3m thick and 10 m square. Many have a pudding-stone appearance, with quartzite pebbles and cobbles protruding from their surface. Examination of these was impeded by the exceedingly dense thorn scrub. The location of each block and examination of its surface was a matter of crawling and weaving

Figure 7.5: Photograph: find spot of R001, lower gritstone/conglomerate horizon

Figure 7.6: Photograph: R001 *in situ*

Face	Length (cm)	Width (cm)	Striking angle (degrees)	L:B Ratio
Face 1				
1)	10.7	10.3	112	1.04
2)	3.2	4.0	-	0.80
3)	>4.6	10.5	110	
Face 2				
4)	6.0	7.8	90	0.77
5)	6.6	6.4	95	1.03
6)	3.5	4.0	100	0.88
Face 3				
7)	5.2	2.6	105	2.0
8)	7.0	7.4	96	0.95

Table 7.1: Details of dimensions of R001

through almost impenetrable thorn scrub. Recording was also made more difficult. In retrospect, we should have numbered each block, and recorded the place from which a flaked piece was removed with a circle of red paint. As it is, all pieces that appeared to have been flaked and protruding from the surface of these blocks were removed, usually with hammer and cold chisel. Out of a total of cobbles that is unknown, but which must have been well over 1000, 23 flaked pieces were removed. (These are discussed in the next section). There is no doubt that the blocks from which these pieces were removed are derived from the lower conglomerate horizon. The blocks all lie within a few metres of the current outcrop of the gritstone/conglomerate, and have identical clast petrology and matrix composition with the outcrop.

As noted earlier, the age of this horizon was unknown in 1983, and it was not until the end of 1986 that Rendell and Hailwood were satisfied with its dating, at ca. 1.9 - 2.1 Ma. The lower gritstone/conglomerate horizon at Riwat is in fact the only conglomerate that we have ever examined for artefacts without knowing its age in advance. The point is not unimportant: if we had known its age in 1983, it is unlikely that we would then have examined it for stone tools; rather, we would probably have continued working back through the early Pleistocene, using the Dina and Jalalpur handaxes as a base line. That is in fact what we did in 1985, when we began to explore conglomerate horizons at Jalalpur and Rohtas that were over 0.7 Ma old; and in 1986 in the first season of field work in the Pabbi Hills, where we deliberately concentrated on deposits between 1.2 - 1.4 Ma because their age fell within the time range for the (then) earliest dates for hominids in Java.

7.5 The Riwat assemblage (R.W.Dennell and L. Hurcombe)

In this section, three aspects of the assemblage from the lower conglomerate horizon will be discussed. The first are its general characteristics; the second, the survey of 1988, and the discovery of a further in situ piece; and thirdly, the replication in 1988 of the socket left by the removal of piece R001 in 1983. We stress that this work is still in progress at the time of writing, and further work is planned for 1990.

Figure 7.7: Photographs of R001

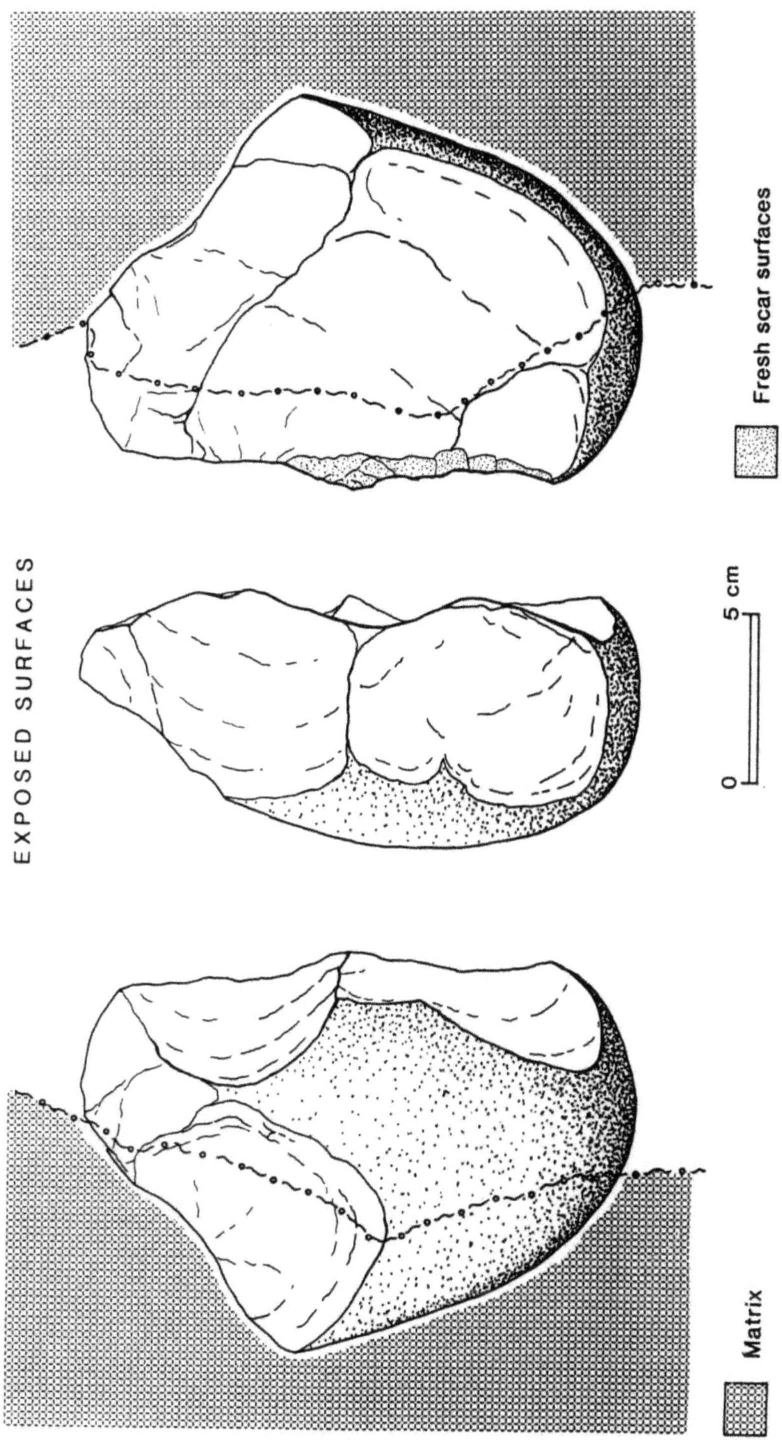

Figure 7.8: Line drawings of R001

Figure 7.9: Photograph of fallen blocks of conglomerate/gritstone on the left bank of the Riwat gully

7.5.1 The overall characteristics of the 1983 Riwat assemblage

Of the 23 items that were found in the lower gritstone/conglomerate horizon at Riwat in 1983, some - as might be expected - were more convincing than others as examples of hominid-struck flaking. For obvious reasons, it is usually difficult to tell whether a flaked stone that is partially or largely embedded in rock was deliberately flaked or not. As it was clearly impractible to remove every single stone protruding from the conglomerate surface, only those that looked fractured or flaked were removed. In many cases, it only then became evident that there was no good reason to regard them as hominid- struck. Nevertheless, they have been recorded so that all the evidence can be considered. Out of the 1983 assemblage, three or four pieces were regarded as convincing examples of hominid-flaking, in so far as they exhibited most, if not all, of the following features: multiple flaking from several directions, a small amount of cortex, clear flake scars, and clear bulbs of percussion. The number of clasts examined to obtain these pieces is broadly comparable to the figures given below for the 1988 survey, as they covered much the same area.

An attempt to grade the Riwat assemblage in more detail was made in 1987, after Rendell had dated the horizon from which they came to 2.0 Ma. By then, the Mission had completed two seasons of fieldwork in the Pabbi Hills, looking for fossil vertebrate remains and stone tools amongst Upper Siwalik strata that were between 0.9 and 2.0 Ma old. One of the authors (LH) of this section had examined the Pabbi Hills material, and recorded for each the following criteria: length, breadth, thickness; scar surfaces with and without clear flake features; number of directions in which flakes were removed; and as a percentage of the total surface area, the amount of surface covered by cortex, positive and negative scars; evidence of retouch; edge roundedness and any obviously post-depositional damage. Given the problems of distinguishing geological- and hominid-flaking of coarse- grained material such as quartzite, each piece was ranked from 0 - 5. A ranking of zero implies that the piece is natural; one of 5, that it is a text-book example of a humanly-struck piece. The critical divisions are between gradings 2 and 3: between "probably natural" with few scar surfaces and little evidence of flake features and "hominid- or naturally-flaked" with a shape suggestive of an artefact and with little cortex remaining but without clear flake features. A score of 4 was reserved for those pieces showing good flake features, with scars indicating removal from several different directions and little cortex remaining. Given the controversies inevitably arising over very early human assemblages - especially outside East Africa - we feel that we have been cautious over claiming that pieces are humanly-struck, and we may have under-estimated the amount of hominid- struck material. In particular we have taken most seriously only those pieces scoring 4 or 5 on our system. We may thus be excluding from our study those pieces made from the coarser quartzite varieties which by their nature cannot score this highly, irrespective of whether they were flaked naturally or not.

The results of this analysis, undertaken independently of Dennell's less detailed assessment are given in Table 7.2. This material has been published previously (Dennell 1988a). Since then a few revisions have been made. These are detailed here together with a very brief account of the reasoning where this is applicable:

R001: originally graded at 4/5; now graded at 5. The lower grade was originally given because of the difficulty of perceiving ripple marks and flake features. These showed up much more clearly on the cast once the piece was stained. **R002**: Still graded at 3. It could be a core but water-rolling and possible edge damge make it difficult to tell. **R003:**

Table 7.2: Details of the 1983 stone assemblage from the lower conglomerate, Riwat

No.	length	breadth	width	(1)	(2)	(3)	Surface			Rank
	(mm)	(mm)	(mm)				+	-	C	
RO01	163	118	74	3	4	5	-	65	35	**5**
R002	170	125	78	2	3	4	-	40	60	3
R003	130	60	60	0	4	2	-	60	40	2/3
R004	100	90	60	0	5	3	-	60	40	3
R005	100	90	60	0	1	1	-	30	70	2
R007	85	64	50	2	?	3	-	80	20	3
R008	96	62	28	5	3	3	40	45	15	**3**
R009	52	50	43	0	1	1		(15)	85	1
R010	52	88	27	1	3	2	40	25	35	**3**
R011	77	160	30	2	1	2	-	55	45	4
R012	110	85	55	0	3	2	-	45	55	2
R013	118	71	56	1	4	4		(100)	0	**3**
R014	132	79	56	4	4	3	40	30	30	5
R015	131	79	47	0	5	1		(85)	15	3
R017	127	135	64	1	1	1	-	35	65	3
R018	121	105	87	6	2	2	-	40	60	3
R019	68	49	36	0	6	3		(75)	25	2
R020	71	36	24	0	3	3		(65)	35	2
R021	51	31	14	3	3	4	45	55	0	3
R022	52	45	20	1	2	3	45	15	40	**2**
R023	84	59	23	0	1	1		(45)	55	1

Notes:(1) No.of flake scars with clearly defined features(e.g. bulbs of percussion); (2) No.of other flake scars; (3) No.of directions in which fracturing/flaking occurred. Surface : +/- = % of surface consisting of positive and negative flake scars respectively; bracketed figures (e.g.85),% of surface covered by positive and/or negative scars; C; % of surface retaining cortex.

Rank: 5:considered wholly convincing as hominid-struck; l: considered convincing as geologically-produced; other values intermediate. R023 is listed as representative of a piece almost certainly natural in origin: only one flake scar is evident, and most of the cortex still remains.

Please note: this table was initially published in Dennell et al, 1988a and 1988b. As indicated in the text, the ranking of pieces R001, R008, R010, R013 and R022 has been revised. These changes are indicated by bold figures in the final column.

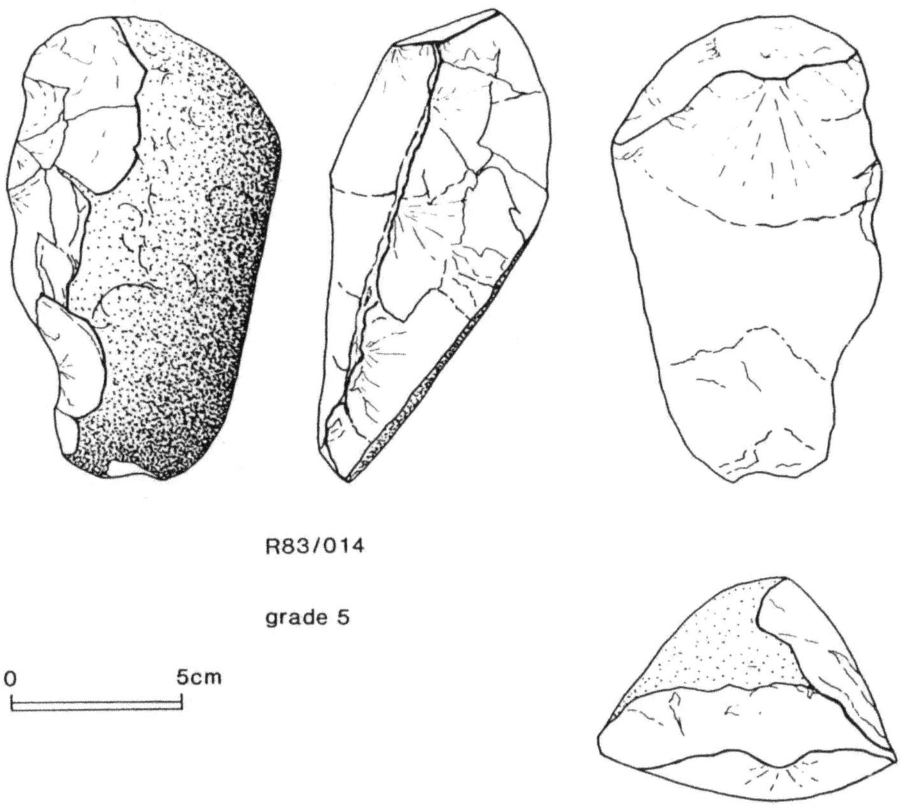

Figure 7.10: Line drawings of R014

still graded at 2/3. No clear flake features but the piece is very rolled. **R004**: still graded at 3. This piece is possibly a core, but there are no clear flake features. **R005** is still graded at 2. It is a very rolled and battered pebble with some fractured surfaces. **R006** was discarded as obviously natural. **R007**: still graded at 3. This is possibly a core but two suspiciously fresh flake scars suggest recent damage. **R008** is now graded at 3 instead of 4/5. It could be retouched or edge damaged and may be a flake but most of the flake features occur on the smaller scars. **R009** is still graded at 1 and is simply a very rolled cobble fragment. **R010** is now graded at 3 but was initially graded at 5. It is a split flake with only one clear set of flake features. Revision downwards reflects the need for caution just in case these features are recent. **R011** is still graded at 4; it is part of a cobble with a flake struck off. **R012** is still graded at 2; it is a fractured cobble, very rolled. **R013** is now graded at 3 instead of 4. There is no cortex; the only scar to show flake features was the largest, and all the other surfaces were rounded. **R014** is still graded at 5. There are clear flake features especially on ventral side (see Figs. 7.10, 7.11 and 7.12). **R015** is still graded at 3. Only 15% of the surface is cortex but no flake features are evident. **R016**, graded at 0, was a very rough cobble, and is included simply as an example of a naturally flaked piece. **R017**, still graded at 3, has 65% cortex and little evidence of flake features. **R018** is still graded at 3. The fractures that show clear flake features look fresher than the other scars, none of which have flake features. **R019** is still graded at 2. It is a very chunky fractured piece with no clear flake features. **R020** is still graded at 2. It is a fractured piece with no clear flake features. **R021** is also still graded at 3. The two scars with flake features could be recent damage. Other scars lack clear features. **R022** is now graded at 2 instead of 3. It is a fragment of an irregular cobble and is probably natural. **R023**, still graded at 1, is included as an example of natural flaking.

Figure 7.11: Photograph: R001, R014 and R88/1

Figure 7.12: Photograph: R001, R014 and R88/1

The overall result of this re-grading is to emphasise two pieces of the 1983 assemblage (R001 and R014) as being incontrovertible examples of hominid flaking, and the rigorous suppression of any pieces about which we have the slightest doubt. The regrading downwards of R008 and R010 is the result of caution being applied to those pieces with flake features that occur on noticably fresher surfaces. We are aware that some bona fide material - notably R010 - may have suffered because of this decision. Examples of the quality of material that we have relegated to our grade 3 are shown in Fig. 7.13, taken from the 1988 survey of the lower conglomerate horizon. The two pieces scoring 5 are discussed in more detail below.

R001

This is a large green/brown quartzite cobble (168 x 118 x 74 mm) with the remnants of eight or nine substantial flake scars (at least three of which show clear flake features) resulting from the flaking of this piece in five different directions. Only ca. 35scar ridges are slightly rounded. The drawing (Fig. 7.8) and photograph (Fig. 7.7) show that at least five scar surfaces were hidden by the matrix when the piece was first seen embedded in the gritstone. The edge that was exposed at the time of discovery has acted as a striking platform for the removal of small flakes (usually less than 10 mm) in length from the face below. The flake scars are clearly fresh and are indicated in the drawing.

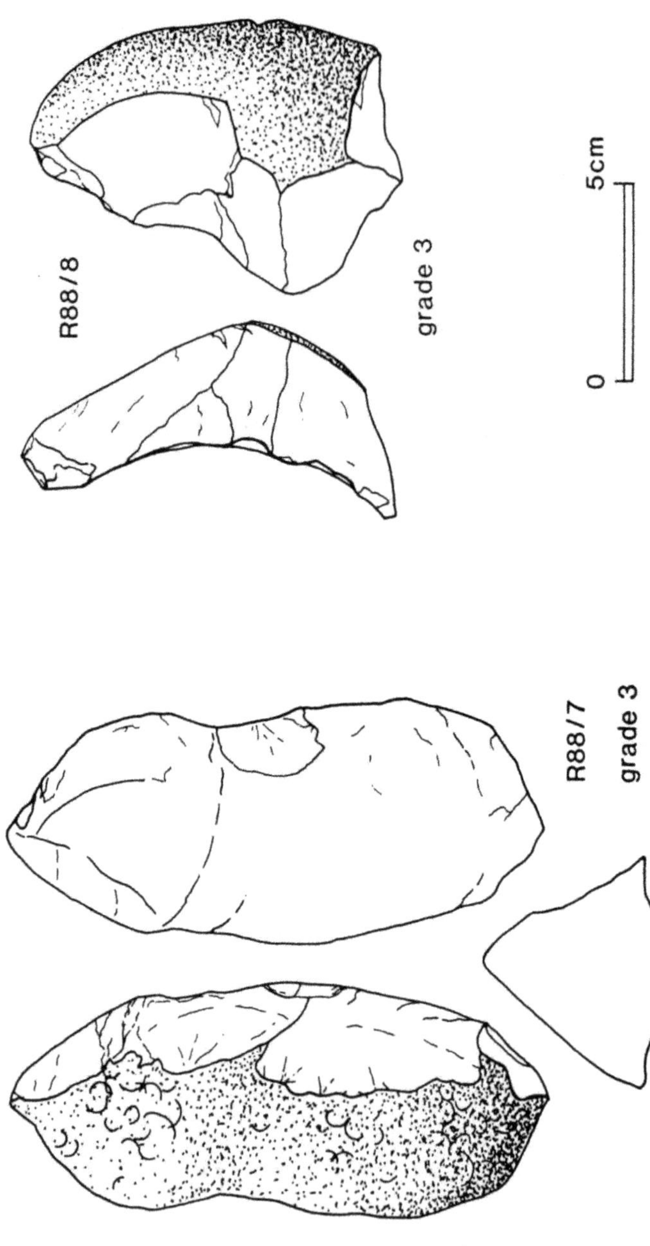

Figure 7.13: Line drawings of pieces graded at 3

Figure 7.14: Photograph of R88/1 *in situ*

R014 (Figs. 7.10, 7.11 and 7.12)
This is a large grey-green quartzite flake (132 x 79 x 58 mm) with a prominent bulb of percussion and associated ripple marks. It was struck from a core, and transversely flaked on one side, creating an edge that is straight in side view. A total of eight scar surfaces can be seen resulting from flaking in three directions. The edges and scar surfaces are all rounded.

7.5.2 The results of the 1988 survey, and the discovery of piece R88/1

At the end of the spring 1987 field season, Rendell demonstrated that the lower conglomerate horizon at Riwat could be traced over several kilometres; in one direction, back to the G.T. road before it outcropped, and in the other, across and down the Kas Dovac. The whole of this horizon was surveyed again in 1988. With one exception, no totally convincing (i.e. grade 5) examples of hominid-struck stone were found in contexts that were felt to be totally secure. The exception was the discovery of R88/1, near where piece R001 was found in 1983.

As noted earlier, the right bank of the gully downstream from the water fall (where R001 was found) grades into a vertical but fairly poorly-cemented or case-hardened section of gravels/clasts up to 2m thick, overlain by ca. 10m of sands and gravels. The face of this section is almost vertical, and is being eroded more rapidly than the better-cemented portions of this horizon by the detachment of individual stones, and of the collapse of large parts of the section triggered by root action, from trees and bushes, and water

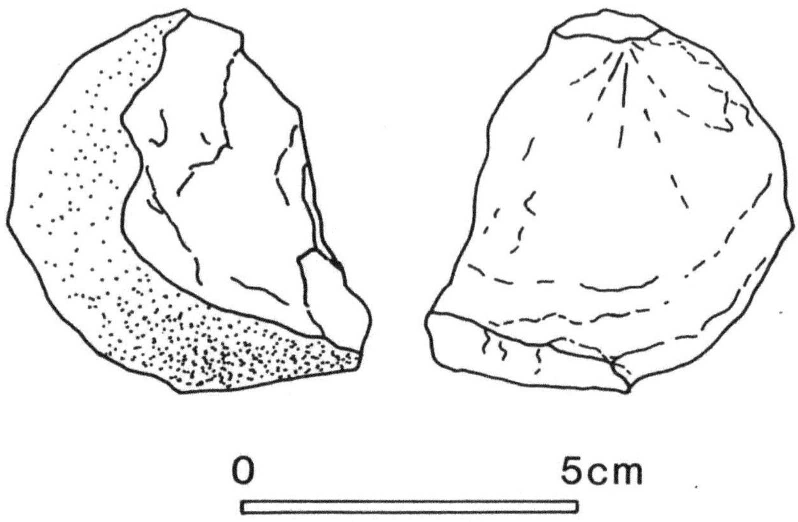

Figure 7.15: Line drawings of R88/1

penetration. The base of this section stands ca. 3 - 4m above the stream bed, and is thus well above its surface, even when in full flood. The section had been examined in 1983 and 1985 without the discovery of any artefacts. In 1988, the survey of the lower conglomerate horizon began here. Some 40m from where R001 was found, a small flake was seen on the face of the section, and at the top of a layer of gravels and cobbles (Fig. 7.14). On inspection, it was seen to be a fresh fawn quartzite flake (59 x 45 x 20mm), with a pronounced bulb of percussion on the ventral face, and with roughly half the dorsal face covered by cortex. However, there is no cortex on the platform. The piece shows evidence of flaking from three directions. The platform and ventral face were still within the matrix of the section and only the dorsal face scar attracted attention to this piece. To demonstrate that it came from the same layer as piece R001 from 1983, the whole 40m section between R001 and R88/1 was drawn at a scale of 1:20, and every stone over 2 cm in size was drawn, and details of each noted. This section is shown in Figs. 7.16, 7.17, 7.18 and 7.19. The piece itself is shown in Figs. 7.15, 7.11 and 7.12.

Seven other pieces were also found during the 1988 survey in that part of the lower conglomerate horizon between the R001 find-spot, and where it outcrops to form a scarp near the G.T. road. These pieces are detailed in Table 7.3, and several are shown in Figs. 7.20 and 7.21. Two pieces, R88/5 and R88/6, scored 3 and 4/5 respectively, and were initially recorded until the reservations over their context expressed by the survey team were confirmed by Rendell, and at this stage they are discounted from the horizon's assemblage. Although both pieces were both firmly embedded in matrix, the clasts were poorly sorted and jumbled indiscriminately, in a way that is typical of the post-Siwalik colluvial fill in the Riwat area. The pieces are therefore currently considered to be in

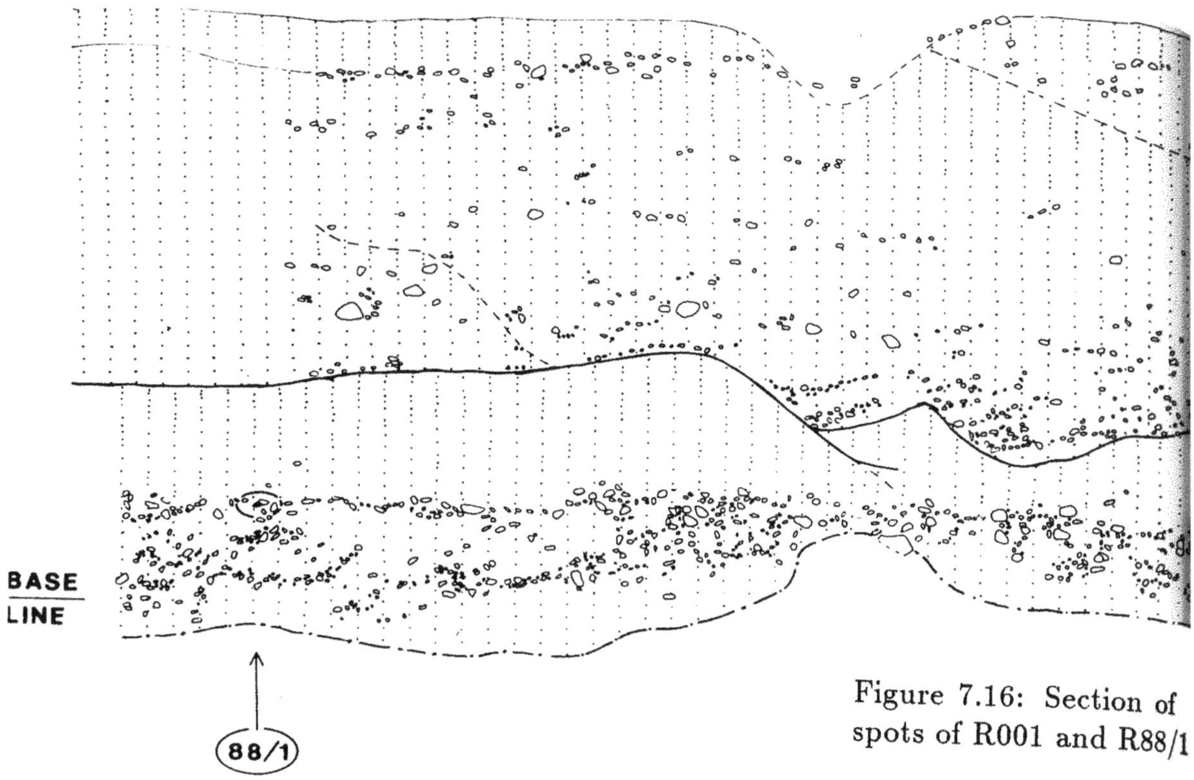

Figure 7.16: Section of spots of R001 and R88/1

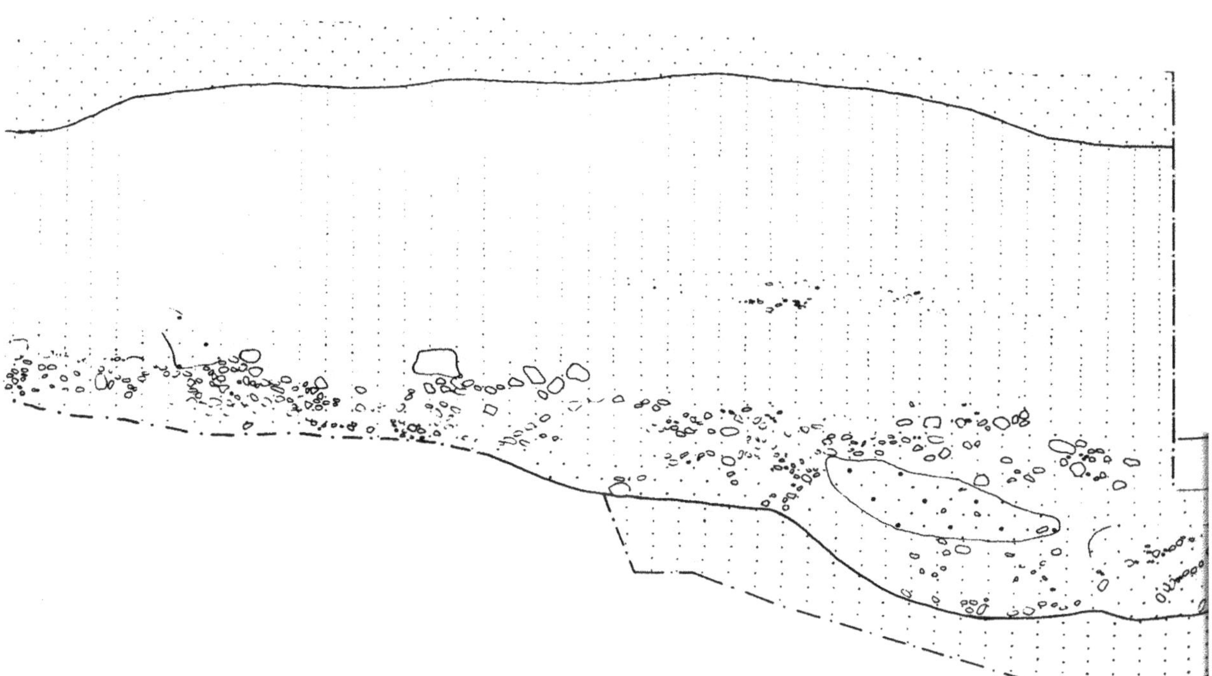

Eastern part of section

Please note that pg 117 and 118 are available to view as a double-page spread from www.barpublishing.com/additional-downloads.html. Please refer to the figure via the download for the original and complete figure as the figure shown on pages 117 and 118 might be cut off or the figure might not be complete in the printed book.

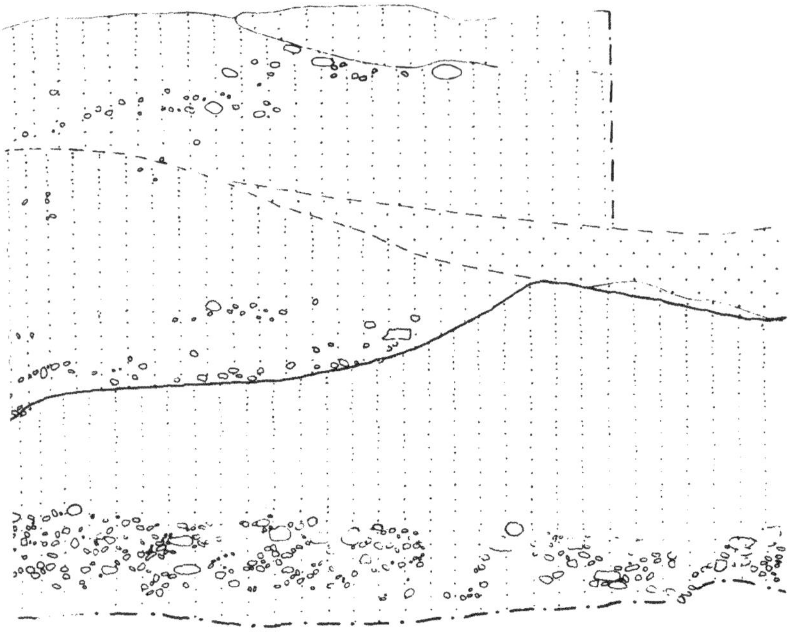

Please note that pg 117 and 118 are available to view as a double-page spread from www.barpublishing.com/ additional-downloads.html. Please refer to the figure via the download for the original and complete figure as the figure shown on pages 117 and 118 might be cut off or the figure might not be complete in the printed book.

the l●wer gritstone / conglomerate horizon showing find-
Western part of section

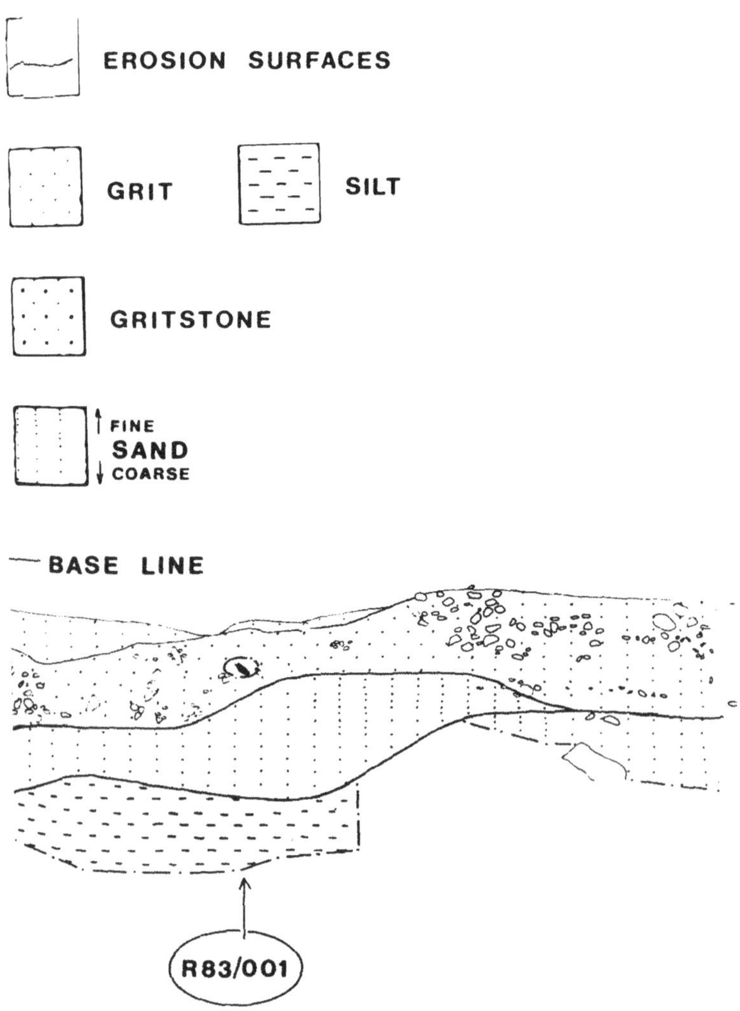

Please note that pg 117 and 118 are available to view as a double-page spread from www.barpublishing.com/additional-downloads.html

Figure 7.18: Photograph of lower gritstone/conglomerate horizon showing the find-spot of R88/1

Figure 7.19: Photograph of lower gritstone/conglomerate horizon showing find-spot of R001

Table 7.3: Details of pieces found in 1988

No.	length (mm)	breadth (mm)	width (mm)	(1)	(2)	(3)	Surface +	Surface -	Surface C	Rank
R88/1	60	50	20	1	2	3	35	25	35	5
R88/2	90	60	40	2	4	6	0	80	20	3
R88/3	40	30	50	1	2	2	45	30	25	3
R88/4	80	80	40	1	5	4	40	55	5	4
R88/5	110	100	70	3	4	0	0	50	50	4/5
R88/6	110	100	6	1	2	2	50	30	20	2/3
R88/7	140	60	50	3	2	2	50	20	30	3
R88/8	90	7-	40	0	6	4	50	25	25	3

(1) Flake scars with clear features; (2) other flake scars; (3) directions of flaking/fracturing. Surface: =/-, percentage of surface consisting of positive and negative scars respectively; C, percentage of surface retaining cortex. Ranking category: 5, totally convincing as hominid-struck; 1, convincing as geologically produced; other values intermediate.

colluvial reworked deposits: further work here is planned for 1990.

Piece R88/5 was in a loose section; and R88/6 was firmly embedded in the conglomerate horizon, but at a point where it grades into a later colluvial deposit.

The brief details of the remaining five pieces are given here which supplements the information given in Table 7.3. R88/2 is graded at 3; the flake features are on fresher scars seemingly more recent, and the older scars show no such features. R88/3 is also graded at 3, and is simply a split flake with evidence of only one direction of flaking. R88/4 is ranked at 4; it is highly suggestive since little cortex is left and the ventral surface is mostly taken up with one large flake scar with clear features. The dorsal face has other scars. R88/3 is graded at 3: the flake features on the large ventral scar are not clear. R88/3 is ranked at 3; it has very coarse fracturing properties, so there are no clear flake features but it resembles a "chunky" flake from a large core.

Two points can be made about the results of surveying the lower conglomerate horizon for artefacts in 1988. The first is that the proportion of pieces thought worth recording in the field is extremely low compared with the total number of clasts examined. In the section (Figs. 7.16 and 7.17) between R001 and R88/1, for example, these two pieces were the only ones with clear flake features out of the 1264 that were drawn in that layer.Only one was fractured but it showed no flake features or any other indications of hominid modification.

There is no reason to see the data from the drawn section as different from the rest of the exposure, so we can estimate that given 30 clasts per metre, over the 4 km of the surveyed exposure, some 120,000 clasts were examined. As each fractured piece was photographed in situ before removal the number of films used suggest that no more than 72 were fractured, and as our data show, only 8 of these scored 3 and over and were collected. Thus, we can estimate that 0.02% of clasts were fractured and only 0.006% of clasts exhibited features that were possibly the result of human activity. Given these kinds of estimates, the recovery of two grade 5 pieces within 40 m of one another is all the more remarkable.

To summarise, it has been extremely difficult to find pieces of flaked quartzite in this horizon that are even remotely like hominid-flaked ones. However, we are still left with

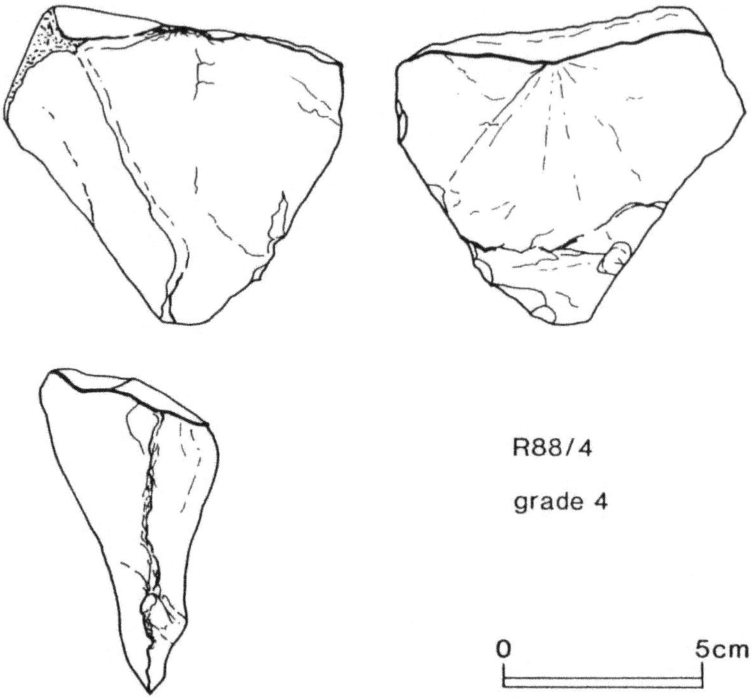

Figure 7.20: Line drawings of 1988 grade 4 piece

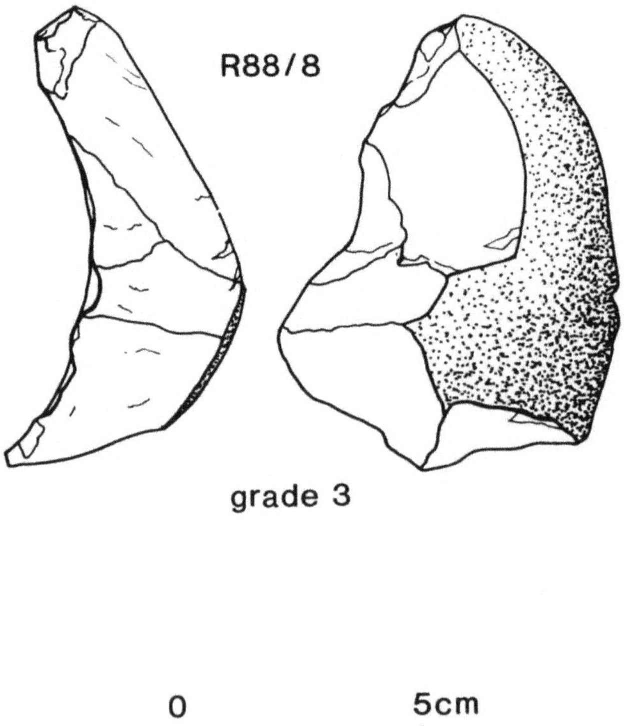

Figure 7.21: Line drawings of 1988 grade 3 piece

three pieces (R001, R014 and R88/1) that we feel cannot be explained away in terms of their flaking characteristics and context. Moroever, all three were found within a short distance of each other.

7.5.3 Replication of the socket of R001

In 1987, we made a cast of the artefact R001 in order to be able to show a replica of it to critics and those interested. (It is forbidden to export antiquities from Pakistan, and the original is now with the head office of the Archaeology Dept. in Karachi). Whilst most of those who saw the replica were prepared to accept the original as an artefact, many raised questions over its context. An attempt to allay doubts over this was made in 1988 by taking a cast of the socket left by the removal of the artefact. Five years after the piece was removed, the socket was still visible (Fig 7.22), and the piece could be fitted back into it without difficulty. Impressions of the flake scars could still be seen. The cast of the socket was undertaken by Dr. Hurcombe as follows: the socket and surrounding area were coated with ca. 20 coats of latex. It was then covered with tissue paper, and layers of wet Gypsona (a 1" and 2" gauze bandage impregnated with Plaster of Paris) were laid over it. This formed a firm support for the latex peel when it was lifted, and when the positive was made up in fibre-glass resin at the Creswell Crags Visitor Centre, in the U.K. Although awkward to carry around, this cast has since proved a useful visual aid when arguing the case for its context.

7.5.4 The sampling of the gritstone around the socket of R001

One obvious point that needed clarification was whether the gritstone around piece R001 had been reworked after exposure, or has later deposits incorporated into it. In 1988, Rendell sampled the gritstone around the socket and well away from it to see if there was any evidence that either had occurred. Her sampling showed that the gritstone/conglomerate horizon is strongly cemented by calcite, thereby forming a far more erosion-resistant horizon than the overlying and underlying sands and silts. The degree of cementation is characterstic of coarser Upper Siwalik horizons. In contrast, the post-Siwalik colluvial sequences show no strong cementation. Hence there is every reason to suppose that the matrix around R001 is integral to that horizon.

These results need to be considered in relation to the local history of this section. As noted already, the lower conglomerate horizon is overlain by ca. 70 m of Siwalik deposits, and also by a series of post-Siwalik deposits including the upper gritstone horizon and a loess sequence. The phase of incision that has formed the present gully has cut through these post-Siwalik deposits, and then down through the Siwalik strata, including the artefact-bearing gritstone/conglomerate. These important spatial relationships are clearly shown in Fig. 7.2.The gully incision is very recent indeed: and on the basis of the TL dating programme of this project, incision must have taken place after the most recent loess was deposited, ie. since 18,000 B.P.

The final point for consideration was the possibility of much earlier exposure and reworking of the gritstone around R001. We consider that this too is unlikely. As noted in chapter 5, there is an erosional unconformity at the base of the lower conglomerate. Above it, however, there is no unconformity until the contact with the unconformable

Figure 7.22: Photograph of R001 socket

post-Siwalik sequence. This major unconformity lies at least 25m above the stratigraphic position of the artefact (see Fig. 7.2).

The incision that has taken place in the last 18 ka is thus the first to have cut into the Upper Siwalik deposits since the uppermost part of these were eroded before the deposition of the post- Siwalik sequence. In conclusion, we see no reason to suppose that there has been any re-working of the lower conglomerate horizon around the area of R001 and R88/1.

7.6 Discussion

Since first publishing our results (Rendell *et al.*, 1987; Dennell *et al.*, 1988a and b), there has been interest in our claim of 2 Ma old artefacts from northern Pakistan, as well as some understandable scepticism. For our part, none of the archaeologists who have seen the replicas doubt that the originals were deliberately struck; conversely, no geologist has yet challenged Rendell or Hailwood on the geological and palaeomagnetic aspects of their work. There has, however, been some criticisms that seem to us to have been wide of the mark, or irrelevant. Before addressing these, a few points should be made.

The first is that the lower gritstone/conglomerate horizon at Riwat has pre-occupied us longer than any other area of northern Pakistan in which we have worked. When we first found piece R001 embedded in gritstone in 1983, the date of that horizon was unknown. However, we persisted with working here for two reasons: i) at least two of the pieces (R001 and R014) that we found that year were so unlike the thousands upon thousands of geologically-flaked pieces that we had previously observed in conglomerate exposures and ii) the context - particularly of R001 - was as good as any we had already seen in 1981 in the upper conglomerate horizon at Riwat, and in 1983 at Dina and Jalalpur. For that matter, it is still one of the best examples we have seen of an artefact in a secure context. Had our confidence in either the context or in the likelihood that we were dealing with hominid and not geological flaking been in doubt, we would not have spent such a disproportionate amount of time and resources here. The eventual dating - laboriously obtained from fieldwork and laboratory analyses in 1985 and 1986 - was considerably older than we had anticipated. Nevertheless, we feel that the reversed polarity of the overlying section, and the rotational data presented in chapter 5, are powerful arguments in favour of the artefact being considerably older than 0.7 Ma.

In our opinion, our conclusions over the dating are at their most vulnerable when it comes to integrating our observations into previous research into the evolution of the Soan syncline. As narrated in chapter 5, the folding of this syncline has been dated to ca. 1.9 Ma at the earliest through the efforts of other geologists, notably G.D. Johnson and Raynolds. Their work, and that of other geologists in the Dartmouth College/University of Peshawar group, is that of highly competent, professional geologists. We see no reason at all to doubt their conclusion that the folding of the Soan syncline dates to the late Pliocene. We are naturally aware that Plio-Pleistocene sections can and often have been re-dated. East Africa has provided several notorious examples in recent years, at Hadar, Olduvai, Olorgesaillie, Shungura, and particularly Koobi Fora. In the case of Riwat, however, we would be surprised if the dating is substantially revised. Nevertheless, we hope to re-check this dating in future fieldwork, preferably in collaboration with other, independent geologists.

As mentioned already, some criticism has appeared to us as ill- founded. One is that the identification of flaked stone as artefacts cannot be sustained without associated fossil evidence, particularly of hominids, or of animals they may have eaten (see Delson, 1989; Hemingway, 1989). It is worth mentioning that the acceptance of the Acheulean in Europe after the 1850's, and of the Oldowan prior to Leakey's discovery of *Zinjanthropus* in 1959 was not contingent upon the discovery of fossil hominids or butchered bone. On a more particular note, the Upper Siwalik deposits of the Soan Valley are notoriously poor in fossil vertebrate remains, and it was largely this absence of fossils that made the Soan sequence so difficult to date prior to the application of palaeomagnetic and radiometric techniques. It was also a reason why we extended our investigations in 1985 to the Jhelum basin, and the Pabbi Hills in particular (Jenkinson *et al.*,in press), where fossil preservation is much better. The lack of hominid remains and animal remains at Riwat, or elsewhere in the Soan Valley does not surprise us in the least. It would indeed have been incredible if hominid remains, butchered bone and artefacts had all been found together in one short section of conglomerate, in an area long recognised as largely devoid of fossils! In any case, it is highly dangerous to identify fractured stones as artefacts simply by their association with allegedly butchered bone. Claims for early hominid butchery have often been based on the association of animal bones with artefacts. However, we know of no instance where flaked stones are accepted as artefacts solely from their association with animal bones. The assumption that animal bones were causally associated with early stone tools resulted in claims that early hominids hunted and butchered animals at sites such as Torralba in Spain, Zhoukoudian in China, or Olduvai and Olorgesaillie in East Africa. As the result of recent re-appraisals (e.g. Klein, 1987; Binford, 1977, 1981), much of the evidence for early hominid hunting and butchery has evaporated; the stone tools remain, accepted as such by their appearance, and not through their association with animal bones. Finally, it is worth noting that vertebrate fossil remains would not have been of major value in dating the lower conglomerate horizon. At present, the biostratigraphy of the Upper Siwaliks is simply not known in sufficient detail for it to provide a check on palaeomagnetic or radiometric age determinations, as happened over the dating of the KBS tuff at Koobi Fora (see Lewin (1987) for discussion). At present and for the foreseeable future, it is radiometric and palaeomagnetic dates that will clarify Upper Siwalik faunal assemblages, rather than *vice versa*.

The main problem over the dating of the Riwat pieces lies less in the field than in their overall context within prevailing theories of human origins. Over the last 30 years, almost all discoveries of early tool-making and early hominids have been made in East Africa. Virtually no major find of lower Pleistocene artefacts or hominids has been made in Eurasia since the 1930's. Consequently, it is now almost axiomatic that Africa is the "cradle of mankind"; according to this view, two million year old artefacts outside Africa are simply archaeologically unacceptable. However, those who maintain that there is no evidence for hominids outside Africa before a million years ago should reflect that there has been precious little research into hominid origins in Asia for more than a generation.

7.7 Summary

i) The upper, post-Siwalik gritstone horizon at Riwat contained two in situ pieces: a hemispherical disc core, and a rolled hand- axe. The gritstone is horizontally bedded, and is overlain by loess dated to 74 ka.bp.

ii) the lower gritstone/conglomerate horizon also contains artefacts. Of these, two (R001 and R014) are still regarded as undoubtedly hominid-struck.

iii) we are equally confident that piece R88/1, found 40m away in 1988, is also an example of hominid-flaking, and that it came from the same horizon.

iv) we are convinced that R001, R014 and R88/1 were flaked before incorporation into the gritstone.

v) we are wholly satisfied that the lower conglomerate horizon is integral to the local section, and not a later bank deposit abutting a series of older deposits.

vi) we are equally satisfied that the 70m of overlying deposits, and the 48m of deposits underlying the horizon have a reversed polarity, and must therefore be older than 0.7 Ma but less than 2.4 Ma.

vii) we are also satisfied that the palaeomgnetic sequence clearly demonstrates the rotation that occured during the folding of the Soan Syncline, and that the artefact-bearing horizon was laid down before the Soan Syncline was folded.

viii) we see no reason to doubt the conclusions arrived at independently by Raynolds and Johnson (1985) over the evolution of the Soan Syncline, and in particular, over the dating of its folding to ca. 1.9 - 2.1 Ma.

ix) we are therefore forced to conclude that at least three pieces (R001, R014, and R88/1) were deliberately flaked, and are at least 1.9 - 2.1 Ma old.

Note: Casts of pieces R001, R014 and R88/1 are available for study and can be purchased for $ 25 U.S. each from either of the British authors. It should be noted that these moulds are slightly smaller than the originals owing to shrinkage of the moulds used in the replication process. The originals are with the head office of the Department of Archaeology and Museums, Government of Pakistan, Karachi.

Chapter 8

The excavation of site 55, Riwat: background and history

"Pity him, I beg you. (The prehistorian) is a surgeon working in darkness and with blunt tools upon that most sensitive of organisms, the human brain. Happily his patient is already dead" (Wheeler, 1966: 72).

"Specialists in proto-human archaeology sometimes have the feeling of groping in a very dark, but not quite pitch-black, room in search of some unknown creature: bits and pieces of information are recovered and subjected to analysis, but a firm image of the total animal is ever elusive" (Toth and Schick, 1986: 1-2).

8.1 Introduction

Any archaeological excavator works on two levels. The first is what might be called the tactical or managerial aspect of his or her work, and concerns the day-to-day running of the excavation: ensuring that all members of the workforce are present and gainfully occupied; that the excavation is proceeding along the intended lines; that finds are properly recorded, catalogued, bagged and so forth; that wages and expenses are accounted for; that vehicles and equipment are properly maintained, that the site and work force are safeguarded against bad weather as much as possible; and all the other hundred and one things that occupy a director's time and attention. This tactical aspect of field archaeology is the one that is the most conspicuous in the eyes of the public, and is one that Wheeler, amongst others, brought to a fine art.

The second level on which an archaeologist works in the field is what might be called the **strategic**. This is essentially conceptual in nature, and concerns the way an excavator perceives his or her fieldwork. All excavators are trustees (or victims) of their generation, and will inevitably select and investigate a site in the light of what they have experienced already, either directly, or indirectly through reading and listening to others. Since we all view the world in ways that we have been taught, it is often only with hind-sight that we realise that our prior experience led us to discover no more than we expected to find. Like most sciences, palaeolithic archaeology can cite a long catalogue of this common human failing. In the early 19th century, many investigators carefully documented the association of palaeolithic stone tools with the remains of extinct animals, yet still denied the antiquity of man. This century, intentional design was unintentionally imposed upon some flaked stone assemblages in England and East Africa, with the result that *"eoliths"* and the Kafuan culture were thought to represent the earliest recognisable stone tools (for references, see chapter 2). Another example relevant to this study are Paterson's drawings of some of the *"tools"* he collected (see e.g. Terra and Paterson, 1939: Figs. 34-35): he was clearly at pains to seperate the *"tools"* from the *"waste"*, and his drawings convey a greater degree of design than was probably warranted. He may also have unintentionally imposed upon the flaked quartzite of the Soan the clear, conchoidal flake scars often seen on the flaked flint and chert that he knew from his studies in England. A more recent example may be the way that a circular configuration of stones in Bed I of Olduvai seemed to represent a hut when first discovered, but may now show little more than tree-root disturbance (see Potts, 1986). This particular example is also a symptom of the habit that Binford (1987) has recently criticised in much lower palaeolithic research, of *"searching for camps and missing the evidence"*.

Given the extent to which the perception of field archaelogists can be shown to have influenced their conclusions, it is remarkable how reticent they tend to be about the factors that influence their fieldwork, and the problems they encounter during it. In most cases, excavations are presented as though they were timeless experiences, independent of the era in which they were conducted, the person who directs them, or the circumstances under which they were dug. It is as though excavations are concerned only with the careful recovery of unambiguous *"facts"*, which are then effortlessly marshalled into an unambiguous account of what the site represented. Few accounts admit to uncertainty and bias on the part of the excavator, still less to mistakes or changes of view. Excavations are often presented less as dialogues between an excavator and his material than as a soliloquy in which all the excavator has to do is to listen to what the site is saying.

This is a preposterous fallacy, since archaeological sites are notoriously mute without their excavators. This excavator cannot be alone in having experienced attacks of doubt, uncertainty and confusion. Such feelings often have a direct result upon the reliability of the eventual conclusions, and are of more than of passing interest to the purveyors of archaeological gossip. There is, for example, little mention in von Koenigswald's academic publications of the way that fossil material was collected in Java in the 1930's, and it is only in his much later, semi-autobiographic account (Koenigswald, 1956) that one realises why there is so much confusion over the biostratigraphy of that region. Similarly, our own work elsewhere in northern Pakistan has been hampered by a lack of information as to how fossil and archaeological material was collected by earlier workers. Field archaeologists should perhaps take a lead from field anthropologists such as Birley (1982) and Turnbull (1974) in displaying more frankness as to how they perceived their own fieldwork, and why they worked the way they did.

So what of this particular excavator, out in the field in 1981 and 1982, overlooking the Soan Valley, and excavating an open-air palaeolithic site with a small team of Afghan labourers? To what extent did the fact that he was working in the early 1980's affect his judgments and decision making? And what particular factors during excavation might have enhanced or detracted from the final results?

8.2 Strategic Factors

Two issues stand out in retrospect as ones that dominated the way site 55 was investigated. These were: the problems of explaining the processes by which a palaeolithic site was formed; and how to infer the function of a palaeolithic site from its lithic assemblages. Each can be considered in turn.

8.2.1 Site formation processes

The main factor that influenced decision-making during (and after) the excavation was the central issue of how palaeolithic sites in general should be interpreted. Most important here was the debate that was then well underway over interpretations of early hominid *"home bases"* and *"kill sites"* at Olduvai and Koobi Fora in East Africa. When the excavation started in 1981, it was already evident that explanations of these would change radically with the realisation that hominid activities were only one of many agencies that could have resulted in the type and distribution of animal remains and unmodified stone found in many of these sites. It was equally evident that these reappraisals had far-reaching consequences upon the rest of the palaeolithic record. In the first place, it was obvious that palaeolithic sites could not be *"read"* as though they represented a direct record of early human activities. Because palaeolithic sites in general had been routinely examined with the assumption that flaked stone and mammalian remains were **causally** linked by activities involving the procurement of meat by hominids, many sites later than those at Olduvai would also need re-appraisal. Similarly, the realisation that the *"hut"* at Olduvai was probably an artefact of tree-root disturbance mimmicing a seemingly artificial configuration cast doubts on other palaeolithic structures. Even where there was no doubt that arrangements of unflaked stones and/or bones were intentional, it was unjustified to assume that these represented the remains of a hut, or other type of a roofed structure, as had often been the case. Gamble's (1984: 72-73) re-assessment of the *"hut"* at Molodova, western Russia, as a windbreak with hearths in its perimeter is one recent example. For these reasons, there was good reason to be wary of assuming too readily that a structure - let alone a *"hut"* - existed at Riwat, and it was difficult to use terms such as *"wall"* or *"occupational floor"* as confidently as would have been possible only a decade or so previously.

Whilst debates of this kind are in progress, excavators are inevitably placed on their guard. Decisions over what to record and what to ignore become more difficult than when there is general agreement as to what one is investigating, and constraints of time, money and labour usually preclude the recording of more than a small fraction of the contents of a site. Given the prevailing uncertainties, the only viable solution when excavating at Riwat was to observe and record both modified and unmodified stone as accurately as was feasible; to separate as much as possible observation from interpetation; and to bring a large measure of scepticism to the site each day. As will be seen, a major part of the excavation in 1982 was concerned with recording as much as possible with a photo-mosaic, and much of this report has been concerned with presenting readers with as much information as possible so that they might draw their own conclusions. If there is a lesson to be drawn from the debates over the early hominid sites at Olduvai, it is surely that none of this debate could have occurred without the high quality of the original observations. Likewise, few would now support the original interpretations of many European open-air upper palaeolithic sites, but the quality of those observations often still allows alternative models (see e.g. Gamble, 1986: 250-268).

8.2.2 Lithic assemblages and site function

The second major area of concern to this particular excavator was how the function of a palaeolithic site such as site 55 could be inferred from study of its flaked stone assemblage. Of critical importance here is our ability first, to identify those items that were used as tools, and secondly, to infer their function.

As many have pointed out, there is little reason to suppose a close relationship between the **form** and the **function** of early palaeolithic flaked stone. Whilst the form of flaked stone can provide a useful basis for describing the major components of palaeolithic stone assemblages, its size and shape is contingent upon several factors (excluding post-depositional damage). These include the form in which stone is found before use (e.g. rounded cobble or angular nodule); the fracturing properties of the stone itself; the techniques used to flake it; the extent to which the shape of a flaked stone is modified during its life-span as an artefact; and the desired shape of the end product. The last two of these features are the ones most likely to be related to the function of a stone tool, and inferences about the function of stone tools are clearly crucial to explanations of the function of a site, and the activities performed there.

Unfortunately, the function of stone tools is usually elusive. Only a small minority of the flaked stone from most palaeolithic sites can be unambiguously referred to as a *"tool"* on the grounds of its form. Conversely, a large percentage of stone flakes could have been used as tools, even though they might be unstandardised in shape and/or size, and at first sight indistinguishable from those that were not used. Consequently, situations often arise when there are no clear-cut ways of distinguishing *"tools"* from *"waste"*. Current inferences on the function of many stone artefacts rest heavily upon their associations with bone debris - as in the case of, for example, the light- and heavy-duty tools at Olduvai and other lower palaeolithic sites - and/or examination of edge modification and polish through microwear studies (e.g. Keeley, 1980). Unfortunately, neither approach was possible at Riwat, since there were no animal remains, and it was not possible to export stone material from Pakistan for microscopic examination in the U.K. Consequently, it was difficult to proceed far beyond categories of shape, and edge damage or retouch in attempting to infer the type of activities performed at the site. This means, in particular, that a basic division into *"tools"* and *"waste"* is difficult, since there is no obvious basis for deciding which were used, and which were not.

Given these limitations, a major source of information was likely to be the spatial distribution of flaked stone. One of the major attractions of site 55 was the possibility of finding relatively undisturbed material resulting from short-term usage. The best way of discovering whether or not this was the case was to look for conjoining pieces of flaked stone, on the grounds that the shorter the use of the site and the less its disturbance, the greater the chance of finding refitting cores, flakes and/or blades. One study that was especially influential to this excavator was the study of the late upper palaeolithic site of Meer in Belgium (Cahen *et al*; 1979), which showed the importance of detailed recording, and the necessity for careful searching for conjoining stone pieces.

8.3 Tactical considerations: the excavation of site 55

As narrated earlier, one of the main objectives of this research has been to find palaeolithic stone tool assemblages in a secure geological context. That task has not been as easy as it may sound. Apart from the problem of finding them in the first place, three additional critical factors in the Soan Valley have been: i) the thickness of deposits overlying potential palaeolithic localities ii) ease of access for a vehicle and workforce and iii) seclusion from unwanted curiosity. As explained earlier, the loess attains a thickness of several metres in the area under study, and this alone prohibited extensive excavation of archaeological horizons buried by more than two metres of loess. This included some potentially interesting localities where fresh flakes and blades were found on a colluvial cobble horizon eroding from under three to six metres of loess. Unfortunately, these localities were also near the G.T. road from Rawalpindi to Lahore, and excavation here would have attracted too much unwanted attention. On the other hand, these considerations did not apply to the area around locality 55: the loess cover here was thin; it could be reached in a vehicle; and was conveniently out of sight of the main road.

This locality was first discovered by Rendell in 1980, whilst making a very brief inspection of the area that we surveyed in much greater detail in 1981 and 1982. Whilst walking behind a low gritstone outcrop (Fig. 8.1), she noticed some conjoining pieces of flaked quartzite and associated débitage (see Fig. 8.2) lying on a stony surface, only partially covered by loess. That season, little more was done to follow up this discovery, but the area around the outcrop was visited frequently during the mapping work of 1981 (see chap. 6). Towards the end of this work, one of us (RD) discovered a very dense concentration of fresh quartzite flakes, blades, core fragments and small chips eroding from an area only 15 by 10 cm in the north side of the gully, immediately west of the outcrop (Fig. 8.1 and 8.3). Of the 85 pieces found, 12 conjoined, and the rest clearly came from the same nodule.

In view of two discoveries of conjoining pieces of freshly struck quartzite, excavation in this area was clearly desirable. Excavation into a narrow saddle immediately behind the 1981 find of conjoining pieces seemed unattractive, because the loess cover there was over a metre thick, and the overall area of deposit severely limited by recent gully erosion. An alternative was to excavate opposite this find, on the south side of the small gully currently parallel to and opposite the edge of the outcrop (Fig. 8.1). If this gully had cut an archaeological horizon, excavation here might have revealed the rest of it. In addition, the loess cover here was very thin (Fig. 8.4), and the original ground surface appeared to have been flat. However, because it lay at the base of a conglomerate ridge to the south, some stone material might have been introduced to this area through colluvial action.

In the end, it was decided to excavate behind the gritstone outcrop, and immediately east of where Rendell had found the conjoining pieces of flaked quartzite in 1980. This decision was heavily influenced by the observation of a line of large cobbles and smashed blocks up to 30 cm long near where Rendell had made her find, and running parallel to the present edge of the gritstone outcrop before dipping under the loess behind the outcrop itself (Fig. 8.1). Although these seemed to have been dispersed, they also seemed to form a convincing line. Some were also split and stacked on one another, and many seemed too large to have been moved by the type of streams seen in the area today.

The interpretation of this observation under-pinned much of what followed. Because this

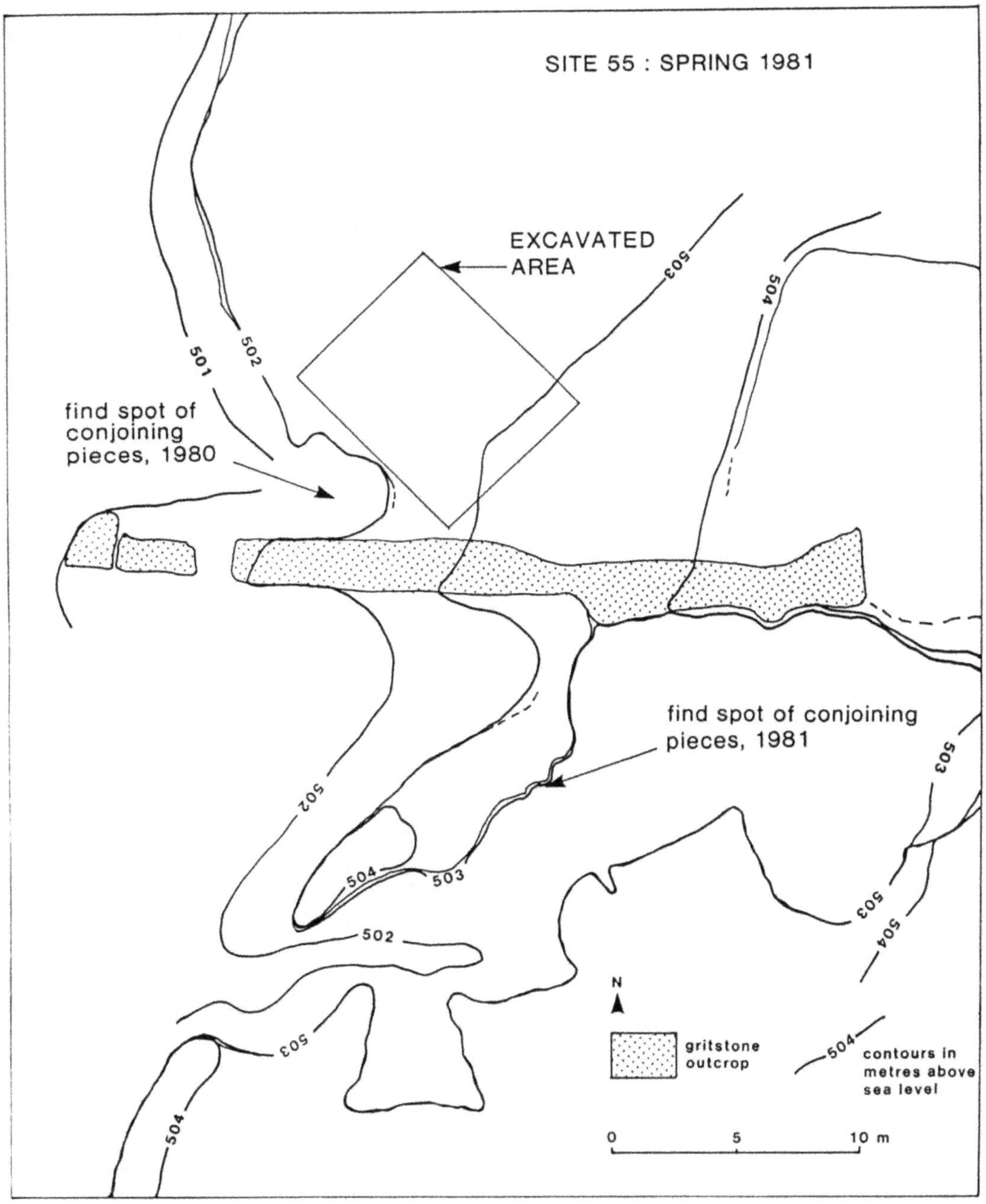

Figure 8.1: Site 55 showing area of excavation, find-spots of conjoining artefacts and the gritstone outcrop

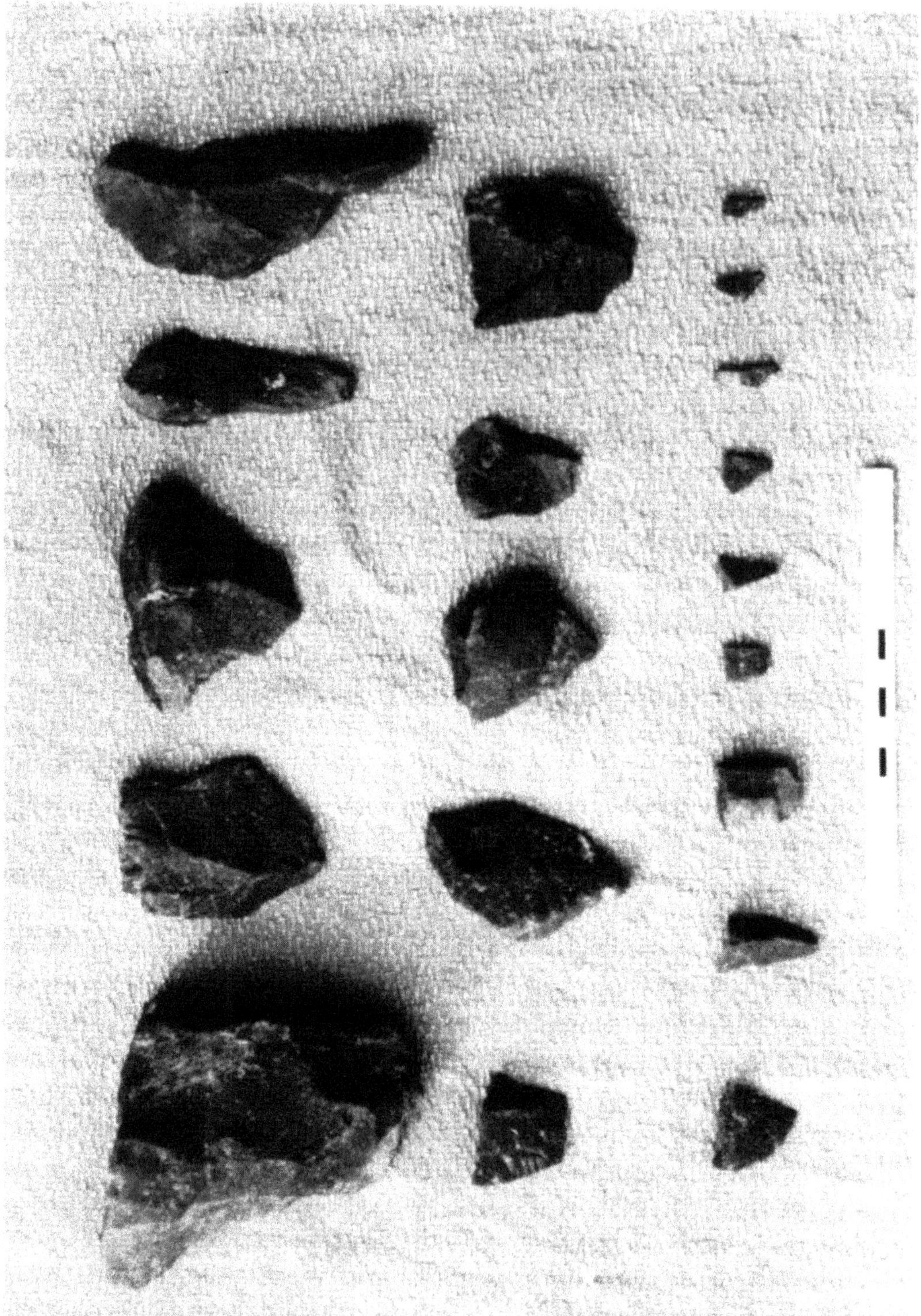

Figure 8.2: Photograph of conjoining pieces, 1980

Figure 8.3: Photograph of conjoining pieces, 1981

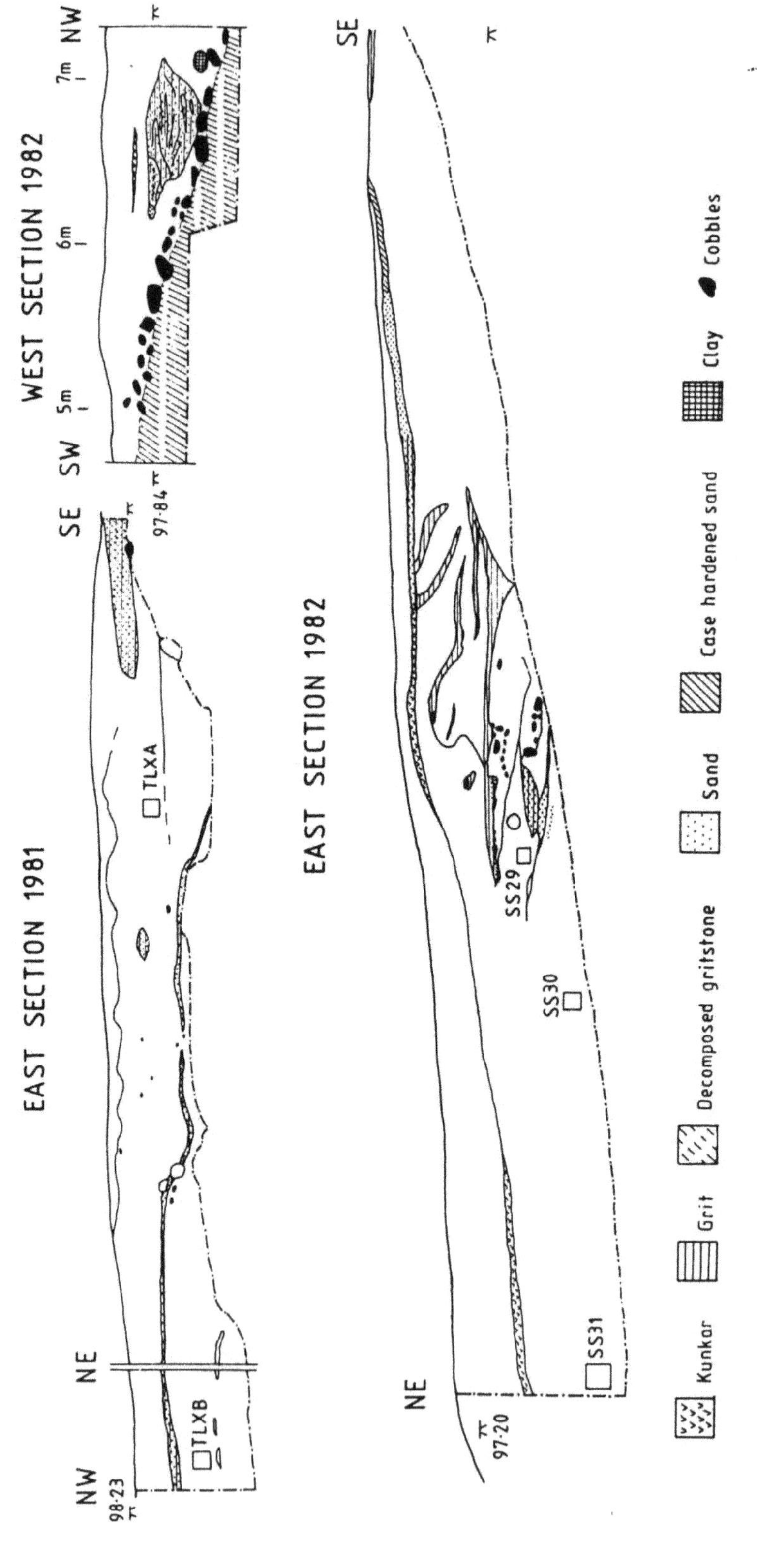

Figure 8.4: Site 55, sections through the site

line of cobbles did not seem to be *"natural"*, in the sense of, for example, a gully bed, or downslope transport of colluvial material, it was seen as artificial. Because large quartzite cobbles of the kind observed here are often used as building materials in the area today, so this feature was thought to indicate the possible remains of a stone structure. Accordingly, it seemed worthwhile to investigate the possibility that:

(i) it was indeed part of a structure

(ii) it pre-dated the loess
and

(iii) was associated with the kind of freshly-struck stone material that Rendell had found nearby.

An additional factor influencing the decision to excavate here was the gritstone outcrop itself. Although small, this is the only such outcrop in what would have been a shallow valley bounded to the north and south by low conglomerate ridges. If, as believed, much of this valley had been filled with loess (see chap.6), this outcrop would have been one of the very few local sources of naturally outcropping gritstone that could have been used for constructional purposes. In addition, it would also have served as a convenient landmark, and acted as a natural shield against any colluvial material transported downslope from the valley side to the south. Consequently, the decision was taken to excavate here.

8.3.1 The 1981 trial excavation

The area selected for trial excavation lay immediately east of the line of cobbles referred to above, and is shown in Figs. 8.1 and 8.5. For reasons of time, the trench was laid out so that its northern edge lay ca. 0.5 m north of the projected axis of the line of cobbles. In 1981, this area was explored with the aid of four Afghan labourers from Riwat, working over a ten-day period under extremely hot conditions in mid- to late-April. This excavation took place in what later became squares A-C 1-3; these areas are indicated in Fig. 8.5

This preliminary excavation revealed five main features, and are shown in Fig. 8.6. An isometric view of the site is shown in Fig. 8.7, and this may help appreciation of the spatial aspects of the site. The highest part of the site was the top of the gritstone outcrop (**feature i**) that has already been mentioned. This dipped into a narrow and level area of loess (**feature ii**) that was bounded on its northern side by a linear feature of cobbles and gritstone blocks (**feature iii**). A second narrow and level area of loess lay to its north (**feature iv**), and this was bounded by an area of densely-packed stones that extended downslope (**feature v**). Each can be taken in turn:

i) the dip-slope of the gritstone outcrop

This sloped at ca. 10-20° into the loess, and its surface comprised several contiguous slabs. In the south-west part of the excavated area, three artefacts (081-083) were found near the top of this slope.

ii) an area of loess, ca. 1 m.wide, and immediately north of feature (i)

Apart from one flake (no. 077), this area was devoid of artefacts, but contained two types of features. One was a gritstone slab, ca. 40 cm long, sloping at an angle of ca. 60-70°, protruding slightly (ca. 2-3 cm) above the loess, and abutting a roughly circular area of

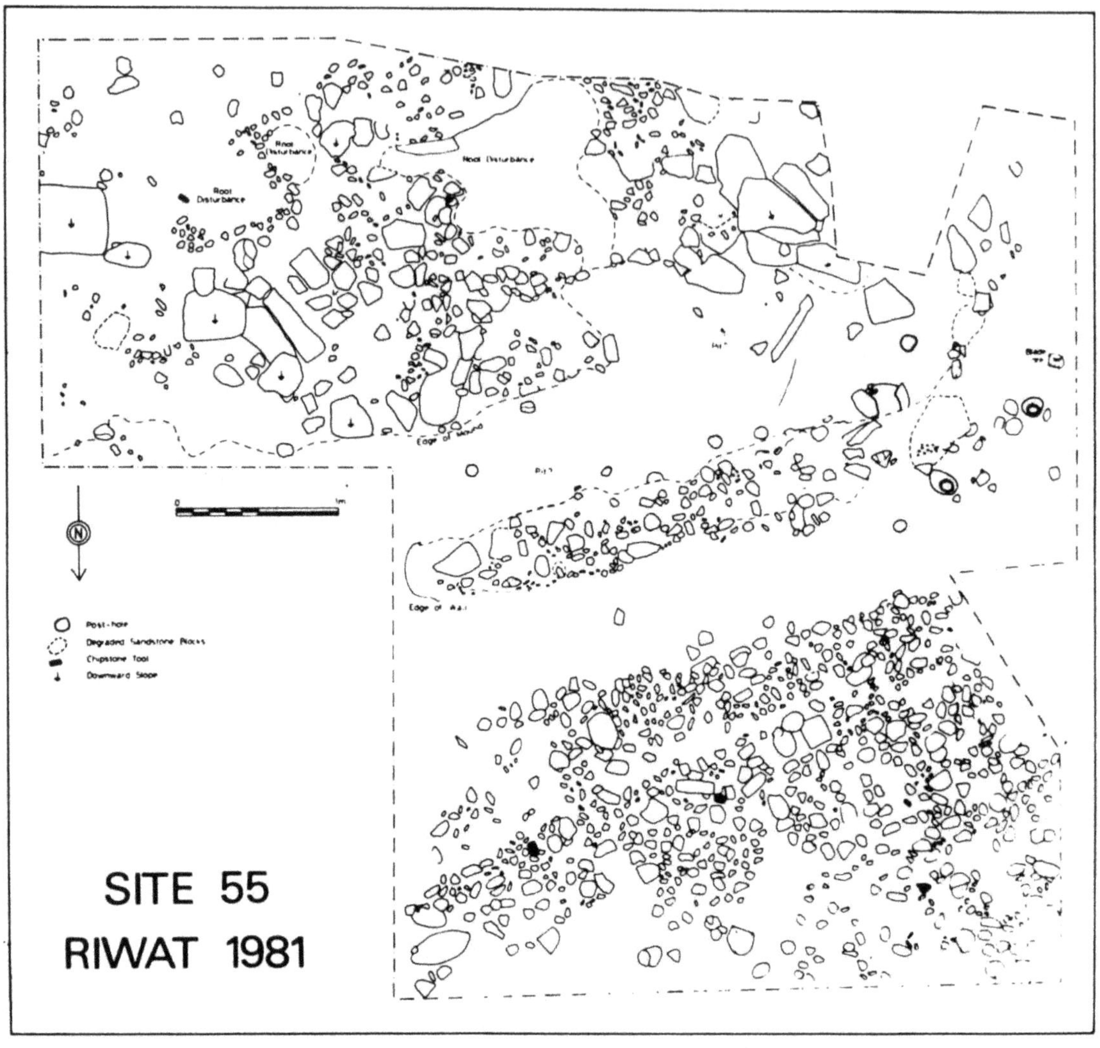

Figure 8.5: Plan of 1981 excavation

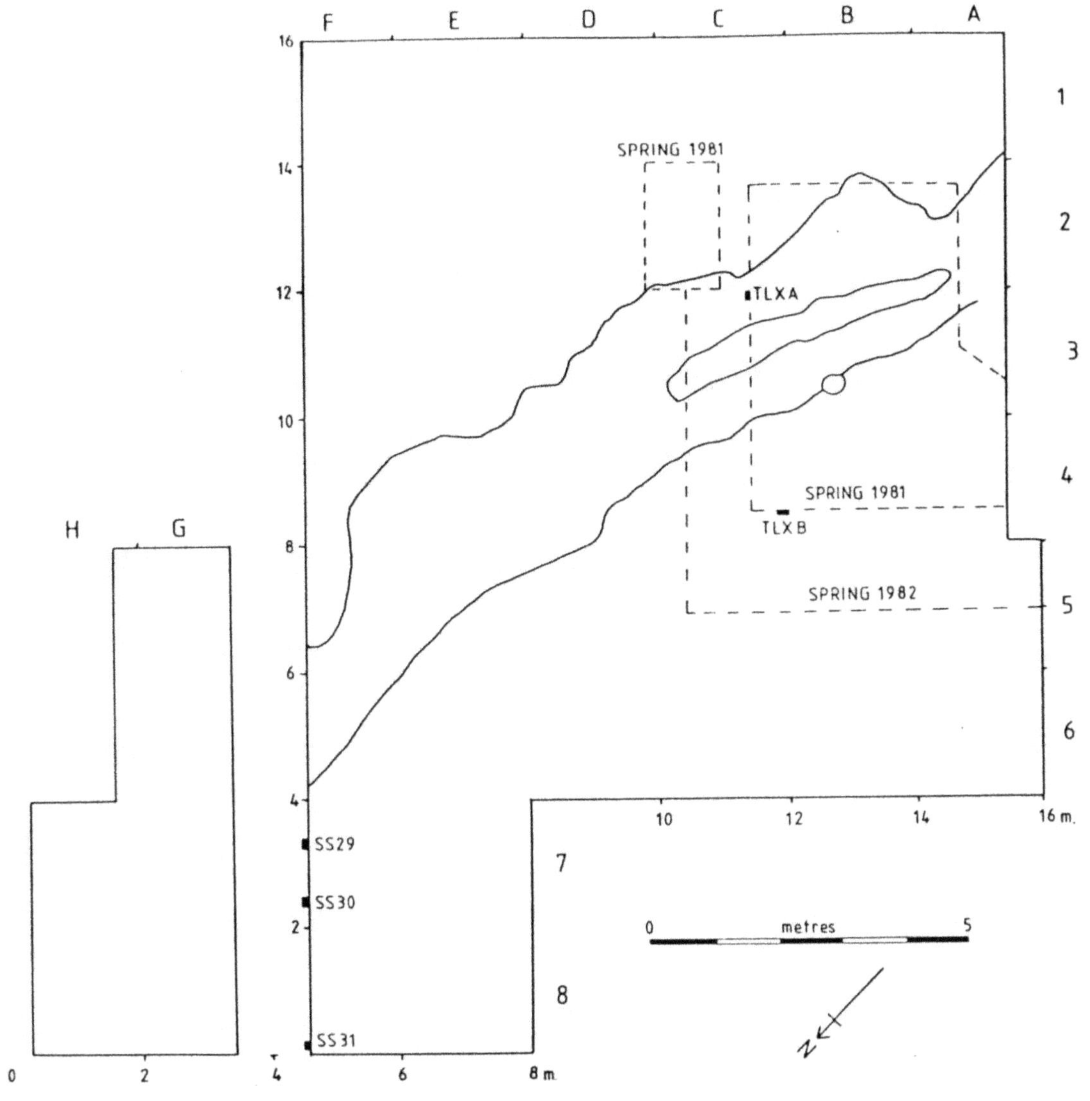

Figure 8.6: Lay-out of trenches excavated in 1981 and 1982

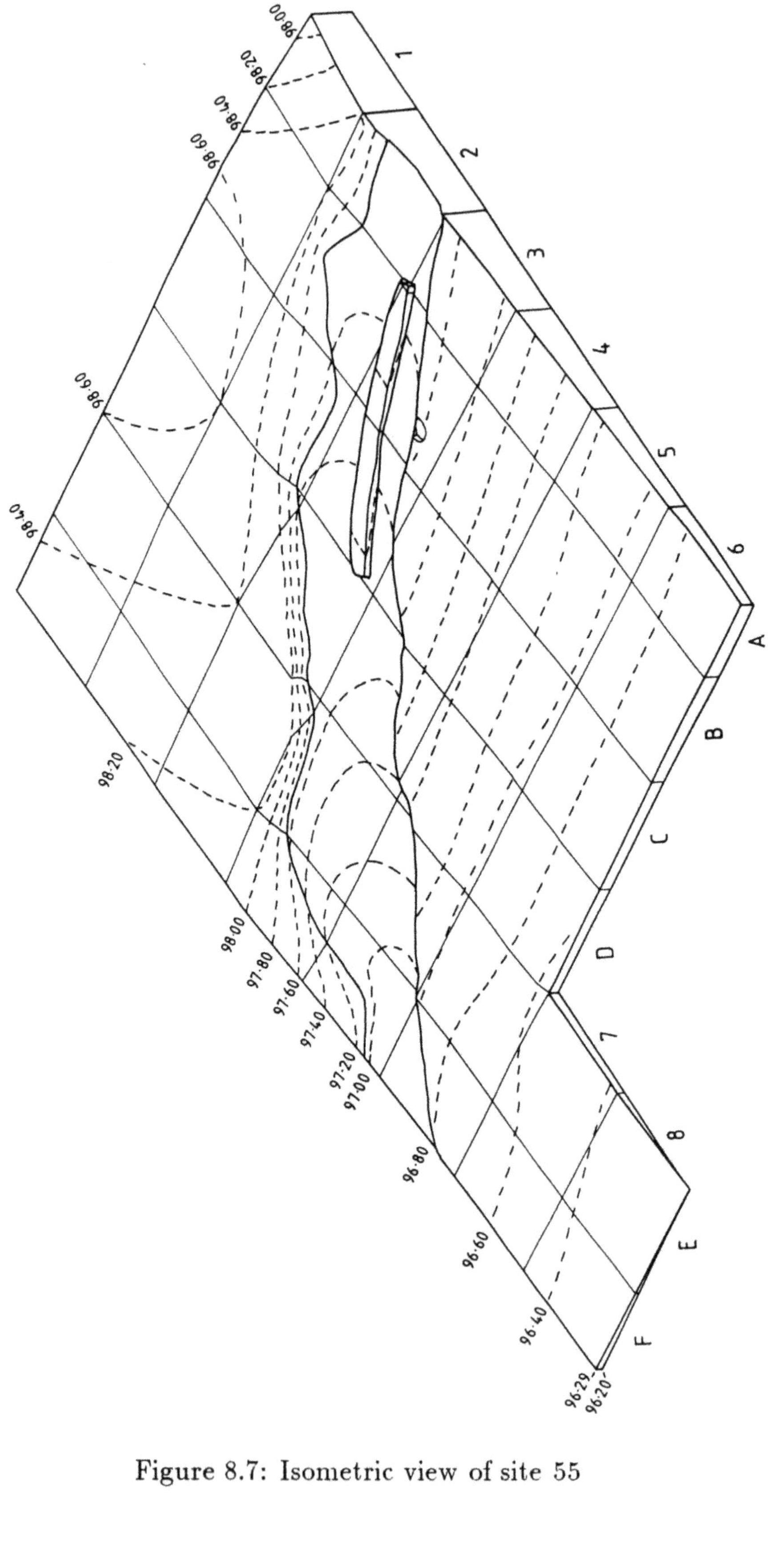

Figure 8.7: Isometric view of site 55

greenish-grey clay. This was tentatively identified as a pit. The other were six circular outlines of reddish clay, each ca. 10 cm in diameter, and reminiscent of what are often regarded as post-holes. These have not been illustrated because they do not show up clearly in black-and-white photographs.

iii) a low, linear feature composed of large cobbles and gritstone blocks

This lay parallel to and ca. 0.5 - 1 metre south of this feature and was covered by loess. These blocks were separated from each other by thin (5- 10 mm) horizontal and vertical cracks that were filled with loess. When first exposed, the edges of the gritstone could be isolated fairly easily from the surrounding loess, although the crispness of the edges deteriorated after heavy rain, and after being twice buried under back-fill at the end of the excavations in spring 1981 and 1982. Its general appearance is shown in Fig. 8.8, and a photomosaic of the eastern end of this feature can be seen in Fig. 8.9, taken towards the end of the third season of excavation in 1982. This feature widened towards its base, which rested on a surface that was gritty and sandy in places, especially on the south side near the gritstone outcrop.

There was no evidence at all that this feature had been cut into the loess, or formed as the result of clearing the area between it and feature (v) of stones and cobbles. To put this another way, the stone surface comprising feature (v) is not likely to have once extended continuously as far as this feature.

iv) an area of loess between feature (iii) and the stone surface forming feature (v)

Apart from two fresh quartzite blades (017 and 080) that were found on a slightly gritty surface coinciding with the base of feature (iv), this area did not contain any other artefacts.

v) a large, sloping stone surface

The southern edge of this was very clearly demarcated on the side nearer the grit-stone outcrop as a line of cobbles. These lay on the same axis as the line of cobbles noticed immediately west of the area chosen for excavation. The northern part of the area opened in 1981 (i.e. downslope from the gritstone outcrop) consisted of a surface which sloped downwards at ca. 15-20°, and which consisted entirely of large cobbles and smaller stones. Many of the smaller stones were split, and there were also a few fresh flakes and blades. The density of both flaked and unflaked stone appeared to be greatest upslope.

At the end of the excavation, the area excavated was covered with polythene sheeting and then back-filled.

In 1981, these features were largely explained in terms of a stone structure. Feature (i) - the rear side of the outcrop - was puzzling in 1981. It (almost) looked as though its pavement-like appearance might be artifical, and the slabs covering much of its surface may have been placed there deliberately. This possibility was tested by making a slit-section ca. 50 cm wide into the back of the outcrop with a cold chisel and hammer. This exercise clearly indicated that this was not the case, as these slabs were integral to the outcrop itself, and the product of surface and sub-surface weathering. The gaps between each block were up to 50 cm deep and filled with loess, in much the same way as mortar separates the bricks in a wall. If nothing else, this exercise usefully illustrated how the weathering of gritstone can mimic human constructions.

The slab and abutting circular area of greenish-grey clay in feature (ii) was regarded as a pit, and thus intentional, but excavation of this was deferred until more time was

Figure 8.8: Photograph of general view of "wall footing"

Figure 8.9: Composite photomosaic of part of the "wall footing"

available. The six reddish clay circles were associated with this, in the sense that they were in the same geological layer. These were viewed as being either post-holes, and thus perhaps part of a structure, or as natural features produced by e.g. rodent burrowing or termite activities. Because of lack of time, most of these were not excavated. However, two were sectioned and not unlike post-holes.

The low, parallel-sided line of cobbles and gritstone (feature iii) was regarded as artificial, and not as the base of a gully. This was because gully profiles observed by us in the area had a V-shaped rather than U-shaped profile; because the gradient was virtually level, and thus inconsistent with the size of cobbles it contained; and because the direction of flow would have been across-, rather than down-slope. As stated already, it was covered by loess, and there was no evidence that this feature had been cut into it. When first exposed as an area only 2m by 0.75m, it was thought to be a deliberate placement that may have covered some kind of pit. However, once it was realised that this feature continued right across the trench, it was interpreted as a second, and subsidary wall, or as a footing for some kind of wind-break or screen made of organic materials. The latter seemed more likely because the loess on either side was largely stone-free, which one would not expect if this feature had been a wall that had collapsed.

The line of cobbles defining the southern edge of feature (v) was viewed as the remains of a stone wall: it seemed too well-defined to be natural, and not unlike what a low wall made of large cobbles with a packing of smaller stones would resemble. This edge also seemed to be the eastward extension into the loess of the line of large cobbles that led to the decision to excavate here. The obvious problem over this interpretation was the absence of a well-defined edge on the north side of this feature. As stated in the field diary, the down-ward sloping stone surface in the north part of the trench was *"viewed as tumble - but (the) problem remains of explaining why all the collapse is to the north."*

In summary, therefore, site 55 was initially thought to contain the remains of a substantial stone wall and a smaller subsidary one. Both were thought to be associated with a stone-lined pit, some post-holes, and a fresh blade and flake assemblage. Two issues were unresolved at the end of this preliminary excavation. The first was whether the fresh quartzite flakes and blades were associated with the the structural remains in a behavioural sense. Although both the structural features and the artefacts were covered by the same loess horizon, it was not clear if they were both residues of one set of human activities. Whilst archaeological associations of this kind are difficult to demonstrate, it nevertheless seemed worthwhile to explore this possibility in a further season of excavation by examining the spatial distribution of flaked stone, particularly if pieces could be fitted together.

As already mentioned, the second problem was feature (v) - the cobbled surface - which seemed disturbingly wide to have been a palaeolithic wall, and too steep to have been a floor. In retrospect, assessments of what this excavation uncovered in 1981 were too heavily influenced by my own prior experience of excavating stone structures, and by the view that was still then widely held that palaeolithic structures were not in themselves particularly unusual, even if the function of many as *"home-bases"* seemed increasingly dubious. At the time, I did not expect the gritstone outcrop to continue downslope as a a conglomerate surface. Indeed, one powerful reason for excavating there was that this area should have been free of the colluvial conglomerates that can be traced to either side. Even less expected was a conglomerate that had a crisp, wall-like edge and was also associated with a freshly-struck archaeological assemblage. At the time, it seemed reasonable to expect that excavation would show whether or not the line of stones on the surface west of the site were the remains of a gully bed or a wall. By showing it was not the former, the case for the latter seemed reasonably strong. The naivety of this supposition was clearly exposed the following year.

8.3.2 The excavation in April, 1982

In April, 1982, excavation was resumed for four weeks, with a larger work-force than in 1981, and a chowkidar (watchman) who camped by the excavation. Once the back-fill from the previous year had been removed, the area already exposed was enlarged to extend downslope and further east (see Fig. 8.6). A subsidiary trench was also laid out on the eastern side, but work here was discontinued after heavy rain (see below).The main aim that season was to see what the cobbled stone surface (feature (v) represented if it was not the remains of a structure such as a wall. A subsidary objective was to see if feature (iii) - the *"wall-footing"* extended further east beyond the area excavated in 1981. It was also hoped that the recording of this surface could be speeded up by making a simple photomasaic.

The results of that season were very mixed. The most enduring result was the realisation that the cobbled surface - feature (v) - was essentially natural in origin, and simply the surface of a gently-tilted conglomerate sheet, extending down-slope from the gritstone outcrop and covering most of the northern part of the excavation. (A general view of this conglomerate surface, taken towards the end of the excavation, can be seen in Fig. 8.10). However, it was also apparent that this surface still retained a considerable archaeological interest. Whilst its extent, gradient, and composition indicated that it was primarily a conglomerate, most of the cobbles on its surface had been flaked, implying that it may have been used as a resource of raw materials. First, freshly struck pieces continued to be found on its surface. Secondly, the northern edge of this surface seemed to have been modified, in that it included in square D4 what appeared to be an intentional placement of a semi-circle of medium-sized cobbles against its edge (Fig. 8.11).

There was also an odd bulge in square B3, where the edge was obscured by stones slumping onto the loess that seemed out of character with the rest of the edge; this had been noticed in 1981, but it was not investigated because of problems with the weather (see below). Excavation of this bulge in the following season showed that it concealed a circular "niche" (see Fig. 8.12).

Other useful results were the discovery that feature (iii) - the stone footing - also extended a little further east beyond the area excavated in 1981, and that one feature at least on the site - the *"stone-lined pit"* was what it was initially thought to be (see Fig. 8.13. When excavated, it was found to be adjacent to a smaller one on its south-west side, containing find 0292: the green clay in the larger one was removed, exposing a pit ca. 20 cm deep, the bottom of which was lined with some cobbles and finds 0296-0298.

On the less positive side, there was considerable anxiety over the circular clay patches provisionally identified as post-holes. Four of those found the previous year turned out to be recent, and perhaps the result of infilling termite nests, rodent burrows, or root holes. Some puzzling examples of possible post-holes also appeared in in the northern-eastern part of the trench, beyond the area excavated in 1981. They seemed to lie in an axis running parallel to the edge of the gritstone, and some 4m behind. More worryingly, they first appeared (along with some stones) some 10 - 15 cms **above** the cobbled surface and the associated fresh flakes and blades. This gave rise to much concern as to whether there had been an archaeological surface within the loess and above the cobbled surface that

Figure 8.10: Photograph of conglomerate surface and Quadrupod, Autumn 1982

Figure 8.11: Photograph of "post-socket"

Figure 8.12: Photograph of "niche"

Figure 8.13: Photograph of "stone-lined pit"

had been missed, or which had not been indicated further up-slope. When excavated, they were found to be shallow, with vertical or sloping sides, and devoid of modern organic material. On these grounds, they were thought unlikely to be part of rodent burrows or termite nests. This particular conundrum was not resolved until the following season (see below).

Various other problems combined to make that month's excavation one of the worst in the author's experience. First, the weather alternated between extremes of aridity and wetness. In the first three weeks of excavation, the main weather problem was one of heat. When first exposed, the loess was a delight to excavate in terms of showing subtleties of colour and texture, but after only a couple of hours in the sun, it would bleach and become as hard as concrete. Attempts to restore colour detail and texture by sprinkling with a watering-can and/or hand-spray were only partially successful, as the surface tended to puddle. As way of a complete contrast, the main problem in the last week was one of flooding, when sustained and torrential downpours completely flooded the lower part of the main trench, and the adjacent one just when the loess over the cobbled surface had been removed. For financial reasons, there was no possibility of erecting a roof over the site, and in any case that would have been a major undertaking. In the end, a metre of water was baled out of the minor trench and further work there abandoned, and a drainage trench was dug in the west part of the site, in what became squares A3 SE and NE. This solved an immediate problem of draining the area that was flooded, and also showed that the stone surface overlay and was embedded in gritstone similar to that in the outcrop further upslope. However, because of the conditions under which it was dug, finds 163-182 cannot be provenanced beyond saying that they came from the area of the drainage trench.

There were other problems. One was the recording of artefacts, and the planning of the cobbles covering the conglomerate surface. At a density of up to 85 stones per square metre, it was clearly impractible under the circumstances to plan these single-handed with a drawing frame whilst directing the excavation and recording the finds. Instead, attempts were made to expedite the recording by using a photographic tripod with a metre-long arm extension to construct a photomosaic of the surface. This did not prove suitable, as the photographs could not be matched accurately enough, and the recording of all stone except for fresh flakes, blades and cores had to be discontinued. As none of the stones on or in the conglomerate surface was removed after exposure, no information was lost when the surface was finally plotted in the autumn. A final problem that season was the 40 mile drive back from Riwat through Rawalpindi to Taxila each day, westwards in to the sun and the evening traffic. By the second half of that season, the excavation was in danger of destroying more information than it was gaining, and its scope was increasingly curtailed to minimise this risk by leaving features for later examination; by halting work until well after rain had stopped; and by restricting workmen to tasks that did not require close supervision.

Upon reflecting on the results of this season when back in the U.K., it was clear that a major change in strategy was required. In particular, the best chance of gaining an overall view of what the site represented seemed to lie in excavating a much larger area. Because palaeolithic sites are often much larger than the area that had so far been exposed, it was highly likely that only a small - and unknown - part of the whole had been examined. However, this strategy would in turn mean major changes in excavation policy. First, the proper recording of the stone surface over a larger area would clearly be totally impossible without appropriate photographic equipment. Secondly, excavation of this type of site during a short spring season was counter-productive: time was too short, and the weather either too hot (as in 1981) or too wet (as in 1982) for excavation. Thirdly, skilled assistance was neccessary to help in the recording of data and the supervision of the workforce. Finally, it made far more sense to base operations in Rawalpindi than to undergo the daily gauntlet of traffic between 'Pindi and Taxila.

For all these reasons, the decision was taken to excavate in the autumn of the same year whilst on sabbatical leave; to stay in Rawalpindi; to take a graduate assistant; and to design a cheap and portable photo-frame for making a photomosaic. The Directors of the Mission were commendable in appreciating these proposals, and in ensuring that they could be implemented.

8.3.3 The autumn 1982 excavation

This last round of excavation lasted for ten weeks, from Oct. 12 to Dec. 22nd, and incorporated the policy changes noted above. Eddie Moth, newly graduated from the Archaeology and Prehistory Dept. at Sheffield, acted as field assistant, and did much to ensure the successful outcome of that season, as well as playing an invaluable role in preparing for publication all the plans and drawings of the site. The "Quadrupod" photoframe (see appendix H; Figs. H.1-3) totally transformed our ability to record the stone surface adaquately. The finds register (appendix A) provides sufficient proof: 25 flakes, blades and cores were recorded in 1981; 73 in the spring season of 1982; and 985 in the last season, as well as almost 1500 cobbles on the conglomerate surface. In addition, a larger workforce of 15-20 labourers (mostly Afghan) was used, and the task of removing most of the over-burden to open up a larger area was accomplished by the use of a tractor with a wide-blade shovel. This saved much time and money, in that the workmen had only a thin layer of loess to remove before reaching the main archaeological horizon. Finally, field operations were greatly facilitated by being based in Rawalpindi instead of Nikra.

The results of the excavation of site 55 are largely those obtained in the autumn season of 1982. At the start of that season, an area of ca. 200 sq.m. was opened up by removing the loess north and east of the previous seasons' excavations (Fig. 8.6). One advantage of this was that rainwater could be diverted into an extension in the north part of the site, well removed from the main area of excavation, and thus minimising damage caused by flooding further upslope. Another was that by now the work-force was accustomed to the type of work that was required: notably, the removal of the loess by shovel until the first stones on the cobbled surface appeared; then the careful isolation of each stone on the cobbled surface with a 4" pointing trowel, the subsequent cleaning with a hand- or paint-brush, and removal of soil with a hand-shovel. The photoframe ensured that each stone could be recorded: each frame comprised four shots covering a 2 m^2; these could be developed and enlarged in Rawalpindi, stuck together, overlain with tracing paper, and then each stone could be recorded by a simple coding system. In addition, individual finds could easily be plotted onto tracings laid over the relevant pictures of each grid square.

Two types of analysis were conducted in the field. Fresh flakes, blades and cores were usually removed when found, bagged and labelled individually, and then cleaned, drawn, photographed (and checked for joins at the end of the field season), alongwith those from the previous two seasons. The vast majority of flaked stone items, however, were the cobbles embedded in the surface itself. Removing and storing these would have been totally impracticable, so once they had been recorded, they were left. Despite increasingly cold and wind-swept conditions by the end of the field season, the excavation was completed on schedule.

The results of the excavation are given in the following two chapters, and its interpretation in chapter 11. Details of the dating of the fresh assemblage at site 55 to ca. 45 ka B.P. are given below.

8.4 A note on the problems of dating site 55

The thermoluminescence dating of site 55 posed a range of problems in sampling, analysis and interpretation. The loess cover over site 55 was relatively thin, and although this was advantageous as far as the excavation was concerned, it posed problems for sample dosimetry. In order to ensure that the environmental radiation dose comes from material of uniform composition, samples need to be taken at least 0.3m from the top or base of the loess. Where the loess cover was thin, with sampling restricted to the "ideal zone" for dosimetry, only one dating sample could generally be taken at each section. The sampling of site 55 was therefore, of necessity, spatially-extensive rather than stratigraphically-intensive. The results of this study are summarised in Fig. refTL55.

A complex picture emerges from this dating exercise. Two distinct areas of loess cover can be identified on the basis of age. The first zone comprises the loess overlying the dip slope of the gritstone outcrop and the spread of quartzite cobbles. The ages for this portion of the loess cover range from 42 ka to 64 ka. These data indicate that deposition rates were low at the edges of the loess sheet. The presence of the small gully, running transversely across the site, further complicates matters since some of the loess sampled for the dating programme may have been reworked older material. One the basis of these dates, a minimum age of 45,000 B.P. was assigned to the loess cover, thus providing a

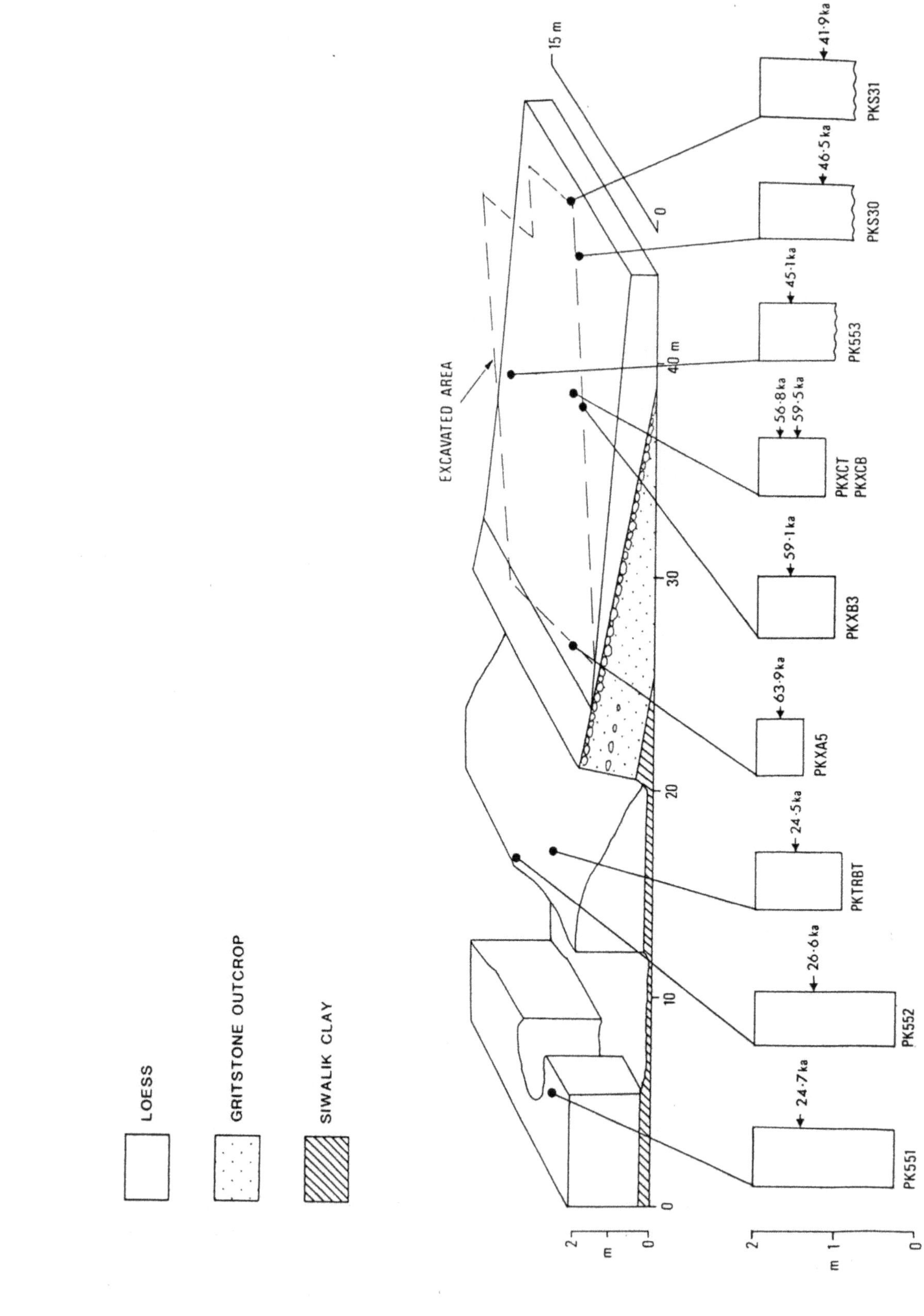

Figure 8.14: TL dates for site 55

minimum age for the materials excavated from site 55 (see Rendell and Dennell (1987) for further details.

The second area of loess cover lies immediately in front of the small gritstone scarp. In this area a much younger loess cover is preserved, dating to the final phase of loess deposition in the Riwat area ie. 27-24 ka. This portion of the loess is currently being eroded by a small gully and it appears likely that during the period 75 ka - 18ka, when loess was being actively deposited on exposed surfaces throughout the Riwat area, the area in front of the outcrop continued to be a focus of local drainage on the slope. Loess was only preserved in this area when surface runoff was less active and therefore when more arid conditions appear to have prevailed.

Chapter 9

The Spatial Aspects of Site 55

9.1 Introduction

In this chapter, we shall consider the spatial aspects of the features and stone material found at site 55. The fresh assemblage of cores, flakes and blades is discussed in more detail in the following chapter in terms of artefact types, and its regional and chronological context. As stated in the previous chapter, site 55 is a palimpsest of natural and human processes. As is common in the study of palaeolithic sites, the residues of human activities are often difficult to distinguish from natural ones, and many of the conclusions reached in this report are qualified to a greater or lesser degree by doubts and ambiguities. Site 55 is additionally unusual as a palaeolithic site in that it is directly on top of a natural conglomerate surface, and it was an interesting, if sometimes difficult, methodological exercise to differentiate between human and natural features, and human and natural flaking of stones.

The excavation uncovered *ca.* 200 m^2 by the end of the final season in 1982. As indicated in the previous chapter, the recording of this area was difficult, but was finally accomplished with a specially-designed photographic frame. The overall plan of the excavated area was drawn at a scale of 1:10, and the resulting site plan was thus 1.5 m^2. Publication of a plan this size is clearly problematic, as the maximum size possible in this series is only A3, and a large pull-out figure would have been prohibitively expensive. In order that the entire plan can be studied in the detail in which it was drawn, adjoining parts of the site are published in Appendix I, Figs.I.1 - I.36. For those who wish to study the spatial aspects of the site as it was recorded in the field, these figures can be cut out of the volume and reassembled. These have been slightly reduced to fit within an A4 format, but without losing any details. A summary plan is shown in Fig. 9.1. For the purposes of photographic reduction, many details have had to be omitted from this figure, notably the location and numbering of flaked material, and stones more than 5 cm in size.

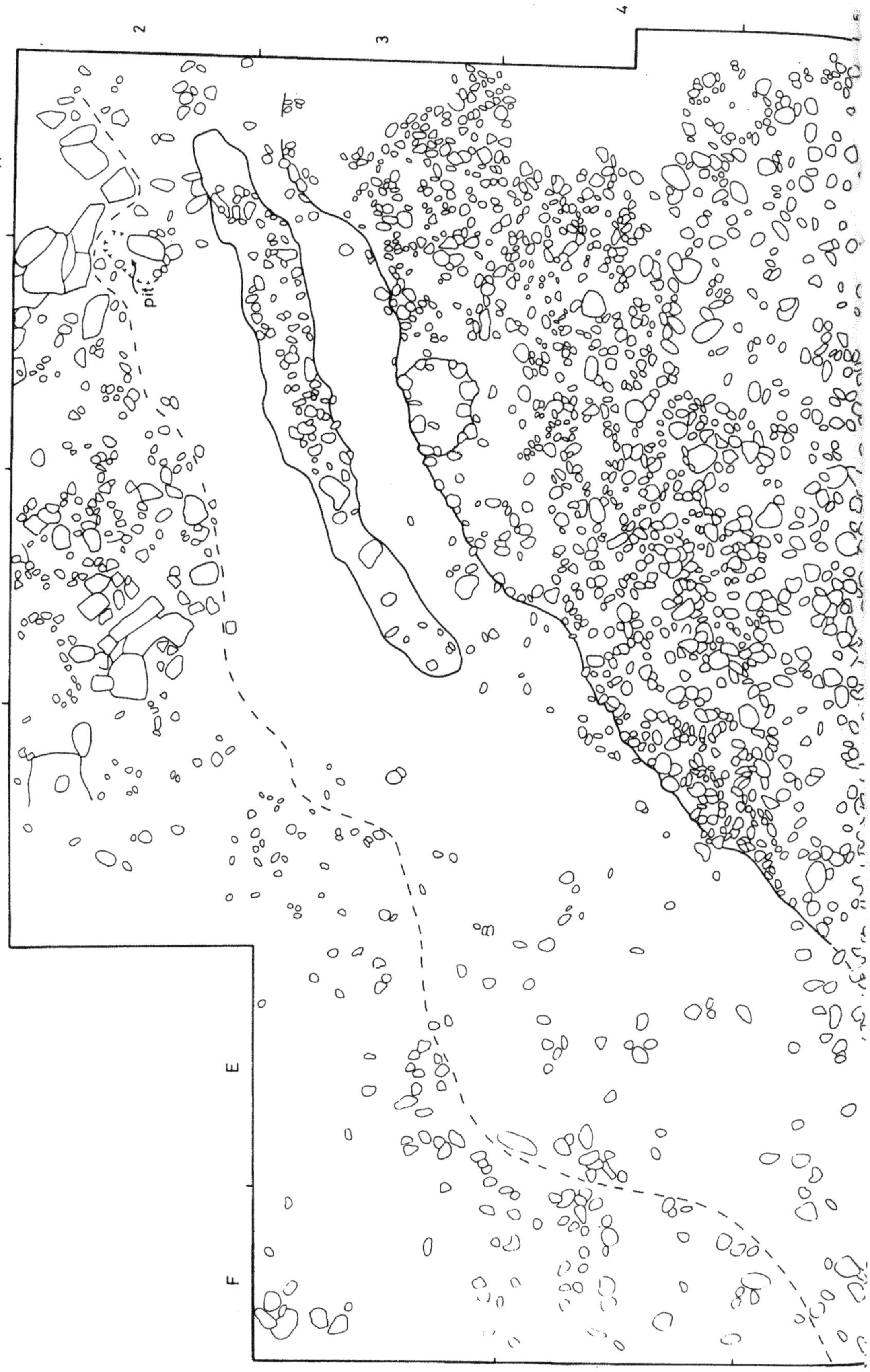

Please note that pg 155 and 156 are available to view as a double-page spread from www.barpublishing.com/additional-downloads.html. Please refer to the figure via the download for the original and complete figure as the figure shown on pages 155 and 156 might be cut off or the figure might not be complete in the printed book.

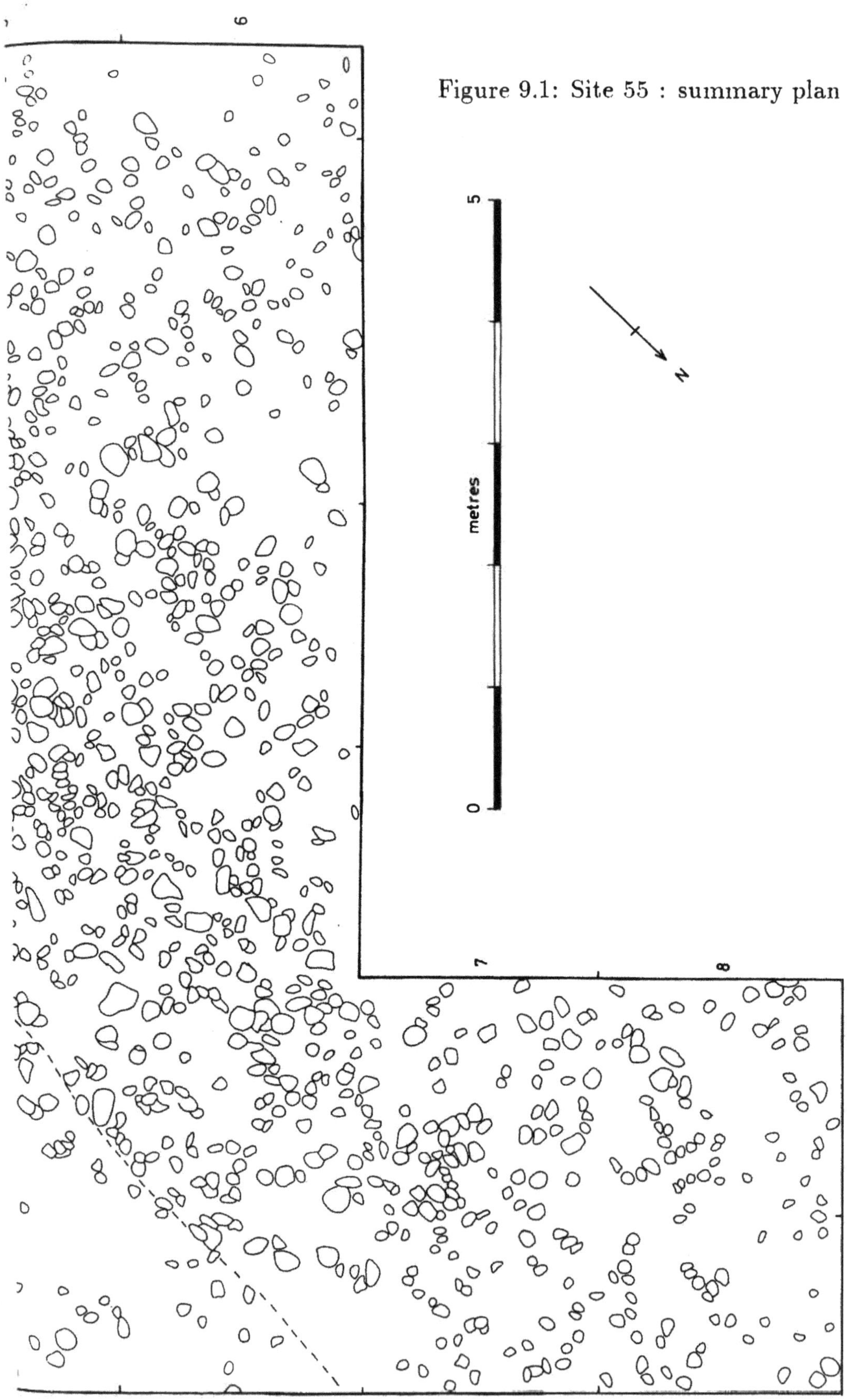

Figure 9.1: Site 55 : summary plan

Please note that pg 155 and 156 are available to view as a double-page spread from www.barpublishing.com/additional-downloads.html
Please refer to the figure via the download for the original and complete figure as the figure shown on pages 155 and 156 might be cut off or the figure might not be complete in the printed book.

Other data from the excavation are listed in various appendices. **Appendix A** is essentially the finds register of the excavation, and lists artefact type, condition and size. **Appendices B** and **C** list the coordinates of all artefacts found over the three seasons of excavation. **Appendix D** lists the coordinates and flaking details of all unlifted, weathered stones on the conglomerate surface from those squares where these were recorded in detail. **Appendix E** gives the spot heights used in contouring the site. Details of the conjoining pieces are given in **Appendices F** and **G**. We can begin by considering the architectural aspects of site 55.

9.2 Human and Natural Features

As in most palaeolithic excavations, some features are more readily identifiable as human in origin than others.

9.2.1 human features

The following features are interpreted as resulting from human activities:

i) the stone footing (Fig. 8.8; Appendix I, Figs. I.22 - I.24 and I.31 - I.33).

This feature was first excavated in spring 1981; and then buried under polythene sheeting until re-exposed in spring 1982; it was then buried again in the same manner, and re-exposed for the last time in autumn 1982. During that excavation it was dismantled. The overlays (Figs. I.31-33) showing "level II" of this feature show it after it had been removed to the level of the loess on either side; the "level III" plans (Figs.I.34-36) indicate what was found immediately below. It is thought unlikely that this feature results from the cutting of a trench through part of the main conglomerate surface. First, the edge of the conglomerate is well-defined, and extends well beyond the end of this feature. Secondly, the base of this "footing" was also well-defined, whereas the base of the conglomerate surface was not.

It is thought unlikely that this feature is a gully-fill. First, it runs across rather than downslope. As Fig. 8.7 shows, the gradient across the slope is almost level, and it seems improbable that a stream flowing over such a gentle gradient could deposit stones as large as those in this feature. Secondly, its cross-section is rectangular, with a flat base and vertical sides, as opposed to the type of V-section seen in small gullies in the area today, and in the gully clearly shown in section at the site (Fig. 8.4). It is also worth remarking that the fill of the gully seen in this section was poorly consolidated, unlike the composition of this feature. Thirdly, almost all the flaked stone was found on its surface, whereas when this type of material is found in gully-fill, it tends to be found throughout its matrix.

In the opinion of the excavator (RD), this feature was probably made from gritstone slabs and large cobbles from the dip-slope of the gritstone outcrop and from the conglomerate surface, respectively, in much the same manner as a low dry-stone wall. As there was no indication of a foundation trench cutting into the loess on either side of this feature, and was completely covered by it, it has to be of late Pleistocene age. This feature is unlikely to have been a "wall", because of its low height (20-25 cm), and because of the absence of material on either side that might have collapsed from it after it went out of use. One possibility is that it was intended primarily to divert rain-water that flowed off the outcrop immediately behind it, thus keeping a comparatively dry space between it and the conglomerate edge. It may also have supported a super-structure of, for example, brushwood, as seen today in many temporary structures in the area.

It is of interest that flaked stone was found on its surface, and in its base. Examples of the latter are the blade fragments and flakes (409-416) in the lower matrix, and two blade tips and a flake (418, 419, 540) found at a slightly lower level. The most reasonable explanation for these finds is that they were discarded before the feature was made; in other words, it overlies some of the fresh blades and blade fragments from this site.

ii) the pit (Fig 8.13; Appendix I, Figs.I.23, I.28).

One side of this feature was formed by a steeply-dipping gritstone slab. Some pieces of flaked quartzite with fresh flake scars (0296- 0298) were found at its base. The fill was a greenish-clay.

bf iii) the "niche" on the conglomerate surface (Fig. 8.12; Appendix I, Fig. I.36)

As related in the previous chapter, the edge of the main conglomerate surface had a slight bulge on its southern side in square B3 formed of stone more loosely packed than elsewhere along that edge. This was excavated in the autumn season of 1982. Removal of loose stones resulted in the finding of some pieces (341, 344, 346, 378, 382A, 382B, 384, 390) and a circular arrangement of stones. It was filled with a reddish clay, but contained no other material. It is difficult to see how this could have been formed naturally, and a human construction seems plausible.

9.2.2 Probable human features

.

The following are regarded as *"probably human"* in that confidence in these is less than in those we have just considered.

iv) post-holes on either side of wall-footing (Appendix I, Figs. I.23, I.24).

Seven were found in 1981. Each was visible as a reddish-brown circle in the yellow loess, and level with the base of the "wall footing" and the top of the pit. These resembled the type of shallow post-sockets often found on later sites. However, they could also have resulted from root disturbance; rodent- or ant-burrows; or the result of clay puddles infilling shallow depressions in the loess surface. However, tree-root disturbances and rodent burrows are typically sinuous, and usually traceable over fairly long distances. In contrast, ant burrows can leave a circular outline with straight sides and can also accumulate a clay fill if flooded before their sides have collapsed. They can also form very rapidly; one example formed overnight in the trench in the autumn 1982 season. On the other hand, if these "post-holes" were natural, they should have been some humus in their fill and been found over a wider area of the site and not just where there were humanly-made features such as the stone-lined pit, wall-footing, and niche.

v) the stone post-socket (Fig. 8.11)

This feature was discovered in spring 1982 against the edge of the conglomerate surface. It was a small semi-circle of stones, and the sides were formed by some stones that projected from the edge of the conglomerate. It might have been a deliberate placement, perhaps to serve as a stone post-socket, or it might be simply a fortuitous arrangement of stones.

9.2.3 natural features

One feature that was initially thought of as artifical can now be regarded as natural. This was found in the spring season of 1982 when the loess was being removed from over the main conglomerate surface in the downslope part of the trench. Eight stones (up to 12 cm long) were found lying ca. 15 cm above the main surface, and in an approximate line across the trench. In addition, some reddish circular outlines were also found, in the same line. No other material was associated with them, and there was no evidence that they belonged to a distinct archaeological surface. Nevertheless, their presence was unsuspected, and raised the possibility that there may have been a subsidary archaeological surface associated with a line of deliberately placed stones and perhaps post-sockets that had been missed.

These doubts were dispelled in the autumn season, when a larger area was being investigated, but with more resources than in the spring. As shown in Fig. 8.4 the east section shows a clear gully section. It is V-shaped, and appears strongly in square F6. This is less pronounced in the (shallower) west section (Fig. 8.4). The base of this gully could also be seen laterally in the course of excavating squares F6. It is probable that during the spring excavation, the workmen had cut through most of the fill of this small gully, and stopped at its base. Consequently, this linear "feature" of stone and reddish clay circles was probably only the truncated base of a small gully channel.

By contouring the conglomerate surface and the base of the gully, as shown in Fig. 8.7 (see also Appendix E), it seems unlikely that the gully cut down as far as the main surface except where it is adjacent to the east section. Consequently, there is no reason to suppose that the distribution of the stone material on the conglomerate surface has been significantly altered by this feature.

In summary, therefore, the man-made features of the site comprise: a pit with a stone slab forming one side, and some flaked quartzite at the base; a low stone footing, 5m long, formed of gritstone slabs and some large cobbles, with a level, flat base and straight sides; and a circular niche in the main conglomerate surface. These features may also be associated with some stake-holes and; a stone post- socket.

9.3 The Stone Assemblage

It was apparent in even the first week of excavation that there were two classes of struck stone material at site 55. One consisted of fresh, sharply edged flakes, blades and cores, and the other were deeply patinated and often abraded flaked cobbles and flakes. By using these simple indicators of patination and abrasion, it was possible to make an initial, and provisional, distinction between artefacts and naturally-flaked material. Although this was useful for isolating the small number of obviously humanly-struck pieces from the much larger body of flaked stone, it is not a satisfactory method of dealing with the latter.

This is because it is not always easy to distinguish between human and geological flaking of quartzite cobbles: at some point a naturally-flaked cobble can grade into a humanly-struck core, and (as the long-running debate over *"eoliths"* testified) some human and geological flaking can be confused with each other. The problem can be stated more precisely by considering the criteria for distinguishing human from geological flaking of cobbles. Clear examples of humanly-flaked cobbles can be recognised by: the removal of several, and probably at least five, flakes from a cobble; evidence of flaking in at least three directions; the presence of deep flake scars, preferably with evidence of bulbs of percussion and ripple scars showing a concentrated and accurate blow; and the removal of most of the cortex. By contrast, geological flaking should – but need not necessarily – be evidenced by: a smaller number of flake scars; flaking in only one or two directions; a low frequency of bulbs of percussion adjacent to the flaked edge; considerable evidence of battering; and the presence of most of the cortex. However, this type of flaking could be produced by human as well as by geological agencies. Examples might be if people flaked a cobble in a cursory manner to see if it was suitable for flaking, or detached a cortical flake that was then used without further modification. There are also other indicators of natural modification of stone, which have been well-documented for flint and chert (e.g. Stapert, 1976), such as sharp-edged ripple scars, thermal fracturing and so on.

In the context of site 55, the weathered and abraded assemblage included three bifaces of the type that we had already noticed on conglomerate exposures near the site, and we subsequently found similar ones in Middle Pleistocene deposits at Dina and Jalalpur in the Jhelum basin (Dennell *et al*, 1985; Rendell and Dennell, 1985). Whilst the bifaces at site 55 need not have been made there, their presence amongst the weathered and abraded material raised the possibility that the site or its environs had been used as a source of raw material before the assemblage of fresh blades and flakes was made, and this in turn implied the need to look for further evidence of human or proto-human flaking.

A further point was that it was obvious from even a cursory inspection that the conglomerate surface of the site contained far more flaked cobbles than flakes. This could indicate the use of the site as a source of raw material before the fresh assemblage of flakes and blades was made, and the preferential removal of flakes that were made into tools elsewhere. From these inferences, it might be only a short step to postulate a stone tool industry based on flakes, and pre-dating the fresh flake/blade assemblage. However, this conclusion might be totally unfounded, as the low ratio of flakes to flaked cobbles might have resulted from geological agencies before the conglomerate had formed. We have seen earlier in chapter 2 that Paterson identified various stages of the Soan industry from collections of flaked stone collected from the type of conglomerate exposure uncovered at site 55, and it seemed prudent to be cautious when dealing with this type of material.

9.3.1 the fresh assemblage

This will be considered in greater detail in the following chapter. Here, we shall consider its distribution with the following two questions in mind:

(i) is it clustered or uniform?

(ii) can it be associated with the structural features already mentioned?

The overall distribution of the fresh assemblage is shown in Fig. 9.2. As can be seen, the highest density of cores, flakes and blades is on the upper slopes of the conglomerate surface. The density otherwise falls off downslope and across slope, and over the loess on either side of the stone footing. The distribution of cores, flakes and blades are shown separately in Figs. 9.3, 9.4 and 9.5. These show an interesting difference between the distribution of cores and flakes on the one hand, and blades on the other. As these figures show, cores and flakes are distributed fairly evenly over the conglomerate surface, whereas most blades and blade fragments are concentrated in the highest part of the surface.

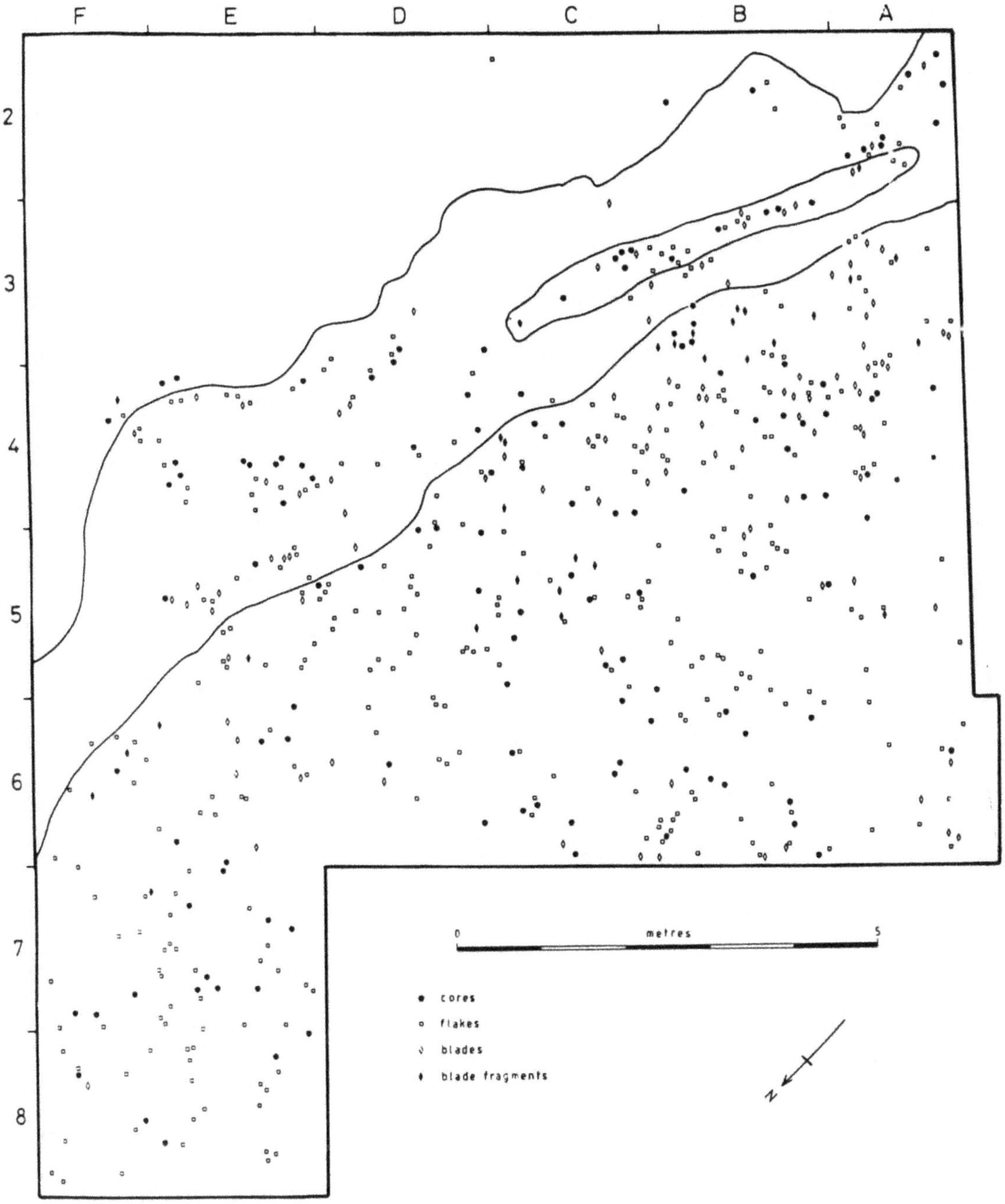

Figure 9.2: Distribution of the freshly-flaked assemblage

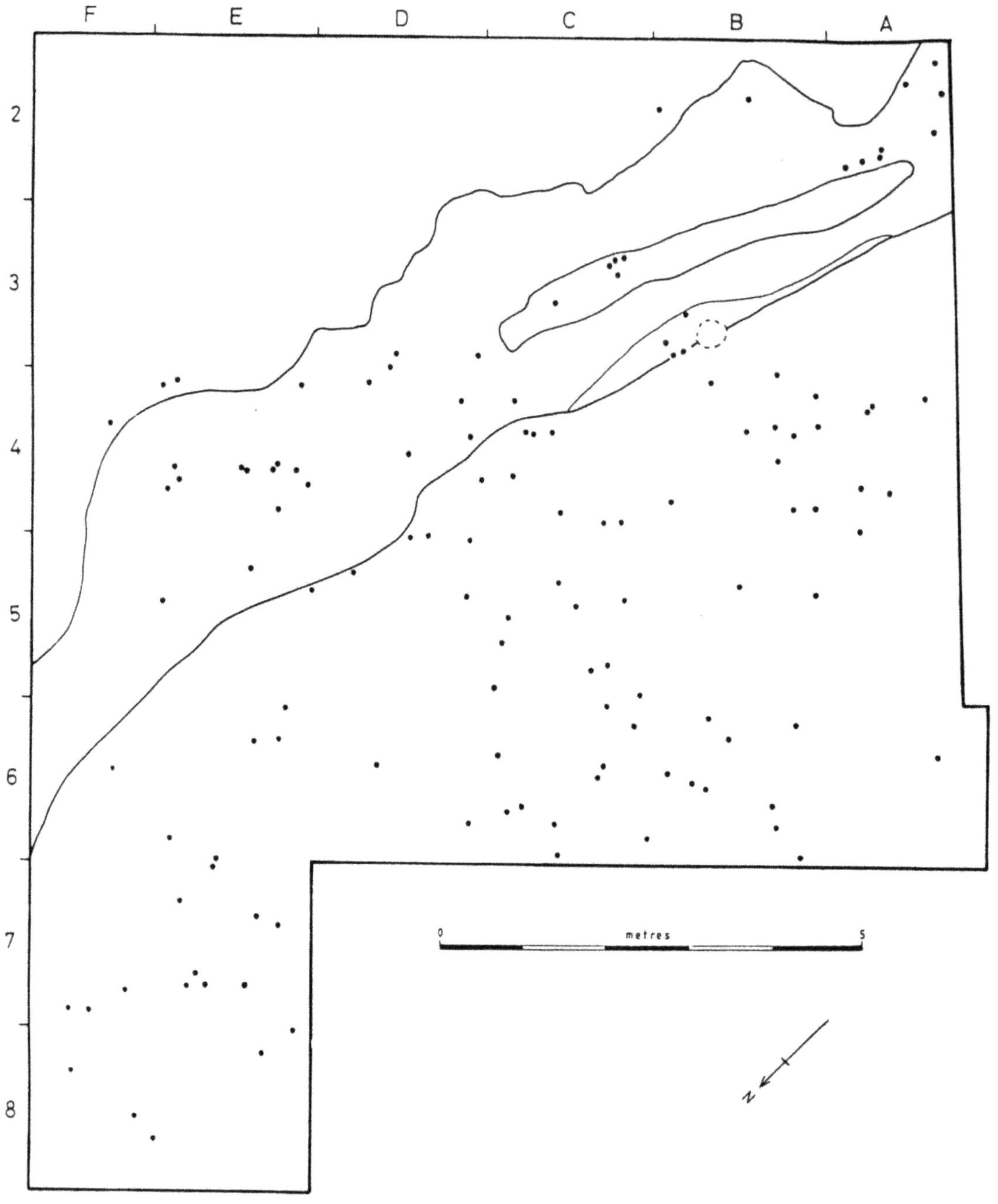

Figure 9.3: Distribution of cores

These patterns may be explained in a number of ways. One is that blades tended to be used, broken and discarded at the top of the surface, but were made over a larger area of the site. If so, many of the flakes found outside the area of greatest concentration of blades may have been waste material produced in making blades. This point is important to assessments of the assemblage: although flakes are more numerous than blades, the latter may have been an important objective of stone flaking. If so, many of the flakes could have been débitage, or regarded as a useful by-product of making blades. It may therefore be misleading to regard the assemblage as largely flake-based.

The distribution of conjoining pieces is shown in Fig. 9.6. This shows several features:

(i) most joining sets are distributed fairly evenly across the conglomerate surface, with the exception of the north-eastern part (squares G5-6, H5-6).

(ii) pieces are usually found within 2m of each other, but occasionally up to 10m apart.

(iii) the overall number of joins per set is low, with only 65 joins in 24 sets, or an average of 2.8 joins per set. This contrasts sharply with the pattern seen in some European upper palaeolithic sites, where very large numbers of fitting pieces are sometimes found within a very small area (see Gamble, 1986: 268-170).

(iv) as appendix F also indicates, some flakes and blades conjoin, thus affirming the integrity of the assemblage.

(v) the wall footing and blade/flake assemblage are clearly associated, in that a blade fragment (no. 0147) found in spring 1982 in the top of the footing matrix joined piece 1356, a flake.

(vi) the colour of pieces was usually a good guide when looking for fitting pieces, with the notable exception of one set (pieces 0121, 1195 and 1278A) in which the colours varied from brown to green and grey. Details of the colour matches of joining pieces area given in appendix G.

(vii) it is noticable that joins were absent from the surface of the gritstone outcrop, and rare in the area beyond the eastern end of the wall-footing and between the edge of the conglomerate surface and the dipslope of the outcrop. As Fig. 9.7 shows, material lying between the base of the outcrop dipslope and the conglomerate surface in this part of the site is most vulnerable to sheet-flow from heavy rains, and it is possible that some material has been swept beyond the excavated area by this means.

9.3.2 the weathered stone assemblage

As noted above, many of the stones making up the surface of the conglomerate encountered in the excavation of site 55 were weathered and abraded, but had nevertheless been flaked. Two abraded bifaces were also found on this surface, and a third was found in the fill of the small gully that ran across and slightly above the main surface. In addition, the distribution and frequency of stones in and on this surface may have been altered by the production of flakes and blades, as these were probably made from cobbles in the

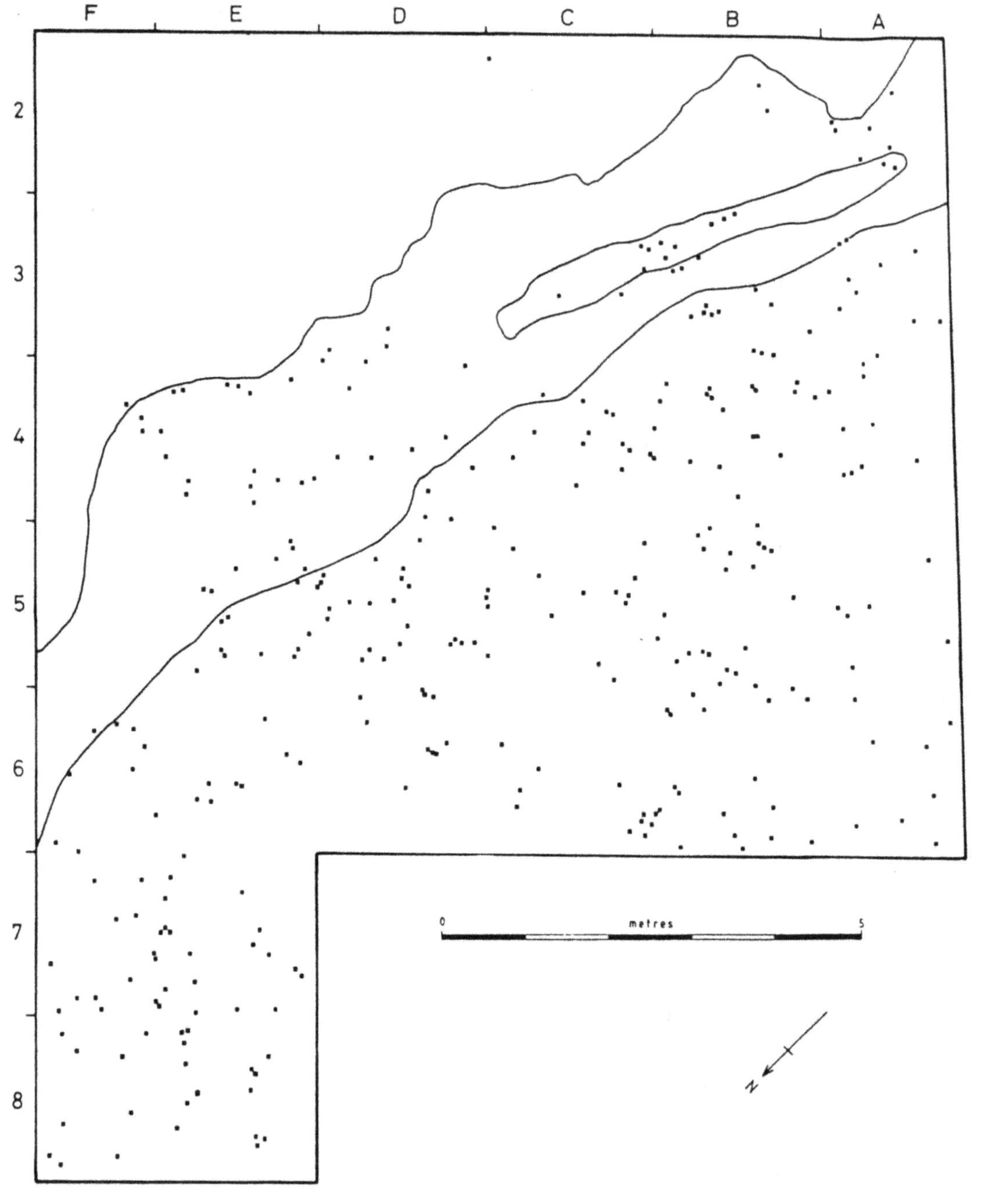

Figure 9.4: Distribution of flakes

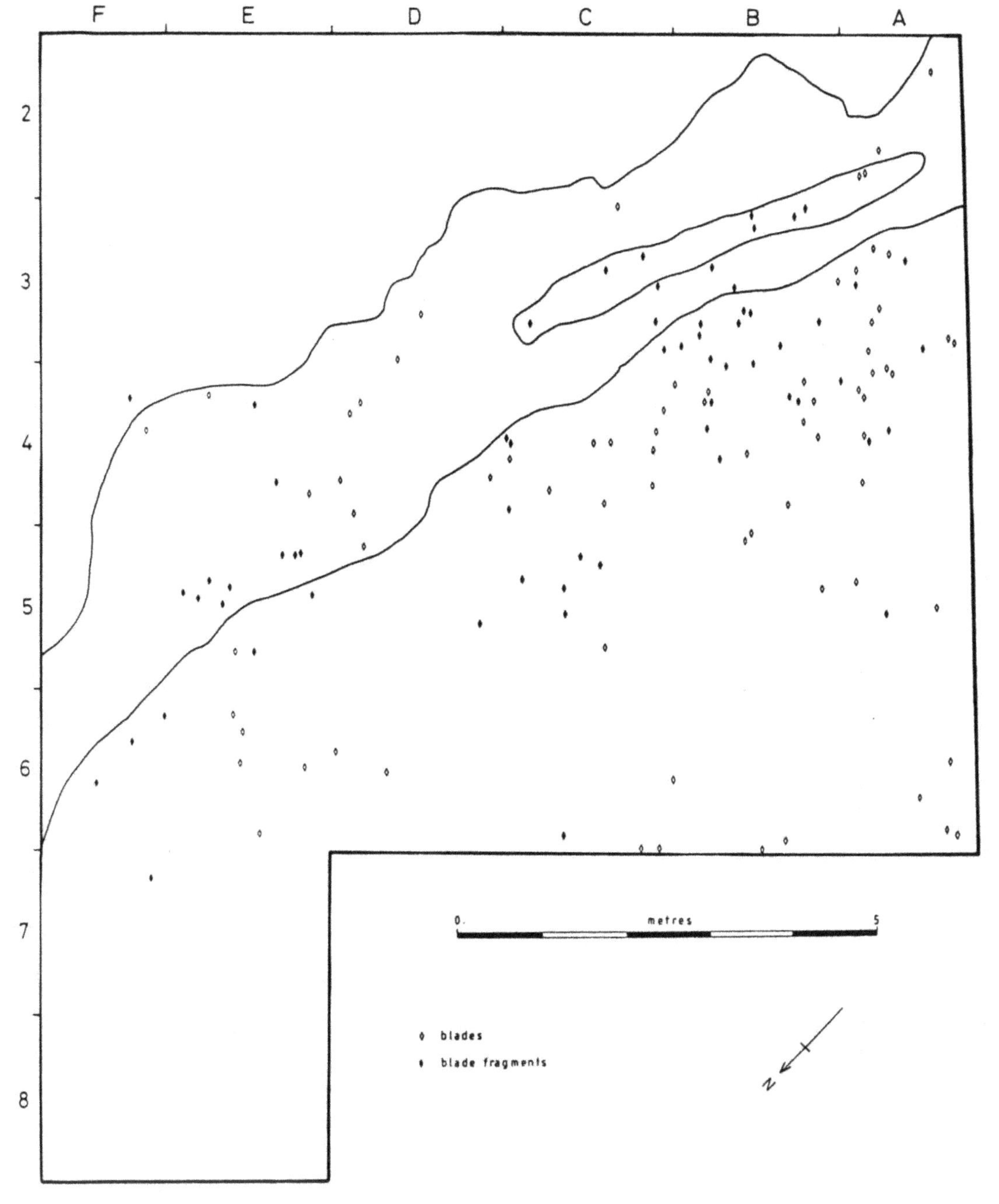

Figure 9.5: Distribution of blades

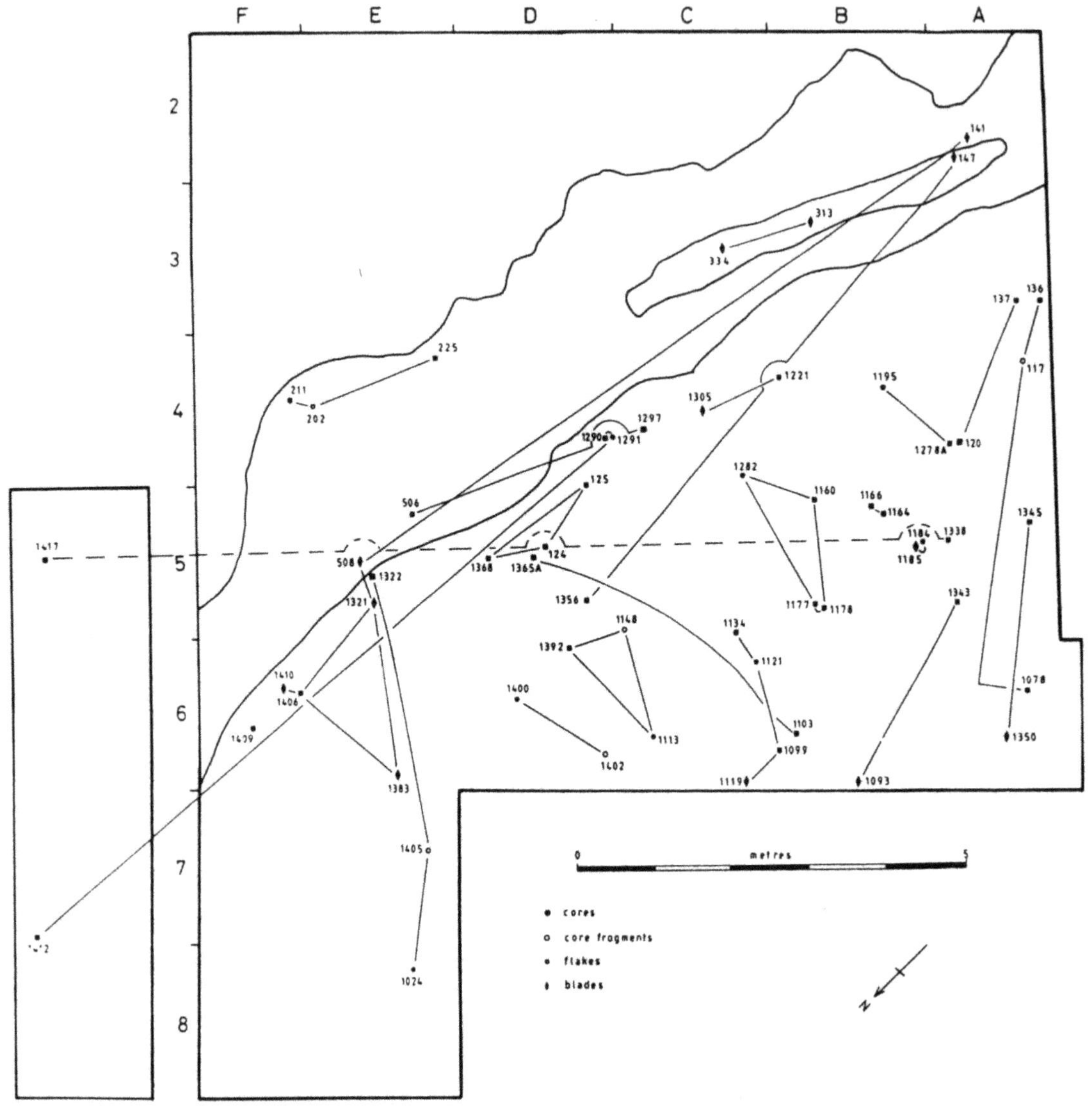

Figure 9.6: Distribution of conjoining pieces

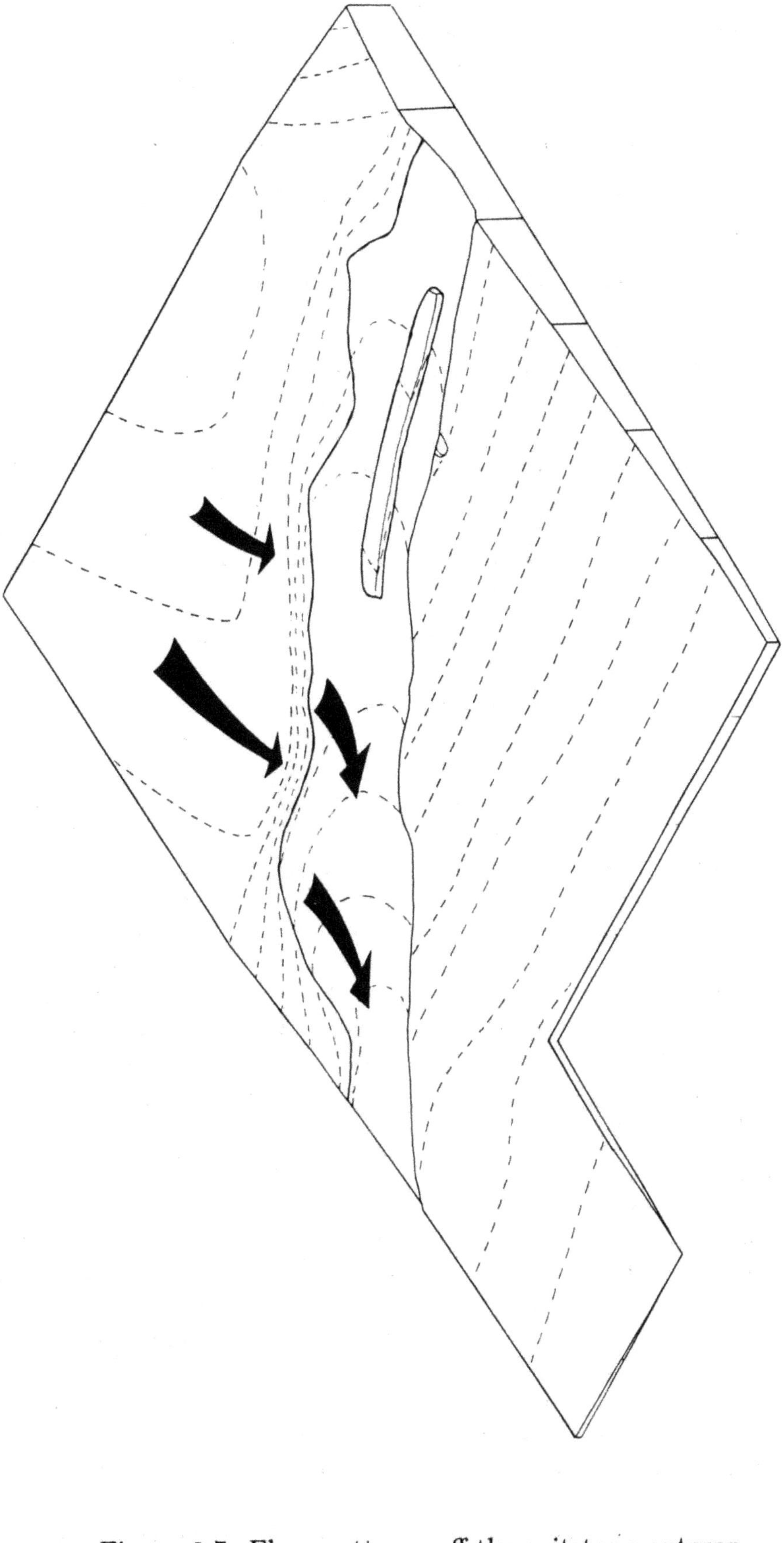

Figure 9.7: Flow patterns off the gritstone outcrop

surface. It was thus possible that the conglomerate surface had been used previously, and contained a mixture of two (or more) episodes of tool-making.

To summarise, the conglomerate surface at site 55 raised the following questions:

(i) were the weathered flaked cobbles and flakes the result of geological agencies?

(ii) if so, how much flaking occurred before the conglomerate was formed, and how much after it was exposed?

(iii) to what extent has this conglomerate surface been modified by humans (or proto-humans)?

(iv) if it was modified by humans, did this occur before and/or when the assemblage of fresh flakes and blades was made?

During the excavation, attempts were made to record the types and distribution of the weathered flaked stone at the site in some detail. At that stage, we had not had the opportunity to examine natural sections of conglomerate in the detail that we have since, notably in the deep section at Riwat, as well as others in the Jhelum basin. Comparisons can also be made now with other conglomerates in the study area.

As indicated in the previous chapter, the problems of recording this surface were finally resolved in the autumn season of 1982, when a photographic frame was used with the help of a graduate assistant. Even so, it was not possible to record every cobble and flake in detail, and a complete record was made of only 38 adjoining one metre squares, 28 of which were completely occupied by the conglomerate surface. Details of the squares which were recorded in detail are shown in Table 9.1. Counts of weathered stone in those squares containing only part of the surface totals have been adjusted in these figures to make their totals represent an area 1m sq. The major categories of weathered material in this surface are summarised in Figs 9.8 - 9.11. The size of each cobble can be seen from the plans (Appendix I.1 - I.36), and the co- ordinates of the 1,481 stones and flakes that were plotted are given in Appendix D. As can be seen from Table 9.1, the commonest element are unflaked cobbles, followed by those that had been split in half, and those split again into quarters. Flaked cobbles are far commoner than flakes. A general indication of the density of stones in the other 12 squares containing this surface can be obtained from the main plans of the excavation. However, detailed counts of each category of weathered stone cannot be made for these squares.

The density of all the weathered material, including unflaked cobbles, in the excavated part of this surface is shown in Figure 9.8. As can be seen, the greatest density of stone occurs in a broad band parallel to, and approximately 1 m. behind, the edge of the surface. It peaks at square C5 NW (at 85 stones/m^2), and then falls off to the east, with neigbouring squares C4 NE and D4 NW containing 59 and 38 pieces respectively. A similar falling-off occurs in the squares west of C5 NW. Downslope, the density of weathered stone also falls off steadily, with the exception of sq. B5 NW., which has a locally high density of 64 pieces/m^2. Apart from this one anomaly, there is little evidence of down-slope transport. This in turn implies that after the conglomerate was tilted and exposed, the stones in this surface have been firmly embedded in it. As there is no reason to suppose that the distribution of unflaked cobbles was significantly altered by human agencies, their distribution can be used to indicate the original, "natural" distribution of stone on this surface, against which other types of struck stone can be compared.

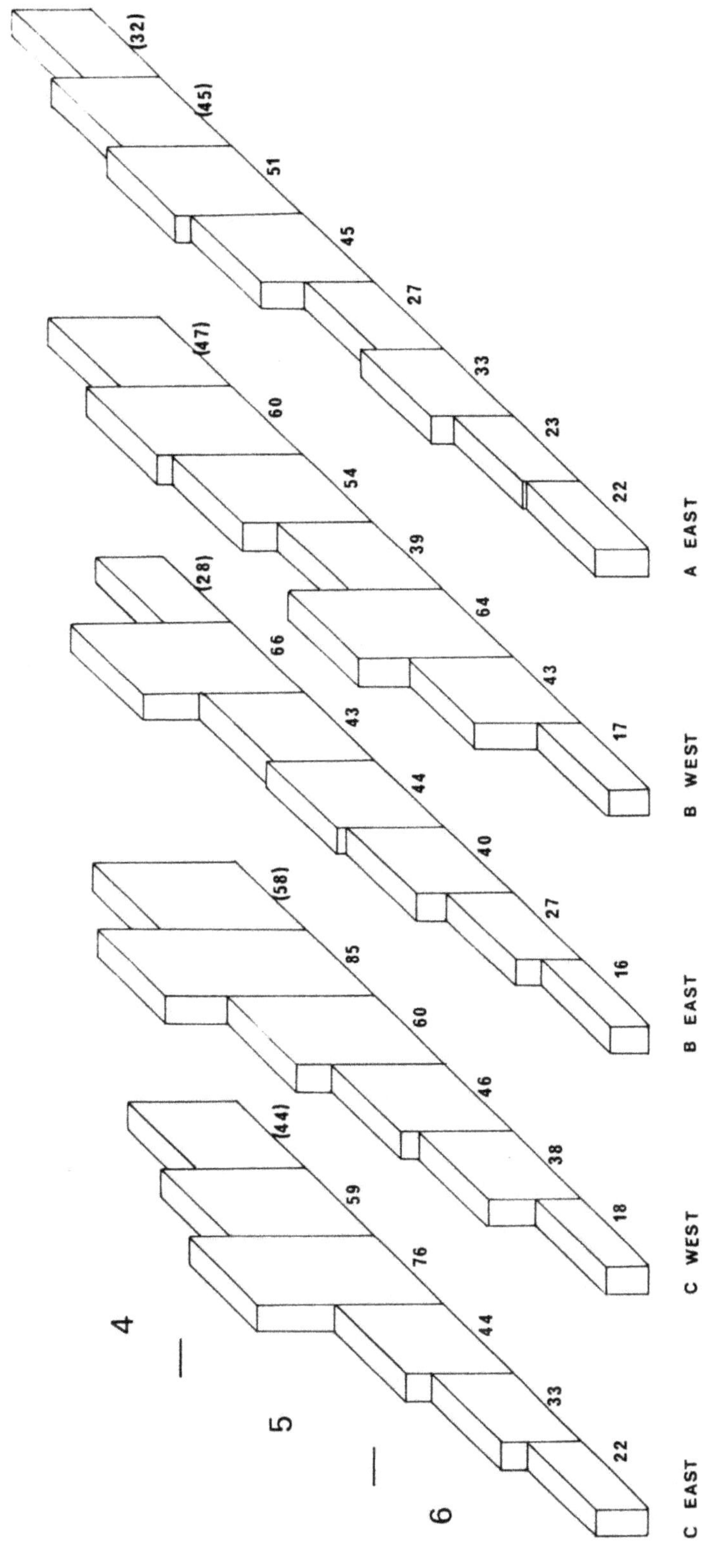

Figure 9.8: Histogram of distribution of all weathered material

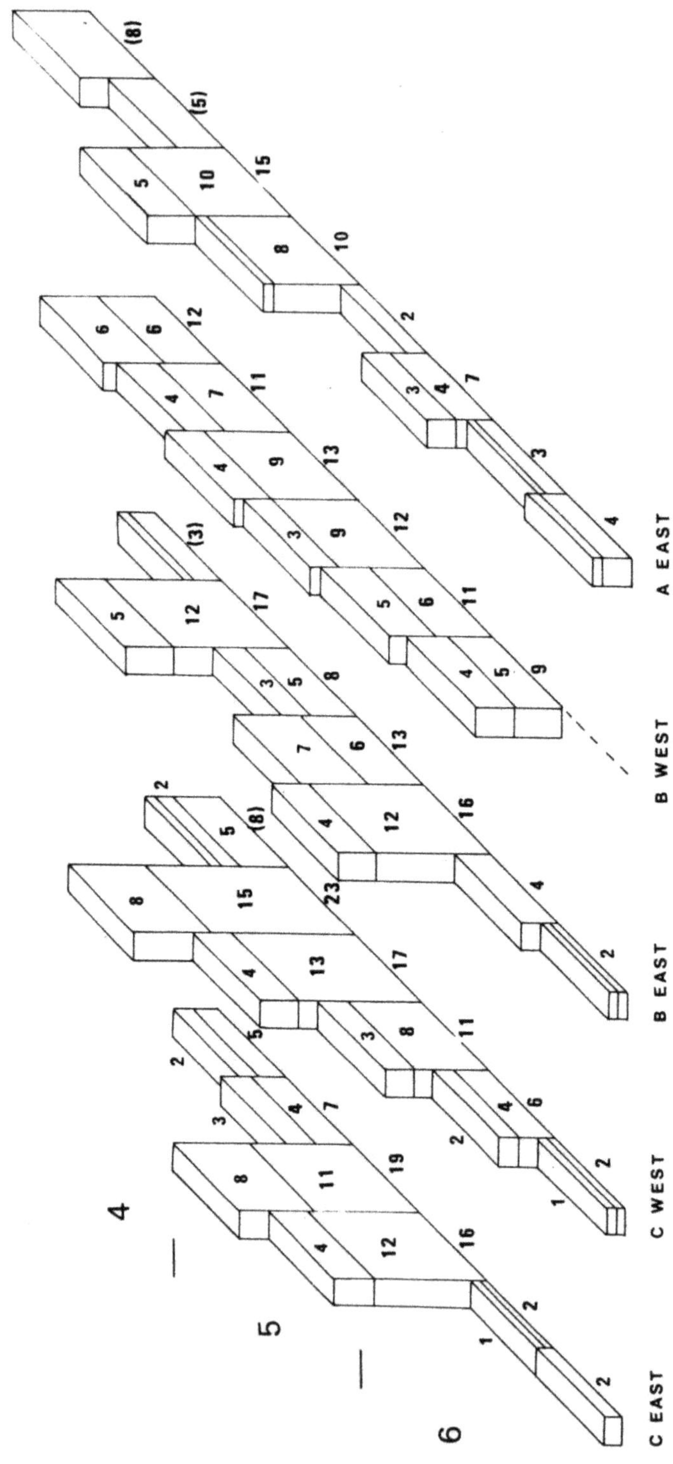

Figure 9.9: Histogram of distribution of split and quartered cobbles

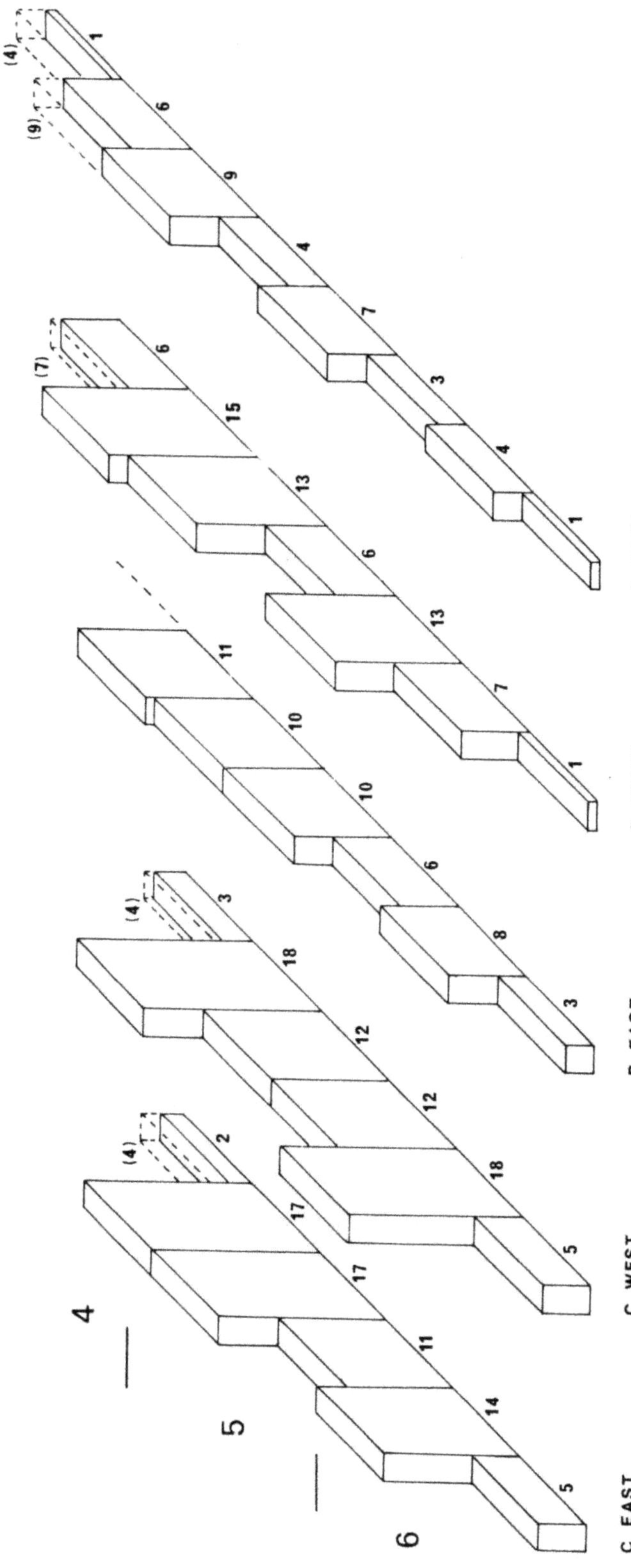

Figure 9.10: Histogram of distribution of flakesd cobbles

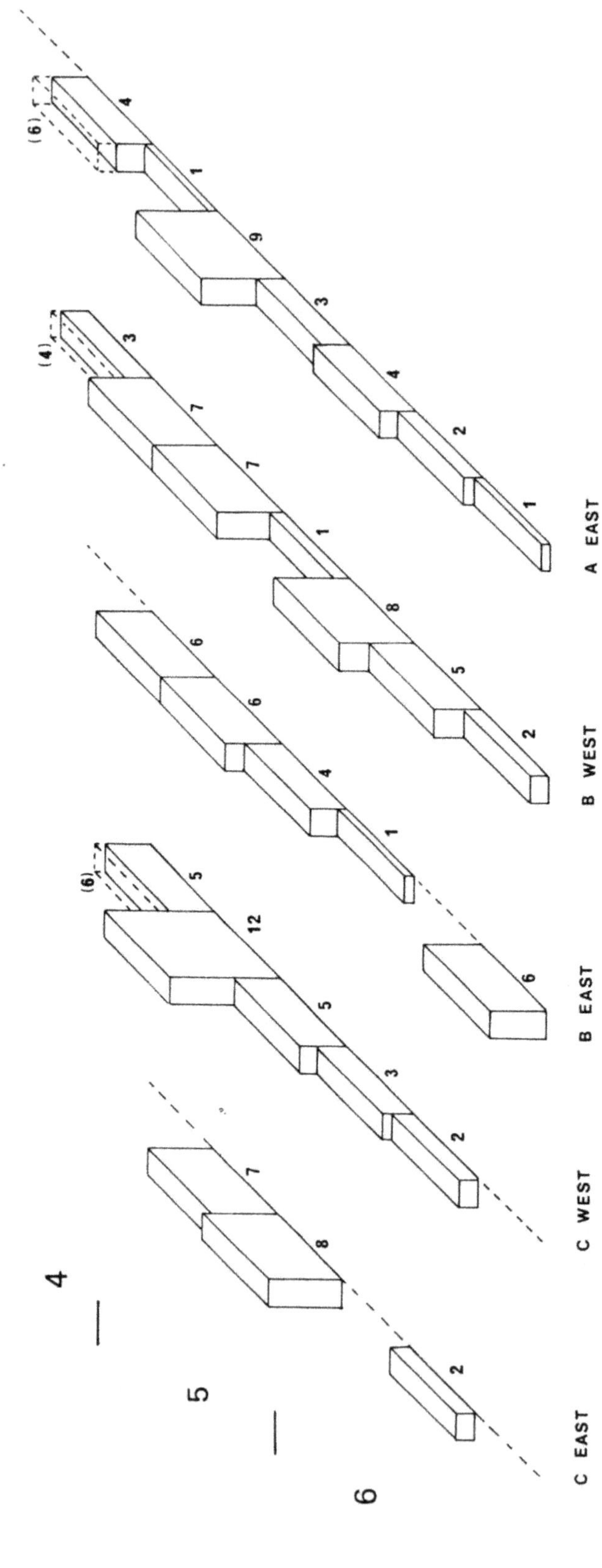

Figure 9.11: Histogram of distribution of weathered flakes

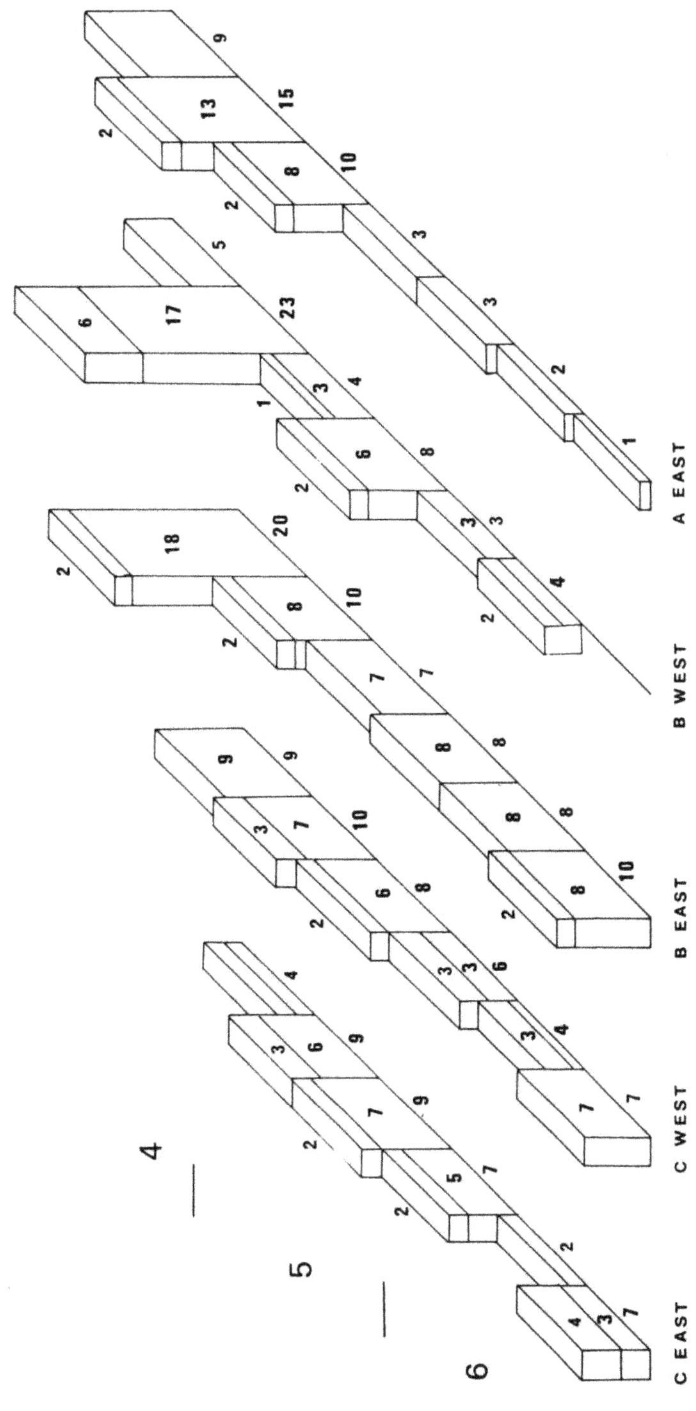

Figure 9.12: Histogram of fresh flakes, blades and cores

Square	Stone Types														Total
	01	30	31	40	02	03	04	05	06	08	20	21	22	23	
A3 SE	5	1	-	1	1	-	-	-	-	-	-	-	-	-	8
A3 NE	16	3	1	-	1	-	1	3	-	1	3	-	-	1	30
A4 SE	26	10	-	5	3	2	3	1	-	-	1	-	-	-	51
A4 NE	22	8	-	2	-	2	1	1	-	-	4	2	3	-	45
A5 SE	16	1	-	1	6	1	-	-	-	-	2	-	-	-	27
A5 NE	19	4	-	3	1	1	1	-	-	-	3	1	-	-	33
A6 SE	13	2	-	1	1	-	3	-	-	-	2	-	-	1	23
A6 NE	17	3	-	1	-	-	-	-	-	-	1	-	-	-	22
	(134)	(32)	(1)	(14)	(13)	(6)	(9)	(5)	(-)	(1)	(16)	(3)	(3)	(2)	(239)
B3 NW	19	5	1	6	2	1	2	1	-	-	2	1	-	-	40
B4 SW	27	5	2	4	4	4	6	1	-	-	6	-	1	-	60
B4 NW	21	6	3	4	2	6	3	2	-	-	5	2	-	-	54
B5 SW	20	9	-	3	1	2	1	1	1	-	1	-	-	-	39
B5 NW	32	6	-	5	4	3	4	2	-	-	5	1	2	-	64
B6 SW	22	5	-	4	-	3	2	2	-	-	2	3	-	-	43
B6 NW	14	-	-	-	-	-	1	-	-	-	2	-	-	-	17
	(155)	(36)	(6)	(20)	(13)	(19)	(19)	(9)	(1)	(-)	(23)	(6)	(3)	(-)	(317)
B3 NE	3	1	-	1	-	-	-	-	-	-	-	-	-	-	5
B4 SE	32	10	2	5	4	3	2	1	1	-	4	1	1	-	66
B4 NE	19	4	1	3	3	3	3	3	1	-	4	1	1	-	43
B5 SE	17	6	-	7	-	1	5	1	3	-	2	1	1	-	44
B5 NE	17	12	-	4	-	4	1	-	1	-	1	-	-	-	40
B6 SE	15	4	-	-	4	2	2	-	-	-	-	-	-	-	27
B6 NE	5	1	-	1	1	-	1	1	-	-	5	1	-	-	16
	(108)	(38)	(3)	(21)	(12)	(13)	(14)	(4)	(5)	(-)	(16)	(4)	(3)	(-)	(241)
C4 SW	34	4	1	2	-	2	-	1	-	-	3	-	2	-	49
C4 NW	32	15	-	8	3	5	5	5	-	-	5	2	4	1	85
C5 SW	26	13	-	4	3	6	2	1	-	-	3	2	-	-	60
C5 NW	20	8	-	3	1	2	4	4	1	-	3	-	-	-	46
C6 SW	12	4	-	2	3	6	6	3	-	-	1	1	-	-	38
C6 NW	11	1	-	1	3	1	1	-	-	-	-	-	-	-	18
	(135)	(45)	(1)	(20)	(13)	(22)	(18)	(14)	(1)	(-)	(15)	(5)	(6)	(1)	(296)
C4 SE	14	3	-	2	-	-	1	1	-	-	-	-	-	-	21
C4 NE	23	4	-	3	7	5	1	4	-	-	-	5	4	3	59
C5 SE	30	11	8	-	2	7	5	2	1	-	-	2	6	-	76
C5 NE	14	12	1	4	1	2	4	2	2	-	-	3	-	-	44
C6 NE	15	1	1	-	5	3	4	1	1	-	-	-	2	-	33
C6 NE	14	-	2	-	2	2	1	-	-	-	-	1	-	-	22
	(110)	(31)	(12)	(9)	(17)	(19)	(16)	(10)	(4)	(-)	(-)	(11)	(12)	(3)	(255)
Total	**642**	**182**	**23**	**84**	**68**	**79**	**76**	**42**	**11**	**1**	**70**	**29**	**56**	**6**	**1348**
D4 NW															
E5 SW															
E5 NW															
E5 NE															

Table 9.1: Distribution of weathered material on conglomerate surface site 55

The distribution of split and quartered cobbles is shown in Fig. 9.9. These are potentially of interest, as they may have resulted from human or geological agencies, or both. They may have split if used as hammer-stones, or as anvils if larger stones were flaked on top of them; alternatively, they may have split before the conglomerate was formed, or after it was exposed through, for example, the action of heat, frost, or trampling by large animals. As shown in Fig. refhist2, their distribution matches that of all weathered material in most respects: the greatest density peaks over square C5 NW; and by and large, tails off downslope and to either side, particularly in files A East, C West and C East. There is, however, a very sharp decrease down-slope in file C East; and a minor peak at B5 NE instead of B5 NW, as was the case for all weathered material. In general, however, split and quartered cobbles are commonest where the overall density of stone is highest.

Fig. 9.10 shows the distribution of flaked cobbles. As before, the highest density occurs in square C5 NW. The most easterly file (A East) mirrors the overall distribution of flaked material very closely, as does file B West, including the secondary peak in sq. B5 NW. There are, however, independent subordinate peaks in squares C6 SE and C6 SW. Apart from these two aberrations, the distribution of flaked cobbles does not seem to differ significantly from that of unflaked and split stones.

The distribution of weathered flakes is shown in Fig 9.11. As shown in appendix A, there are also four weathered flakes (1168, 1298, 1299, and 1312) and one weathered blade (1304)included in the artefact count which could be transferred to this group; however, this exercise does not significantly alter the basic pattern. Although the sample is small, it nevertheless follows the pattern shown so far, in that the greatest density is in sq. C5 NW and then declines to either side, as well as downslope. The main differences are that there is a subordinate peak in sqs. B6 NE; and unexpectedly low frequences mid- slope in squares B5 SW and B6 SE.

We can finally consider the distribution of fresh flakes, blades and cores (Fig. 9.12). This resembles that seen so far, in that the density of material in general is highest mid-way and immediately behind the southern edge of the conglomerate, and then declines to either side, and to some extent, downslope. There are some noteworthy differences, however. In the first place, the greatest density of material is slightly to the west of that for weathered material, with a peak over square B4 SE. Secondly, there is a local high density of material at the bottom of one file (B West), which is not found elsewhere. Interestingly, there are some differences in the distribution of fresh material and flaked cobbles. As Figs. 9.10 and 9.12 show, the **high** densities of flaked cobbles in squares B5 NW, C6 SW and C6 SE are matched by correspondingly **low** densities of fresh material; conversely, **high** densities of fresh material in squares B6 NW and B6 NE. are countered by **low** frequencies of flaked cobbles.

Before proceeding further, there are three aspects of the data on flaked cobbles and flakes that are worth commenting on:

(i) the incidence of flaked cobbles
The incidence of flaked cobbles appears to be far higher than in the 2 Ma conglomerate at the base of the deep gully on the other side of the GT road. As seen in chapter 7, the conglomerate exposed here was almost entirely composed of intact clasts, and the frequency of flakes and flaked/split cobbles was negligible. Similarly, qualitative assessments of other conglomerate horizons that we have investigated elsewhere (at e.g. Dina, Rohtas and Jalalpur) suggest that unflaked cobbles are usually by far the commonest

component, and make up >95% of the total.

(ii) the multiple flaking of cobbles
Table 9.2 summarises the numbers of times weathered cobbles have been flaked. These data are interesting in that one might expect the incidence of multiple flaking to decline steadily if it had been done under conditions of fluvial transport. That is to say, cobbles with only one flake removed should be commoner than those with two, and those in turn should be commoner than those with three flakes removed. In fact, this does not occur, and the commonest type of flaked cobble were those with **three** flakes removed. Thereafter, there was a very sharp fall-off, and only one had more than seven flakes detached.

(iii) the low incidence of flakes
The low frequency of flakes relative to cobbles is even more evident when one considers the ratio of flakes to flake scars. The results of counting the numbers of flake scars on cobbles and also on those weathered flakes struck more than once is shown in Fig. 9.13 relative to the number of all flakes/blades, both weathered and fresh. As seen in Table 9: only ca. 20% (127) of the expected number of weathered flakes are present, and even if the 211 fresh flakes and blades in those squares are added, the total is still only 49% of that expected.

Each of these features can be taken in turn.

i) the incidence of flaked cobbles

The high frequency of flaked cobbles at site 55 (Table 9.2 relative to that in the lower conglomerate horizon at Riwat could be a result of the way each was examined. Whereas the conglomerate at site 55 was examined as a **surface**, the others were inspected in **section**. It is only by removing each stone in turn from its matrix that it is possible to ascertain how much flaking has occured. When sections are inspected, often only a small part of each clast is exposed, and one side is always – by definition – embedded in the section. As there is at most a 50/50 chance of a flaked side being exposed, one would expect the frequency of flaked material to be at least half of that estimated when each cobble is fully inspected. Nevertheless, this factor fails to explain the wide discrepancy in the frequency of flaked cobbles in these two conglomerates

It is useful at this juncture to consider some data on the incidence of flaked cobbles on conglomerate exposures. As De Terra and Paterson noted (see chap. 2), humanly-flaked stone can usually be found wherever conglomerates are exposed in this area, and it is probably impossible to find exposed examples of conglomerates that have not at some stage been used for raw material. Despite this obvious potential bias, they might still offer a useful perspective on site 55. In the spring season of 1981, conglomerate exposures under a thick layer of loess were examined along the side of a gully within the study area, and ca. 500m from site 55. Artefacts and freshly flaked material had been noticed amongst the surface of the colluvial conglomerate the year before, and some concentrations were tentatively identified as quarry sites. The purpose of this sampling experiment was to see whether or not worked quartzite was localised, and if it seemed common relative to unworked stone because it was more conspicuous, or because it was actually so. Two five metres squares were laid out ca. 100 m apart, near known concentrations of worked quartzite. Ten of these squares were selected using random numbers. It is likely that some artefacts had been removed from the sampled areas by prior collecting, as these localities are very close to the main road, and some artefacts were collected nearby in 1980.

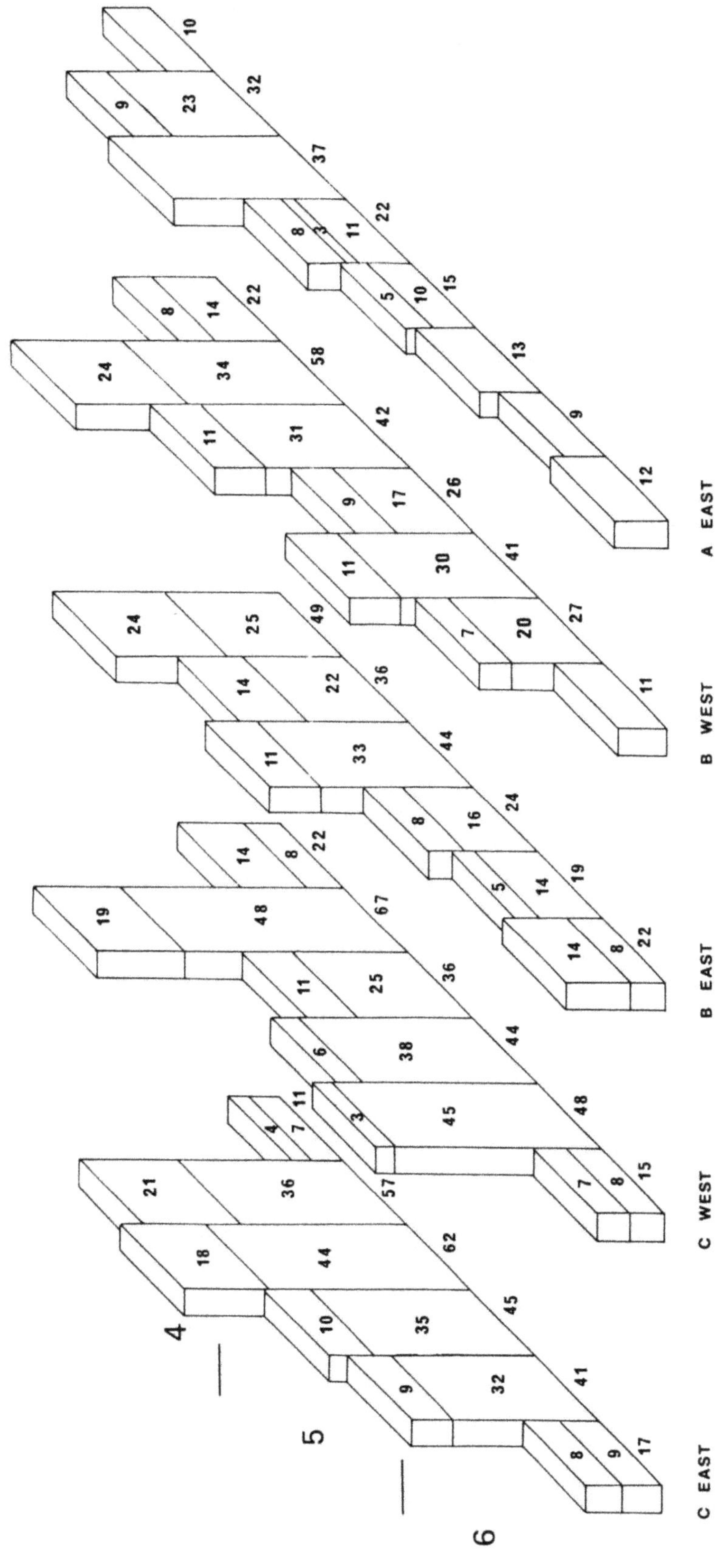

Figure 9.13: Histogram of ratio of flake scars to flakes

Square	Flaked			cortical flakes	fresh flakes	split/ quartered	Total
	once	twice	3 times				
AA1							3
AA3	1						4
AA4	2			1	4	1	3
AA8	11	1	1	2	2	5	19
AA9	3	1				3	33
AB4	5					10	83
AB6						1	28
AB9	1					3	40
AB16	1		1	1			31
AB24				1		1	66
totals	24	2	1	4	6	24	310
%	8.7			1.3	1.9	7.7	

Table 9.2: Flaking frequencies on colluvial exposures near site 55

Counts were taken of the total number of intact, flaked and split/quartered cobbles, as well as the numbers of weathered and fresh flakes. The results are shown in Table 9.3. These show first, the very low frequency of worked material, and this in turn suggests that worked quartzite occurred in patches, and was very localised. Secondly, the incidence of both flaked and split/quartered cobbles is very low, and more comparable with the proportion seen in the section of the lower conglomerate at Riwat than at site 55. This in turn implies that the high incidence of both flaked and split/quartered cobbles at site 55 is significantly different from that in Upper Siwalik conglomerate sections, and on conglomerate exposures near where quartzite was worked.

ii) the incidence of multiple flaking of cobbles

Earlier, we observed that the commonest type of flaked cobble at site 55 was one from which three flakes had been removed. This is not what one would expect of naturally-flaked clasts in a fluvial conglomerate. On common sense grounds, one would expect that when cobbles are transported in rapidly flowing and turbulent water, some will have one flake detached, a smaller number two, and so on. This type of pattern is evident in Table 9.3, which shows the degree of flaking of cobbles on the colluvial conglomerate exposure that was sampled near the excavation. The low incidence of multiple flaking on these exposures, and in observed sections of fluvial conglomerates suggests that site 55 is unusual in this respect.

Whilst it is tempting to invoke human or hominid agencies to explain the high incidence of multiple flaking of cobbles at site 55, other explanations should also be considered. One is that it resulted from fluvial agencies before the conglomerate at site 55 was formed. If the clasts in this deposit had been extensively reworked, the chances of a stone that

cores				
02	1 flake removed	77	flake scars	77
31	1 flake removed	12	flake scars	12
03	2 flakes removed	83	flake scars	166
04	3 flakes removed	88	flake scars	264
05	4 flakes removed	45	flake scars	180
06	5 flakes removed	11	flake scars	55
07	6 flakes removed	–	flake scars	–
08	7 flakes removed	1	flake scars	7
	total	317	total	761
flakes				
20	resulting from 1 blow	89	flakes removed	89
21	resulting from 2 blows	31	flakes removed	62
22	resulting from 3 blows	26	flakes removed	78
23	resulting from 4 blows	6	flakes removed	24
24	resulting from 5 blows	1	flakes removed	5
	total	153	total	258
ratio : flakes/cores ... 48.3% ratio : flake scars/flakes ... 33.9%				

Table 9.3: Ratio of flakes to flake scars at site 55 (all squares)

had already been flaked once being flaked again should have been the same as those for an unflaked one. Indeed, their chances of being flaked again might have been higher, as their angularity would have increased through already having one flake knocked off. Thus the high incidence of flaked (and split) cobbles at site 55 might simply indicate that a large proportion had been re-worked and often rolled against each other before the conglomerate was formed.

A second possibility is that the flaking of cobbles on this surface is post-depositional, and results from the exposure of this surface. Stones could have been flaked by a variety of means, such as thermal splitting, or by trampling of animals; by previous colluvial transport and the pressure-flaking of cobbles against each other. Resulting flakes would tend to show shallow flake scars.

iii) the low frequency of flakes relative to flaked cobbles.

This can be explained in several ways. One is that flakes could have been broken down into increasingly small fragments by fluvial action, and/or deposited further downstream before the conglomerate was consolidated. We can note in support of this idea the extremely low frequency of flakes relative to flaked cobbles in the artefact-bearing conglomerate section (chap.7), and on the colluvial exposures near the site, where 34 flake scars were recorded, but only 10 flakes. Even here, however, the ratio of 3:1 is considerably higher than the 5:1 recorded at site 55. Another possibility is that after the conglomerate was formed and exposed, flakes could have been comminuted by, for example, trampling by large animals, and worked either downslope by rain, or into the interstices between cobbles. As seen already, however, there is little evidence of downslope transport from the general distribution of material on the site. Human agencies are, of course, another factor that may have been responsible: stones might have been flaked, and the flakes removed for use elsewhere.

9.3.3 Discussion

We have seen that whilst little of the weathered assemblage provides unambiguous examples of human flaking, the composition of this assemblage is unlike that in conglomerate sections, and colluvial exposures near where stone has been worked. We have also noted that the distribution of fresh flakes and blades does suggest some negative correlations with the density of flaked cobbles. This in turn implies that where flaked cobbles are rare, freshly worked material is common, and *vice versa*.

It is probably wise to refrain from reaching a firm conclusion on these observations at present. More data are needed on a wider variety of conglomerate exposures and sections than we have been able to study. Nevertheless, whilst the flaking of the weathered cobbles and flakes could have been natural, the frequency of flaked and split cobbles, and of stones flaked several times was far higher than that observed in the conglomerate section at Riwat, as well as on colluvial conglomerate exposures. Furthermore, the density of flaked cobbles at site 55 seems to be inversely related to the density of fresh flakes and blades away from their area of maximum concentration. This last point implies that at least some cobbles on the site were used for making fresh blades and flakes. We do not, however, have any firm basis for concluding that the conglomerate at site 55 was used intensively or frequently as a source of raw material before the fresh assemblage was made, and it seems reasonable to conclude that most of the flaking of this surface is post-depositional in origin, and the result of exposure to surface conditions.

What this study suggests is that conglomerate exposures are likely to have been modified in three ways. The first is when they were originally formed under conditions of swiftly-flowing water. As indicated in the deep section at Riwat, this is unlikely to result in extensive flaking or modification of clasts; indeed, our observations suggests that very little modification takes place at this stage. The second, and much more substantial, modification is when a conglomerate is exposed as a surface. The data from site 55 would imply that sub-aerial exposure to a wide range of biological, physical, and chemical processes results in extensive and often multiple flaking, as well as extensive splitting of clasts. Finally, stone-knapping produces a third set of changes, by the breaking of stones, and the reduction of cobbles into cores, flakes, blades and débitage.

9.4 Conclusions and summary

The main conclusions reached in this chapter are as follows:

i) the architectural features of the site comprise a wall-footing; a pit; a circular niche; and perhaps some post-sockets.

ii) the distribution of conjoining stone pieces indicates that the wall-footing at least is associated with the fresh assemblage of flakes, blades and cores.

iii) the spatial distribution of flakes and blades implies that whilst flakes are more numerous, many may have been by-products of making blades.

iv) the weathered stone assemblage is different from that observed in sections of fluvial conglomerates, and on other exposures of conglomerates seen near the site. Although it has been modified to some extent by stone-working, its main features are probably best explained as the result of re-working before incorporation into the conglomerate, and surface attrition after exposure.

Chapter 10

The Fresh and Abraded Flaked Stone Assemblage from Site 55

10.1 Introduction

In this chapter, we consider first some of the attributes of the predominantly fresh assemblage from site 55, and then attempt to place it in its chronological and geographical context.

As indicated already, the main source of raw material for making stone tools in the Soan Valley are quartzite cobbles. The shape of cobbles, and the fact that they are quartzite, places severe constraints upon the type of material produced in flaking it. Quartzite is less ideal than flint or obsidian as a raw material, as it is coarser-grained. The outcome of flaking can be hard to control, and quartzite can easily shatter if struck in the wrong direction. Repeated evasive flaking is also more difficult on quartzite than on finer-grained materials like flint. In general, one would expect quartzite tools to look more "primitive" or crude compared with their counterparts in flint, in much the same way as Jones (1979) showed that the "crudeness" of some East African bifacial implements probably stems more from being made on basalt than any incompetence on the part of their makers.

In the Riwat area, tortoise- or Levallois-cores and flakes are occasionally found as surface finds. However, none was found in the excavation of site 55, and the flaking of quartzite cobbles appears to have taken place with minimal core preparation. The likely sequence of flaking is as follows. First, the cobble must be inspected by detaching a flake from the cortex to see if it is the right quality (and perhaps colour), or by tapping the nodule to see if it has the right resonance. Secondly, cortical flake(s) need to be removed from the end(s) to prepare a striking platform. This will usually result in the detachment of further cortical flakes from the side of the cobble. These flakes are often sharp, and could be used as tools. Some of these are likely to be elongate, and could be classed as blades if their length is more than twice their breadth. Large cortical flakes might also have been used as cores from which further flakes were detached. Thirdly, as flaking progresses, flakes (or blades) will have little, or no cortex, although their proportions might otherwise be similar to those of cortical flakes. Finally, a core is discarded either because no more flakes or blades can be detached, or because no more are needed. Even after discard, however, a core may still be used as a tool in its own right.

This simple sequence allows one distinction to be drawn in the types of flakes or blades produced from cobbles. Primary flakes are those struck early in a flaking sequence, and still retaining a substantial amount of cortex. Secondary ones are those with little or no cortex. These are formed either by removing the cortex from from primary flakes, or by detaching flakes or blades from a core after the cortex had already been removed. On common- sense grounds, one would expect some primary flakes to show signs of use (through e.g. retouch or edge damage inflicted in tool use, or being snapped). Secondary flakes or blades should exhibit more frequent evidence of use than primary ones, unless the aim was to produce a core tool, in which case these too could - but need not necessarily - be waste material.

From personal (and usually unimpressive !) attempts to experiment with flaking quartzite cobbles, it is much easier to flake ones that are already angular or "blocky", and have distinct facets. A cobble can then be split more easily, or reduced into two or three large flakes which can then be further flaked. Core preparation can be minimal, but the flakes produced tend to be highly irregular in shape, especially if (as in this case), the knapper is none too proficient.

If quartzite cobbles had been knapped at site 55, one would thus expect to find several types of flaked objects. These could include: primary, cortical end- and side-flakes that were detached in preparing a core and then either discarded or used; elongate (blade-like) and squat (flake-like) secondary flakes removed either as artefacts or in refining a core; some cores made from large cortical flakes; cores from which flakes and blades could have been detached; cores that were fully worked-out and from which no more flakes or blades could be removed; and flake spalls, produced in flaking, and probably of no use as artefacts. There might also be evidence of core-tools, such as bifaces and cleavers. There will, however, be a continuum between each category. For example, at some, arbitrary, point, a flaked cobble becomes a core; likewise, long, thin flakes and short, squat blades can be hard to differentiate; large cortical flakes from which further flakes were detached can grade into flake- cores.

There is no general agreement as to how such a gradation of different types of flaked stone should be classified. One could decide, for example, to divide flaked material into two basic categories of cores and flakes/blades. Alternatively, each shape could be regarded as a separate type in its own right. Both are potentially dangerous. A simple classification may mask important and "real" categories of artefacts, whilst one that is more elaborate may give the impression that stone tool production was more complex than was actually the case. Much depends upon the aims of the analyst. In this report, a cautious attitude has been adopted: categories of flaked stone are kept to what are considered to be a reasonable minimum, on the grounds that too little is yet known about the function of different types of early palaeolithic flaked stone to devise complex categories on the basis of their shape.

10.2 The drawn part of the fresh assemblage from site 55

226 pieces of this assemblage, or roughly 20% of the total, were drawn, and are shown in Appendix J, Figs J.1 - J.30. Decisions as to what should or should not be drawn were fairly arbitrary, but the underlying aim was to draw a representative sample of the whole. All pieces with evidence of edge modification were drawn, along with most blades, a smaller proportion of flakes, and some of the cores. Unless otherwise mentioned, all the

material is fresh. The term "abraded" was used to indicate unusually clear indications of mechanical damage to the edge of a specimen. For reasons that should already be apparent, the material was not always easy to categorise, in particular, blades, flake-like blades, blade-like flakes, and flakes grade into each other; some flakes may have been used as cores, and some cores as tools. These subtleties aside, the general features of the fresh assembage should be clear from the illustrations, and a basic division of blades, flakes and cores can be retained as a general guide.

10.2.1 blades and blade-like flakes (Appendix J, Figs. J.1 - J.7)

Fig. J.01

Some pieces in this figure have indications of retouch or edge damage: 210 and 1230 along one side, and with a slight hint of a shallow notch on the latter; along both sides and one end of 1141 (which is better described as a flake); and along both sides of 239 (including a possible notch), 1142 and 1247.

Fig. J.02.

1255 retains much of the original cortex on one side, and may be simply a small and delicately-struck cortical flake. Nos 065 and 1198 seem to have been snapped, as might have 1310. There is a slight indication of retouch or edge damage on the sides of 1198. No. 1310 is weathered, whereas the others are fresh.

Fig. J.03.

Most of the pieces shown here appear to have been broken: 163, 299, 303, 327, 1241, 1389, 1398, 1410, 1415, and possibly 303 and 1119. Some retouch or edge damage is evidenced down one side of 303, 1415, and perhaps 1119; and on both sides of 1398 and 1410. Piece 299 is abraded, but the others shown are fresh.

Fig. J.04.

As in the previous figure, most of the pieces shown are broken: 102, 109, 135, 258, 1102, 1191, 1207, 1219, 1285, 1298, 1325, 1377, 1418, 1420, 1423A, and perhaps 268. Retouch or edge damage can be seen on the converging sides of 1219 (similar to that on piece 1198 shown in Fig. 01) and perhaps on the end of 258, and on one side of 1418 and 1191. Pieces 1304 and 1298 are weathered and 1420 is abraded.

Fig. J.05.

Pieces 040, 074, 141, 1196. 1259, 1263, 1321 and 1383, are broken. There is a notch and some retouch along one side of 1383, and some retouch or edge damage on one side of 1196 and 1278B. Pieces 163 and 1278B were abraded; all the others are fresh.

Fig. J.06.

Most of these have been broken. The following, proceeding from top to bottom of the page, have been broken once: 1120, 1424B, 1375, 1376, 1350, 1342, 306, 1311, 1153, 1251, 1399, 1137, 214, 126, 113 and perhaps 351. Numbers 1346, 1424A and 508 have been broken twice. Retouch or edge damage can be seen on one side of 1120; 1424B, 1346. 1376, 1350, 306 and 351; and on two sides of 1342, 508 and 1153. Number 214 has a notch on one side. Number 1311 was slightly abraded on its sides; the other pieces shown were fresh.

Fig. J.07.

Proceeding from top to bottom of this figure, the following specimens were broken: 1162, 1265, 1289, 1351, 1093, 105 and 133. Retouch or edge damage can be seen on one side of 175, 1162 and 1289; in the last two of these, it takes the form of a small notch. Numbers 1354, 1305 and 1093 have a slight indication of retouch on two sides. Numbers 1185, 17 and 1154 are very small indeed, and seem to have been intentionally made; the last-mentioned shows some edge damage on one side. 1354 is slightly reminiscent of a burin.

10.2.2 flakes and flake-like blades Appendix J, Figs.J.8 - J.20

Fig. J.8.

1368 is one of those pieces that could be classed as a blade or blade-like flake. Edge damage or retouch can be seen on the end of 1221; the side of 1084; the end and one side of 1152 and 1368; part of one side of 1322 and 1365A. Shallow, evasive flaking ioccurred over both sides of 1221, and most of the cortex was removed from the others, apart from 1368. Numbers 1084 and 1136 were abraded.

Fig. J.9.

Numbers 1409 and 136 are somewhat blade-like and snapped. Numbers 376 and 1189 are trimmed cortical flakes. Some retouch or edge damage can be seen on the side of 1409, 383, 1426 and 1100; and the base of one side of 1134. Specimens 376 and 383 came from the gully in square F6 overlying the main surface of the site, and are thus need not be the same age as the rest of the assemblage.

Fig J.10.

Some retouch/edge damage can be seen on the base of 1103; the side of 1143 and 137; and two sides of 1179. Number 1338 is not unlike piece 1304 (Fig. J.04), described above as a blade. Number 1338 may have been snapped.

Fig. J.11.

A very slight suggestion of retouch or edge damage can be seen on 1382 and 1233, but more clearly on 1046 and 506.

Fig. J.12.

All these are fresh except 1124 and 1088, which were abraded. Some retouch or edge damage is apparent on 1124 (base and side); 161 (notch); 1088 (around one side); 1417 and 1175 (side); 1190 (base of side). The last-mentioned is similar to the "blade" specimens 1310 (Fig. J.02) and 126 (Fig. J.06); note also its broken tip.

Fig. J.13.

All these were fresh except the abraded piece 1278A. Retouch/edge damage can be seen on: one side and perhaps also the base of 1000; the base of 1180; and the side of 1278A and 1406.

Fig. J.14.

Some retouch is visible on the side of 010, 125 and 1334. There is also a pronounced notch on piece 216.

Fig. J.15.

All these are fresh except for the abraded blade-like piece, 1224. Some edge damage or retouch can be seen on the side of 082 and the base of the side of 1224. 130 is a good example of a cortical flake from which two flakes have been detached to remove most of the cortex.

Fig. J.16.

There is some retouch or edge damage on the side of 1108, 1361 and 1385; and on the base and side of 124.

Fig. J.17.

There is some retouch on the side of 121; a little on the side of 1087; and some evasive flaking over 1297, like that in 1221 (Fig.J.08).

Fig. J.18.

Some retouch or edge damage is visible along the side of the blade-like 1011; and on the base of the 1176. Specimen 1011 had been snapped.

Fig. J.19.

The largest pieces shown here, numbers 1089 and 1144, are abraded. There is some edge modification to the side of 1144 and 1181.

Fig. J.20.

All these are fresh, except for the abraded pieces, 1267 and 1091. Edge modification can be seen on the sides of 1239, 1357 and 1369, and the base of 1212. Specimen 1217 is similar to 1191 (Fig. J.04) and 1354 (Fig. J.07)

10.2.3 Cores and core fragments (Figs. J.21 - J.29)

Fig. J.21.

181 and 1328 are examples of cores that have been worked out; some edge modification on one side of 1328 may indicate it was then used as a tool. 1282 retains ca. one-sixth of its cortex after having 7 or 8 flakes detached.

Fig. J.22.

There is some cortex and some deep flake scars on both 1116 and 1064.

Fig. J.23.

Both these retain ca.one-third of their cortex. 1291 is a cobble that was truncated and split before being used as a core. Number 169 has some signs of battering.

Fig. J.24.

Very little cortex remains on 1055 and 1283. There is some on 1184, which is also one of the few examples from the excavation of centripetal flaking.

Fig. J.25.

Number 1121 may be a core retaining some of its cortex and worked on both sides, or a core-tool (e.g. a cleaver) with a deliberately made chopping edge. Judging from the obliqely-angled bulb of percussion on 1296, this example may have been detached from a larger block before being further flaked.

Fig. J.26.

Number 1113 still retains a fair amount (perhaps one third) of its cortex. 1425 is a long cortical flake detached from a core and then flaked to remove much of its cortex. It can be compared with the similar but smaller flake number 1000 (Fig. J.13).

Fig. J.27.

A lot of cortex (perhaps 40%) remains on 1122; only some on 1195, and very little on 1237. One edge of 1237 has some indication of modification. Example 1122 is reminiscent of a "chopping-tool" and may have been used as a core-tool as well as a source of flakes.

Fig. J.28.

Six flakes and ca. two-thirds of the cortex have been removed from 1148. Core preparation was fairly minimal, and comprised simply the removal of one cortical flake to create a striking platform. 1421 is abraded, and may be an example of a bifacially- worked core tool. There is some indication of centripetal flaking and edge modification on the flake specimen 1333.

Fig. J.29.

These specimens (117, 202, 1402 and 1405) are fresh, and probably examples of worked out - or at least heavily-used - cores in that they have little cortex and several flake scars.

Fig. J.30.

Specimens 206 and 1220 are good examples of lower palaeolithic- type of bifaces that can be found on the surface in the Riwat area. 1220 in particular is like the bifaces that we have found in secure early Middle Pleistocene contexts in the Jhelum basin (Rendell and Dennell, 1985) at Dina and Jalalpur, ca. 80 - 100 km to the east.

Fig. J.31.

Number 1427 is less weathered than the pieces shown in the previous figure, and indicates the type of centripetal flaking already seen in specimens 1184 (Fig. J.24) and 1333 (Fig. J.28).

10.3 Discussion

Features of this assemblage which merit comment include core preparation; the manufacturing of flakes and blades; and the question of retouch or edge damage. Each can be taken in turn.

10.3.1 Core preparation

In most cases, this appears to have been minimal. Overall, site 55 contains little evidence of disc or prismatic cores, or of Levallois (tortoise) cores. The best evidence for systematic flaking to a pre-determined pattern is the cleaver-like object (1121, Fig. J.25), and the two instances of centripetal flaking (1184, Fig. J.24 and 1333, Fig. J.28), and perhaps 1427 (Fig. J.30). In the case of cores such as 1296 (Fig. J.25), 1113 (Fig. J.26), 1122 and 1195 (Fig. J.27) and 1148 (Fig. J.28), irregularly-shaped flakes were detached from several directions around part of the surface, thus leaving some of the cortex.

As noted above, examples such as 1122 (Fig. J.27) could be regarded as a core-tool rather than (or as well as) a core; if so, it indicates that core-tools did not require elaborate prior preparation.

10.3.2 Flake and blade production

Flakes

As might be expected from the meagre evidence for core preparation, few flakes indicate formal patterning. Most of those shown can be regarded as cortical flakes that have been flaked from the non-cortical surface to remove some or most of the cortex on the other side, and/or to sharpen the flake edge by reducing its angle. Good examples are: 1141 (Fig. J.01); 1368 and 1365A (Fig. J.08); 1134 (Fig. J.09); 1143 (Fig. J.10); 1370 (Fig. J.11); 1000 and 1026 (Fig. J.13); 71 and 130 (Fig. J.15); 1361 (Fig. J.16); 1132 and 1160 (Fig. J.17); and 1147 (Fig. J.18). The fact that five of these (1141, 1000, 1143, 1368 and 1365A) were retouched or had edge damage implies that trimmed cortical flakes were used as tools.

Other flakes have no cortex, and indicate that this was removed before or after flaking. These tend to be smaller, and more frequently retouched. Examples of flakes with no cortex are: 1136, 1084, 1221, 1152 (Fig. J.08); 1409, 136, 1078, 1100, (Fig. J.09); 75, 137 and 1103 (Fig. J.10); 506, 1046, and 1370 (Fig. J.11); 1175, 1190 and 1417 (Fig. J.12); 1180 and 1406 (Fig. J.13); 010 (Fig. J.14); 081 (Fig. J.15); 1083 and 1356 (Fig. J.16); 1087 and 1297 (Fig. J.17); 1408 (Fig. J.18); and 1212, 1314, 1369 and 1357 (Fig. J.20). Several of these have signs of edge damage and may have been used as tools. Listed in their order of appearance from Fig. 10:08 onwards, these are: 1084, 1152, 1409, 137, 1103, 506, 1046, 1175, 1417, 1180, 1406, 010, 1087, 1212, 1369 and 1357.

Finally, there are a few pieces that seem to have been more carefully made than the others. Examples are the small "thumb- nail" flakes (1084 and 1152: Fig. J.08; 1175 and 1417: Fig. J.12; 1406: Fig. J.13).

Blades

As with flakes, some of these are primary pieces that still retain their cortex, whilst others are secondary and struck after the cortex had been largely removed from that part of the core. An example of a primary piece is number 210 (Fig. J.01), which has some edge damage. However, most blades from this site have only a little cortex. Examples are: 239 and 1247 (Fig. J.01); 1389, 1415, 1410, 1398 (Fig. J.03); and 1311 (Fig. J.06). Of these, numbers 239, 1247, 1415, 1410 and 1398 have modified edges. This may indicate that most of the used blades came from well within a core.

Some pieces stand out as unusually small and delicate: 1285, 1423A, 1377, 1191, 1325, 1298 (Fig. J.04); 141 and 1321 (Fig. J.05); 1120, 1424A and 1424B, 1346, 1375, 508, and 1342 (Fig. J.06); and 175, 313, 1289 in Fig. J.07, along with the tiny pieces 1185, 017 and 1154. Most of these are broken. If nothing else, these pieces indicate that blades were an integral part of the fresh assemblage, and when occasion demanded, the quartzite from which these were made could be worked in an extraordinarily delicate and skillful manner. They also help considerably to offset the predominant impression of "crudeness" in the rest of the assemblage, and in surface collections. There is no doubt about the context of most of these pieces. As can be seen from the site plans, almost all were well away from the gully-line and collection from F6, and most were found on the main archaeological surface, up to 2 metres below the present gound level. Of those quoted above, the following came from the conglomerate surface away from the areas likely to have been disturbed by slope-wash from the outcrop (see Fig. 9.7); 175; 1120; 1154; 1185; 1191; 1285; 1289; 1298; 1321; 1325; 1342; 1346; 1375; and 1377. Piece 313 came from the top of the wall footing, and piece 017 from the space between that and the main conglomerate edge. Only 508, 1423A and 1424A/B came from where slope-wash and stream-flow may have transported material.

10.3.3 Retouch and edge damage

As noted in passing already, many pieces have small flake scars on one or more edges, and this type of modification may have been deliberate, or unintentional. If we assume that it resulted whilst a piece was being used, it provides some indication of those which were artefacts. Details of retouch or edge modification are as follows:

11 blades had some edge modification, but were unbroken: 210, 1230, 1141, 239, 1142, 1247 (Fig. J.01); 1117 (Fig. J.02); 1278B (Fig. J.05); 175, 1354 and 1305 (Fig. J.07).

26 blades had some edge modification, and were broken: 1198 (Fig. J.02); 303, 1398, 1410, 1415, perhaps 1119 (Fig. J.03); 258, 1219, 1191, 1418 (Fig. J.04); 1196 and 1383 (Fig. J.05); 1120, 1424B, 1376, 1350, 1342, 306, 508, 1153, 214, 1346, and 351 (Fig. J.06); 1162, 1289 and 1093 (Fig. J.07)

46 flakes (excluding number 383 from the gully in square F6) had some edge modification: 1221, 1084, 1152, 1368, 1322 and 1365A (Fig. J.08); 1409, 1426 and 1100 (Fig. J.09); 1103, 1143, 137 and 1179 (Fig. J.10); 1382, 1233, 1046, 506 (Fig. J.11); 1124, 161, 1088, 1417 and 1175 (Fig. J.12); 1000, 1180, 1278A and 1406 (Fig. J.13); 010, 125, 1334 (Fig. J.14); 082 and 1224 (Fig. J.15); 1108, 1361, 1385 and 124 (Fig. J.16); 121, 1087 and 1297 (Fig. J.17); 1011 and 1176 (Fig. J.18); 1239, 1357, 1369 and 1212 (Fig. J.20); 1333 (Fig. J.28). This total includes one flake (1190, Fig. J.12) that had some edge damage and was also broken.

Three of the drawn cores showed some indication of edge modification: specimens 1328 (Fig. J.21); 1121 (possible cleaver; Fig. J.25)), and 1237 (Fig. J.27).

10.4 Regional comparisons

Unfortunately, few useful comparisons can be drawn between the assemblage from site 55 and other sites in the same region, even if the term region is used to denote the whole of the Indian sub- continent, Central Asia and most of the Near East. As seen in chapter 2, the early palaeolithic of this region is plagued with problems over the dating and context of relevant assemblages.

The paucity of evidence for Levallois-type core preparation in both the core and flake component at site 55 precludes regarding this assemblage as Middle Palaeolithic, still less Mousterian, in the sense that these terms are used in Europe and the Near East. From the little that is known of the late Mousterian assemblages at Shanidar in Iraq, the Levallois and Mousterian components there are noticably far higher than at Riwat (see Solecki, 1963; Akazawa, 1975). Similarly, if a Mousterian can be recognised in Afghanistan (Dupree, 1972), it cannot here. Nearer to Riwat is the site of Sanghao, a rock-shelter in the North West Frontier Province that is claimed to have a long, three-metre sequence that spans much of the Middle and Upper Palaeolithic. It is unfortunate that this material has not been published, for reasons discussed by Allchin and Allchin (1982: 51-52). Nevertheless, the few *"belle pièces"* shown (see Allchin and Allchin, 1982: 53) do not encourage one to regard the Riwat and Sanghao assemblages as similar. Further south, the Middle Palaeolithic or Middle Stone Age assemblages summarised by Agrawal (1982: 28-55), and by Allchin and Allchin (1982: 47-49) from Central India and the Deccan indicate tortoise cores and flakes, what appears to be extensive edge retouch, and an absence of the type of thin blades found at Riwat. However, too little material has yet been published from India to allow more than the suspicion that it represents something different. Moving northwards into Central Asia, Ranov and Davis (1979) have reported late Middle Pleistocene assemblages, but again, the information published so far precludes detailed comparisons.

Comparisons between early upper palaeolithic assemblages in the region and Riwat are similarly frustrated. Whilst blades, often backed and retouched, and burins are commonly reported from the Thar desert (Allchin *et al.*, 1978) and south and central India (Agrawal, 1982, *ibid.*), few details are avaliable on assemblages older than 30,000 b.p. Further west, Baradostian assemblages from western Iran have thin blades at ca. 40 ka. and further work between here and the Punjab may show closer parallels than can be seen at present. An indication of the results that may be expected from further work in Pakistan may be seen from recent studies in Egypt and the Levant. Research here, where the data base is now fairly good, suggest that there is no sharp transition between the middle and upper palaeolithic. Instead there seems to have been a long transition in these areas spanning several millenia, and characterised by industries that can be called *"transitional"* or *"initial upper palaeolithic"*, in which bladelets and Levallois-type flakes were sometimes produced from the same core (Bar-Yosef and Belfer- Cohen, 1988; Marks and Reid Ferring, 1988; Vermeersch and Peer, 1988). It may consequently be suggested that blades featured in lithic assemblages between Europe and India during the equivalent of the late Mousterian (i.e. from after 45 ka), but prior to the emergence of a *"pure"* upper palaeolithic.

10.5 Conclusions

The assemblage of fresh flaked material from Riwat is characterised by:

- a virtual absence of prepared core techniques
- a virtual absence of centripetal flaking of either cores or flakes
- the production of large numbers of cortical (primary) and non-cortical (secondary) flakes
- a smaller proportion of blades, with or without most of their cortex
- the occasional production of thin, narrow blades, and of very small and delicate pieces that could be called microlithic
- by a small amount of systematic edge retouch and evasive flaking

At present, detailed comparisons cannot be drawn between this assemblage and others in the region, and it is difficult to place it within even a general grouping of assemblages. This is particularly the case when only a tiny - and almost certainly atypical - part of other relevant assemblages have been illustrated in publications. The paucity of evidence for the Levallois techniques of core preparation at site 55 implies that this assemblage should not be called Middle Palaeolithic, in the sense used in areas further west, or Middle Stone

Age, as defined in peninsula India. As one blade - or even a few - does not make an assemblage Upper Palaleolithic or Late Stone Age, neither term seems appropriate here. In any case, it seems prudent to avoid terms that have connotations as to whether the assemblage at site 55 was made by Neanderthals or by anatomically modern man.

In addition, the assemblage from site 55 cannot be regarded as Late Soan, in the sense defined by Paterson (De Terra and Paterson, 1939: 285, 295, and Plates 39-42). Paterson's *"late Soan"* was poorly dated, and *"late"* largely on typological grounds; it was also identified from surface collections that may have been composite and almost certainly excluded the small items seen at site 55. Paterson's *"late Soan"* also included large numbers of keeled scrapers, high-angled burins (De Terra and Paterson, 1939: 277), and large numbers of items made by the Levallois technique (ibid: 285) that are not found here.

On account of these factors, and in view of what has been said earlier (chapter 2), it is recommended that the assemblage from site 55 should be regarded as either a *late Early Palaeolithic* one that also includes a blade component, or an *initial Upper Palaeolithic* one, characterised by the production of blades and bladelets. More detailed comparisons with contemporaneous assemblages in neighbouring regions should meanwhile await the publication of more excavated material than is available at present.

Chapter 11

Interpretation of site 55

11.1 Introduction

This chapter attempts to synthesise and interpret the results of the excavations at site 55. As indicated in chapter 8, debate over the function of palaeolithic sites has been lively in recent years, and the interpretation of such sites is no longer the simple matter it once seemed to be. We can begin by summarising the main points of site 55 from chapters 8 - 10:

(i) it contains a low parallel-sided stone feature that has been interpreted as the footing of a low wall. It is thought unlikely that this was made simply as a place for sitting (although it may well have been used for that purpose). It could also have served as the base of a wind-break made from wood and/or hides. However, its main purpose may have been to divert rain-water flowing off the gritstone outcrop behind it so that it left a comparatively dry area of loess between it and the conglomerate surface.

(ii) between this footing and the outcrop there was a small pit, which was lined on one side with a gritstone slab, and packed with struck stone at its base. No charcoal or burnt material was observed, and the pit is explained as a post-hole. Judging by the way it drained quickly after heavy rain, it is unlikely that it was used for retaining water.

(iii) the loess on either side of the wall footing was largely clear of both unflaked and flaked stone. This implies that artefacts were made and discarded elsewhere on the site.

(iv) the conglomerate surface was modified near its edge by the construction of a circular stone-lined "niche". The top of this may have been raised a little with stones, as it was obscured by stones when excavated. Alternatively, it may have been deliberately buried once used. There is also a possible post-socket in the edge of this conglomerate.

(v) a far higher proportion of the stones on this conglomerate surface appears to have been flaked than we have observed in sections of conglomerates, and in other colluvial exposures. There is also some indication that the incidence of flaked cobbles is inversely related to the density of freshly-struck material, implying that some of this assemblage was made on the spot.

(vi) the assemblage of fresh flakes, cores, and especially blades, is concentrated on or near the highest part of the conglomerate surface. It may be significant that this part would have been exposed the longest as the loess continued to accumulate down- slope. In addition, the fresh flaked assemblage was found directly on top of the conglomerate, rather than in the overlying loess.

(vii) the distribution of conjoining pieces shows a similar pattern, and shows also that the fresh assemblage is probably associated with the use of the structure. However, conjoining pieces did not cluster in the manner resulting from the intensive knapping of one piece of stone, and the number of joins per set was generally low.

(viii) no animal remains or evidence of fire was found at the site. The absence of bone may be explained by either the poor circumstances of preservation, or indeed by a genuine absence, as it has not been observed in loess sections nearby. Ash may have been removed by wind and rain, but the absence of any hearth, charcoal, burnt stones, or burnt clay suggests that fire was not used at the site.

(ix) if the white quartzite used for making some artefacts was imported, then some tools were not made from immediately available materials.

(x) a high proportion of the blades - especially those with edge damage - was broken.

(xi) it seems unlikely that the site was disturbed by geological and biological agencies after it was abandoned. Even the gully running across the main surface does not appear to have disturbed the main archaeological horizon. Overall, there seems to have been little post-depositional disturbance in the area of the wall footing and the main conglomerate surface.

11.2 Preliminary considerations: the resolution, integrity and grain of palaeolithic sites

Before we attempt to infer the function of the site from its contents and location, it is useful to consider some general aspects of the evidence, as summarised above. As noted in chapter 8, one of the main lessons learnt in palaeolithic studies over the last decade is that most palaeolithic sites are palimpsests of several agencies and activities, of which human ones may be the least important, even if the most interesting. One way of assessing the quality of data from a palaeolithic (or for that matter, any prehistoric) site is to consider its degree of *resolution* and *integrity* (Binford, 1981: 19-20).

Resolution refers to the homogeneity of events and activities that created the archaeological remains of a site. A site with low resolution is one where individual activities cannot be readily distinguished; instead, the evidence for any one activity is swamped by the evidence for several different activities carried out many times. In contrast, a site with high resolution is one that (ideally) contains evidence of only one activity, performed once. *Integrity* refers to the variety of agencies that were involved in the formation of the assemblage. If integrity is low, numerous agents and agencies were involved in the distribution and accumulation of material – a good example might be a cave used throughout the year by humans, bears, foxes, birds and bats, each leaving its own particular refuse. In contrast, a site with high integrity is one where only one agent was involved (see Binford (1981: 19-20) and Gamble (1986: 23-4) for relevant discussion of these points).

The combination of resolution and integrity allows a measure to be made of the *grain* of a site. If the evidence is *coarse-grained* - because either the degree of resolution, or integrity, or both - are low, *"the correspondence between an event and the archaeolgical record it generated is poor"* (Binford, 1980/1:). However, if the evidence is *fine-grained*, *"the materials deposited reflect more precisely the activities that were carried out at those locations and in relation to the immediate environment"*. In terms of inferring the function of a prehistoric site, it is clearly the fine-grained sites, with both high resolution and high integrity, that offer the most rewarding prospects of inferring site function.

Site 55 appears to be one with both high resolution and high integrity. As noted, there is little evidence over most of the site for disturbance during and after the use and discard of the fresh assemblage. The evidence we have from the site indicates that its human occupants were the main agents responsible for the structural and artefactual remains. The incidence of conjoining pieces linking fresh stone material with the wall footing suggests that this assemblage can be associated with a structure. All the evidence suggests that the site was occupied for only a short period, and not re-occupied later for a different set of purposes. Accordingly, it seems worthwhile to attempt to infer its role from its residues, without fear that it is a mish-mash of several human and non-human agents, each performing different activities over a long period of time.

At this stage we need to consider the type of palaeolithic sites that have been identified.

11.3 The classification of palaeolithic sites

Over the last twenty years, there have been numerous attempts to classify the types of sites evidenced, or likely to be found, in the palaeolithic (see Gamble, 1986: 300). Most of these have been relatively simple, and drawn a major distinction between *residential* and *non-residential* sites. Binford and Binford (1966), for example, suggested that the archaeological record of hunter-gatherers would include *base camps*, *work camps* and *transient camps*. Hole and Flannery (1967) recognised *seasonal base camps*, *butchery sites* and *transitory stations*. Isaac (1978), in his classic paper on early hominid subsistence, recognised *living sites*, *workshop*, *transitory* and *butchery sites* on the basis of different ratios of stone to bone. Price (1978) devised a more complex scheme based on the area and density of artefacts as well as site contents, and recognised *small extraction camps*, *small base camps*; *medium short-term base camps*; *medium long-term base camps*; and *large aggregation camps*.

Although these schemes differ in terminology and often overlap, they agree in some important respects. Most rest on the assumption that the habitation site, or base camp, are an inherent part of human and hominid behaviour. Whilst this seems valid for the archaeological record after 30,000 years ago, doubts are now voiced about its applicability to the earlier palaeolithic record (e.g. Binford, 1987), so this concept might not be valid for sites of the age-range of site 55. A second area of agreement lies in the basic distinction between *residential* and *non-residential* sites. The former may be used by part of a group, an entire group or even several groups; it may be used for a few days or several months; used once, or repeatedly. Non- residential sites are usually associated with specific tasks, such as butchering of animals, procurement of raw materials, or specific functions such as places from which to monitor game. Ceremonial sites might also be included in this category. However, it is likely that residence and function were not always kept distinct. A site might have been *non-residential* in the sense that its primary function was for a specific task or set of tasks, but whilst those tasks were being performed, part or even all of a group might have stayed there, and created some of the archaeological detritus evidenced on residential sites. A third area of agreement in these schemes is over so-called *transit* sites, although this term is perhaps often used as a heuristic device to explain why some sites have so little evidence. Like *living- sites*, they are perhaps more commonly assumed than demonstrated. Finally, most discussions of palaeolithic sites assume that they represent areas where artefacts were both used and discarded: in other words, the spatial patterning of discarded artefacts is supposed to reveal information as to how those artefacts were used **within** the area of discard.

The central issue implicit in all these schemes is that specific types of sites should leave specific types of archaeological "signatures" that we should be able to identify. Unfortunately, we are far from being able to do this. There is no neat checklist, and there is not even agreement as to how many types of sites are, and were, left by hunter-gatherers; there is no ready equivalent of the Munsell soil chart for indentifying the function of a palaeolithic site from its archaeological equivalents to colour, chroma and hue. Interpretations of palaeolithic sites have often been characterised by a combination of vivid imagination and intellectual poverty: imaginative often, because of the detail with which structures and activities materialise from so little evidence; poverty-stricken because so few break free from stereotypes of animal dismemberment on the one hand, and the domestic serenity of the home base on the other. In the time over which this investigation has taken place, the number of kill sites and home bases allegedly evidenced in the palaeolithic has dwindled remarkably. Increasingly, it seems likely that our classifactory systems have yet to encompass the full range of the localised residues of stone and bone debris that constitutes the recent and archaeological record of hunter-gatherers. We need also to bear in mind that extinct types of humans and hominids may have behaved very differently from ourselves, and that pre-upper palaeolithic sites may have no direct equivalent today.

A beginning can be made by considering what site 55 was probably **not**. The complete absence of animal remains and evidence for hearths implies that the site was not a *"base camp"* or *"habitation site"*. This point is made by even a cursory look at the evidence from those few open-air palaeolithic sites that do seem to have been base camps, such as Kostienki in the Ukraine (Soffer, 1985). If doubts are currently (and in my view, reasonably) expressed over whether base camps exist in the European palaeolithic (see Gamble, 1986: 299), there are certainly no grounds for regarding site 55 as an outside contender. The predominance of flakes and blades over recognisably patterned core tools implies also that heavy duty tasks such as butchering animals are not represented: there is thus no reason to assume that site 55 represents a butchery site where the bones have not been preserved. We can note also the absence of any artefact that could be called a projectile, unless one includes the small number of fragile (and usually) broken blade tips. Topographically, the site is not well- placed to serve as a place from which game could be observed, as Gamble (1984: 72) has suggested for Molodova in western Russia along the lines proposed by Binford (1978a) for the Mask site in Alaska. If people wished to monitor game, they could have done so much more easily and ably from the top of the conglomerate ridge a few hundred metres to the south. For obvious reasons, the site cannot be regarded as a burial place. Site 55 also seems to contain too much dbris for a transient camp; and in any case it is hard to see why a group should prefer an overnight stop on a sloping and stony surface, when there was so much flat ground nearby.

If we dispose of explanations of site 55 as a base camp, a butchery site, an observation stand, a burial or ceremonial site, or a transit site, other possibilities have to be explored. Three can be considered: that the site is a discard area; a lithic procurement site; and/or a skin-processing site. Each can be considered in turn.

11.4 Discard zone or activity area?

One possible explanation of site 55 is that it does not represent a "site", in the sense of a residential or activity area, but is simply an area where material from a site nearby was discarded. A good example of how the two might be confused in given in the account of a modern skin-working in Ethiopia by Clark and Kurashina (1981), who documented a skin-worker using obsidian tools. What was interesting was that the area where he worked had very little discard, and most of the refuse of finished artefacts, débitage, as well as the ash and bone from his own domestic refuse was dumped elsewhere on middens. If through time, the ash were blown away, and the more perishable organic materials decomposed, the remaining flaked stone and bone in such a midden could easily be interpreted archaeologically as an activity area where scrapers were used and discarded after processing bone, the most commonly associated type of organic remains. As they noted (*ibid.*, 315): *"From Pleistocene times most of even this slender evidence would not be available"*. Judging from the distribution of scrapers in the midden surface that they plotted, artefacts show a distinct clustering in one part of the midden. Although they did not look for conjoining pieces, there seems no reason why some should not be found under such circumstances. Moreoever, one would expect joining pieces to be dispersed,and there to be only a small number per set.

An explanation of site 55 as a discard area can account for the low number of conjoining flakes, blades and cores per set, and the high proportion of broken blades. The spatial distribution of these and other types of flaked stone would obviously be fortuitous on this explanation.

However, we have noted that the fresh assemblage from site 55 is associated with a structure, and it thus seems unlikely that the flaked stone was used elswhere but then discarded on the site. This may have happened if the flaked assemblage post-dates the structure. We have already noted that this structure **overlies** some of the fresh assemblage; however, one flaked piece on the wall-footing conjoined another piece on the main surface. There thus seems no reason to infer that the site contains an assemblage-less structure and also a structure-less assemblage.

11.5 Lithic procurement site?

We noted earlier (chapter 2) that palaeolithic sites in the Soan Valley tend to be found on conglomerate exposures, as these were readily accessible sources of quartzite cobbles, the material most commonly used for making stone tools. This explanation would also explain why people focussed their attention on one of the stoniest parts of the area.

It is useful at this point to consider some characteristics of lithic procurement sites in a well-documented area, such as the Ukraine. Soffer (1985) has some valuable comments on how these might be identified, from broadly analogous open-air sites from the loessic landscape of western Russia. As she (*ibid.*, 364) notes:

> *"I suggest that we can anticipate lower blade-nuclei ratios at lithic workshops and higher ratios at sites with a wider array of maintenance activities. I assume that blades are manufactured at lithic workshops, not used extensively there, and removed to other types of maintenance sites. At base camps, on the other hand, the wide array of maintenance activities performed should necessitate a far more extensive use of blades. We should therefore anticipate blades in far greater numbers at this class of sites".*

(Ratios of blades to cores are considered low if ¡25:1).

Soffer (*ibid.*, 388) further states:

> *"I identify sites exhibiting the presence of only lithic assemblages, predominantly made of locally available materials, as well as high densities of nuclei, low blade-nuclei ratios, and low end-scraper and burin densities ... as lithic workshops. These sites lack grinding stones, exotic lithics and decorative objects, worked bone, hearths, and small pits".*

An example that makes the point is Pogon where:

> *"general low densities of tools and end scrapers and burins, a high nuclei densitiy, a low proportion of nuclei, and a low blade-nuclei ratio all indicate extractive activities associated with lithic procurement"* (Soffer, 1985: 390-1).

However, this site also contains exotic lithics, ochre, and bone charcoal, and may therefore (like so many other sites) be a palimpsest of several activities.

Site 55 accords with some of Soffer's criteria of a palaeolithic open- air lithic procurement site. There is a low ratio of blades and flakes to cores: at site 55, the fresh assemblage contained 399 flakes and 210 blades and blade fragements, compared to 172 cores and core fragments. The ratio, of 609 flakes and blades to 172 cores, is thus low, and only 3.54: 1. The site also lacks evidence of hearths and exotic items (apart from a few items made from white quartzite that may have been taken there from nearby). If the site was used for lithic procurement, there is no reason to assume it was anything but small-scale. It is most unlikely that the quartzite at site 55 was any different from that easily obtained elsewhere in the Soan Valley, and there is no reason to assume that it was valued further afield.

However, it seems unlikely that site 55 can be explained solely in terms of lithic procurement. If it had been simply a lithic workshop, the following aspects of its evidence are hard to explain:

(i) why then retouch or damage the edges of so many?

(ii) why import white quartzite nodules to a lithic procurement site to make stone tools when there were at least 3000 other quartzite cobbles to choose from?

(iii) why were most of the blades broken?

(iv) why build a structure that includes a wall footing, a niche, and a stone-lined pit?; and also:

(v) one would expect the patterning of conjoining pieces of flaked stone to show distinct clusters of where stone was worked, and a high number of joins per set. Instead, there is no evidence of clustering, and the number of joins per set is generally very low.

11.6 A skin-processing site?

As shown already (chapter 6), the area around site 55 was largely loess-covered when the site was used, and almost all the valley between the conglomerate ridges on either side of the site was stone- free. We should thus consider why a sloping conglomerate surface might have been attractive to those who used it. We can start with the premisses that stone may have been made at site 55 for immediate use, there or nearby; and that the conglomerate surface itself was a major reason for the use of the site. At this point, we can consider the possibility that the conglomerate surface was a useful area to dry animals skins.

11.6.1 Skins and skin-working

Animal skins of deer are in their best condition for making clothing in autumn and early winter. Spiess (1979: 29-30) notes that reindeer skins are at their best in winter, and their worst in summer at the annual moult. At end of summer, warble flies hatch subcuteaneously and eat their way through the skin. Thereafter, the holes heal quickly. The best time to procure reindeer skins is in the early autumn, in September and October.

The preparation of deer (and most other types of herbivore) skin involves several stages. They have to be cleaned of fat; depending on their intended usage, loose hair may have to be scraped off (sometimes after soaking); it may be important to make, or keep, them supple; and they may also need to be carefully dried (see Reed, 1972 for detailed discussion). Whilst drying, it is often the practice to hang them so that both sides can dry simultaneously, or to let them dry on the ground and turn them over repeatedly. Stony surfaces are attractive for this because they heat up faster than flat, densely-packed and smooth surfaces during the day, and allow air currents to circulate between the skin and the ground.

There are numerous and often lengthly accounts of non-industrialised methods of skin-preparation in the ethnographic literature (e.g. Mason, 1891 and Rogers, 1973) which show the variety of methods that can be used to convert animal hide into bedding, clothing and other items. A good description of these processes is found in Clark (1952: 219). He suggested that hides were often soaked before the hair was removed with scrapers. Details of tanning sheep-skins – using roots of tormentil *Potentilla sp.*– in the Faroe Islands in recent centuries are also given:

> *"The roots were gathered by the women who prepared them for use by grinding them in natural depressions in the rocks by means of stone pestles. The powder was mixed with water to form a paste, which was rubbed into the hairy side of the skin after the wool had first ben removed. After a couple*

of days or so the skins were washed in sea water and dried." (*ibid.*, 219 footnote).

Some illuminating descriptions of skin-processing are also found in Binford's account of the Nunamiut Eskimo in Alaska. Binford (1978b: 285) observed that:

> *"Skins taken at that time of year (i.e. later summer and fall) are frequently destined for use as clothing. Such choice skins are placed on the ground, hair side down, and weighted around the edges with medium-sized stones to prevent their blowing away and curling around the edges as they dry. In this manner they are dried and therefore reduced in weight prior to being rolled into bundles".*

One skin-processing site was Little Happy New Year, which consisted of two hearths, a windbreak, and two small skin-weight circles where fall skins had been weighted down for drying (Binford, *ibid.*, 352). This site is also a hunting-stand, so contains more than one type of residue. Other skin-processing sites observed by Binford lacked wind- breaks. One was Kongumuvuk, another autumn hunting stand where there were: *"four 'eating' hearths on the site as well as seven well-defined small stone rings used in drying fall caribou skins"* (*ibid.*, 355). Another illuminating pointer from this ethnographic study is that rings of stones for drying hides are usually found with scrapers and chipping debris from sharpening scrapers (*ibid.*, 402).

Another excellent description of reindeer skin-working is given by Gronmow *et al* (1983: 34), from their work on Eskimo sites in Greenland:

> *"The skins are treated in two different ways, according to what they are to be used for. If they are to be bed skins, part of the legs is first cut off and sewn to the head so (the skin) is rectangular. The fat and things are scraped off with an ulo (woman's knife), if that has not been done before. It is stretched out and washed. One day when it is dry, it is moistened and folded. It is then scraped with a blunt ulo and when it is quite dry, rubbed and scraped with a stone. If the clothes are to be made from caribou skin, you do not sew onto the head. It is stretched out and washed. When it is dry, it is rubbed in the hands and treated with fine sand and a stone, without moisture".*

From these descriptions and from consideration of the inherent properties of ungulate skins, some predictions can be made as to the character of hunter-gatherer skin-processing sites:

(i) they do not need to be associated with a habitation site, as skins can be processed whilst other autumn activities take place. In fact, because skin-working could be undertaken almost anywhere, the location of skin-processing sites will probably have been determined by other activities, such as the need to monitor game, or obtain raw materials.

(ii) the main artefacts used in skin-working are scrapers, and weights to keep the skin from blowing away. According to Seminov (1973: 82-93), the commonest purpose of the upper palaeolithic end- scrapers that he examined from Soviet sites was *"for treating skin, for scraping and softening skins after they had been taken off the animal"* (*ibid.*, 87). He further commented that *"In all probability during upper palaeolithic times there were no special scrapers for cleaning the flesh side of the hide. Cleansing the skin of flesh, fat and muscle fibre which usaually occur on freshly removed skin could have been done with end-scrapers with a wider and sharper working edge. The blunted ones were intended for rubbing the skin to make it soft, that is currying it".* (*ibid.*, p.98). Interestingly, he also observed that irregular flakes, cores, and so-called "pebble tools" could also be used as scrapers for skin-working.

(iii) Evidence for skin-drying facilities such as racks, and containers for water, powders, pastes and other chemicals is likely to be elusive, particularly on pre-industrial sites.

(iv) it seems very unlikely that a site would be used only for the processing of animal skins. After all, watching skins dry must be only marginally more exciting than watching paint dry, and usually some *"boredom component"* should be evidenced, such as the servicing, and manufacturing of other tools, if only as a way of passing the time.

Site 55 fits reasonably well with these descriptions of skin- processing. First, the structure is compatable with the supposition that a small group may have spent some time there on a specific task. Secondly, the *"niche"* at the top of the conglomerate surface can be seen as a way of storing water that may have been needed whilst processing hides; thirdly this explanation accounts for the high proportion of broken blades better than the suggestion that the site may have been only a lithic procurement site. Finally, skin-processing explains why the conglomerate surface was the main focus of attention at this site, and not the flatter, loessic surfaces nearby.

There are a number of weaknesses in this explanation, however. The most obvious is that there is no positive evidence that any of the fresh assemblage was used for processing skins. There are two reasons why: the first was that at the time (1981 - 1982), it was not possible to examine any of the material microscopically for edge-damage and - modification. It was not then possible to have artefacts sent abroad for this type of detailed study; and even if it had been, it would have been very difficult to find a person with the relevant experience of quartzite wear analysis to conduct the study. Hopefully, this type of research may be undertaken at some future date. A second weakness is that the low ratio of cores to flakes and blades implies that many of the latter items were discarded **off** the site; and this in turn may imply that the site was also used as a lithic procurement site in the sense defined by Soffer. Finally, if the *"niche"* was used for storing water, it is unclear how, or from where, it could have been filled.

11.7 Conclusions

The main points reached in this chapter are as follows:

1) Site 55 is unusual in the sense that it has both *high resolution* and *high integrity*; its evidence is thus *fine-grained*, and probably relates to one main activity, undertaken for only a short period of time.

2) Current difficulties of classifying and identifying different types of palaeolithic site are noted, as is the need to widen the frames of reference for dealing with palaeolithic sites beyond simple categories of living sites, butchery, workshop and transit sites.

3) Site 55 does **not** support explanations that it was a residential, a butchery, a ritual, or a transit site.

4) It is considered unlikely that site 55 repesents an area where refuse was dumped from a neighbouring activity area. The most obvious type of site to which site 55 can be compared are *lithic procurement sites*, such as those identified by Soffer in a broadly analogous area of western Russia. However, this explanation does not seem wholly satisfactory, and it is consequently suggested that it may also have been used as an location for processing animal skins in the autumn, with the procurement of flakes and blades as an ancillary activity.

Chapter 12

Summary and Conclusions

This volume presents the results of six years of investigations into the Pleistocene and Palaeolithic history of a small area of the Soan Valley. We began this report with a review of De Terra and Paterson's (1939) *Studies of the Ice Age of India and Associated Human Cultures*, published 50 years ago under academic and logistic circumstances that are virtually unrecognisable today. In evaluating their work, we have often been critical of their methodology and results, and we have shown that almost all of their conclusions have been untenable for some time. In De Terra's case, he was at fault in assuming that the geological record of the Soan Valley and Kashmir showed an essentially climatic rather than tectonic record; he made serious errors in the field in imposing a sequence of river terraces on the Soan valley landscape, and in assuming that this sequence would provide a complete record of changes in Pleistocene climate. He made a major error in following Wadia's (1928) mistaken view of the relationship between the northern and southern limb of the Soan syncline, and consequently of the relationship between Siwalik and post-Siwalik strata on the southern limb of the fold. His attempts at correlating glacials and interglacials, deduced from his supposed terrace sequence, with a four-fold sequence of world-wide glaciations are also untenable, not least because of what has since been learnt about the glacial record of the Pleistocene. Unknown to him, his main error was in assuming that the Upper Siwalik "Boulder Conglomerate Stage" represented a single time unit within the Pleistocene, which he then correlated with the "Great Glaciation" (i.e. the "Mindel" glacial) of the Alps. Recent work has demonstrated that the Upper Siwalik Boulder Conglomerate is time-transgressive, and cannot therefore be correlated with any single event elsewhere. De Terra compounded his mistake within the Soan Valley by mapping the post-Siwalik Lei Conglomerate as Upper Siwalik, an error that sowed seeds of confusion for subsequent studies because the former is around a million years younger than the latter. Paterson's work was no less faulty in several important respects. We have shown how his approach to the Soan Valley in particular was pre- conditioned by his main research on the Clactonian assemblages of southern England, and how he was influenced by current French views of gradual, "progressive" evolution in the palaeolithic that have been obsolete for many years now. We have shown that few of his assemblages can be dated, either relatively to one another or absolutely, and that most are probably mixed and derived from long and perhaps intermittent use of conglomerate exposures as sources of raw material. Most importantly of all, we doubt his identification of a pre-, an Early and a Late Soan, and his conclusion that the Soan Flake Tradition existed alongside, and independently of, the Acheulean bifacial tradition.

These are harsh conclusions, but ones that we feel are fair. This is not to belittle De

Terra and Paterson's achievements, provided that their work is judged in the context of its time. We have tried to tease out the reasons why they worked they way they did; to identify the ingredients that shaped their view of the past, and to understand why they used the methods that they did. To that extent, our appraisal of their work is also an enquiry into the history of Pleistocene geology and Palaeolithic archaeology. If their work was at fault, so too was most of that of their peers and contemporaries, including figures that were, or later became, more important, such as the Abbé Breuil and Louis Leakey. Had we been working in the 1930's, we doubt whether we would have approached the Soan Valley very differently; similarly we hope that putative judges of our work will attempt to set it in the context of the 1980's. Having lived in De Terra and Paterson's shadow for several years, we cannot but admire their breadth of vision, their energy and self-confidence, however wrong their conclusions may now appear.

What we have attempted to do in the second part of this monograph is to put something in place of Studies of the Ice Ages of India. As is evident, we have worked on a totally different scale, with totally different field and laboratory techniques, and wholly different perceptions of the past. Instead of attempting to cover the whole of northern Pakistan and India in one five-month field season, we have spent twelve months between us, over ten seasons, in one tiny part of the Soan Valley, examining an area that De Terra and Paterson would probably have exhausted in a matter of days. Unlike De Terra and Paterson, our main geological concerns have been with sequences rather than with landforms; and unlike them, we have had to grapple with a glacial-interglacial sequence that is far more complex than the appealing simplicity of only four world-wide glaciations. Thanks to palaeomagnetism and radiometric dating techniques, we have been able first to date absolutely sections and exposures that were undatable to them; and secondly to appreciate the recent nature and extent of tectonic changes in the Soan Valley. Unlike them, we can accept and demonstrate that major geological units are time-transgressive, and we do not need to be unduly pre-occupied with large-scale regional correlations as a way of establishing a chronological framework. Archaeologically- speaking, we have concentrated on finding small amounts of well-dated material, and paid scant attention to the type of surface collections that constituted the bulk of Paterson's evidence. Absolute dating has relieved us of the need to devise typological sequences based on notions of gradual, progressive change; and research into the early palaeolithic in the last few decades has shown how inter-assemblage variability within a given period frustrates most attempts to show simple, uni-directional change through time. This is not to say, of course, that change did not occur in the lower palaeolithic; merely that merely that others factors such as the circumstances and location of tool use and discard, the type of raw materials used, and the taphonomic history of each assemblage need to be considered before such changes can be identified.

Not surprisingly, our results are very different from those that De Terra and Paterson published in 1939, but we hope, no less fascinating. As we hope to have shown, the Plio-Pleistocene history of the Soan Valley is just as enthralling and exciting as it must have appeared to them, even if it must now be seen in terms of tectonic as well as climatic change. As reviewed in chapter 3, the Upper Siwaliks have been investigated longer and in greater depth than any other Neogene sequence, and in the last decade, our understanding of it has advanced enormously through the application of palaeomagnetic and radiometric work. Our debt to that work, particularly that undertaken by the Dartmouth – Peshawar group is immense, and much of what we have attempted would not have been possible without their programme of palaeomagnetic dating. In particular, their provision of a chronological framework for the Upper Siwalik, and their demonstration that the"Boulder

Conglomerate" is not a "stage" but is in fact time- transgressive were major breakthroughs in developing an understanding of the Pleistocene geology of the Potwar.

The post-Siwalik record tends to be overshadowed by the fascinations of the Upper Siwaliks, but should be treated as being of equal interest. Much of our work has been on the Upper Pleistocene, and particularly on the so-called "Potwar loessic silt". We have shown that much of this is true loess, and that much of it was deposited during the last glaciation (i.e. between 75 ka and 18ka), and not during the penultimate ("Riss") glaciation, as supposed by De Terra and Paterson. Where we have been fortunate is that our work coincided with the development of TL as a way of dating loessic deposits. This technique was developed during this current project, and the opportunities of sampling extensive and thick loess deposits at Riwat also led to some innovative and useful contributions that have helped establish TL as a useful and viable dating technique. In methodological terms, one of the most rewarding parts of this study for the principal geologist was the opportunity to participate in the development of a laboratory technique that could be used to date the deposits under investigation in the field.

Our archaeological results from an area less than 3 km by 1 km fall far short of replacing Paterson's sequence with another of equal breadth, but hopefully provide a series of starting points for future workers. Three sets of discoveries have been presented in this volume, of very different ages and type. Each can be briefly reviewed in turn:

i) the initial upper palaeolithic or late early palaeolithic

By far the largest amount of archaeological evidence presented in this volume is from site 55, dated by TL to ca. 45,000 years b.p. As shown from the numbers of conjoining pieces, this assemblage has a high integrity, and probably relates to a very short period of use and discard, perhaps in the order of weeks rather than years. By excavating a large area of up to 200 m^2, spatial variability in the distribution of different types of flaked stone could be shown. In particular, it was demonstrated that a smaller excavation might have inadvertently over- or under-estimated the importance of blade manufacture. What was of interest was the presence of small blade elements in an assemblage otherwise dominated by flakes, but made without elaborate core preparation. Unfortunately, the regional status of this assemblage cannot be assessed until more is published on comparable assemblages from elsewhere in Pakistan, northern India, Afghanistan and Iran. In following recent trends, we have avoided terms such as Middle Palaeolithic or Middle Stone Age; for reasons evident in chapter 2, we cannot retain Paterson's concept of a late Soan, as that was poorly defined and is largely undated, and perhaps even undatable. Future work may clarify whether the site 55 assemblage is classed as *"initial upper palaeolithic"*, or as *"late early palaeolithic"*; however, the distinction between the two is likely to be blurred, and a lengthy terminological debate over this point at this stage is probably of little value.

As is evident in chapters 8 and 11, explanations of the distribution and type of material found at site 55 have been cautious, given ongoing and long-term uncertainties over the interpretation of palaeolithic sites in general. The most definite conclusion that can be reached is that it does not represent a *living site*, as normally defined. Other suggestions are inevitably more tentative. A structure of some kind existed there, but we have resisted temptations to foist yet another hut on the palaeolithic from the evidence at hand. The suggestion that the site was used for a specific function such as skin working is one that needs further enquiry. One line is through wear analysis on some of the artefacts; another is to compare site 55 with other palaeolithic examples in better documented areas. A third approach that could not be tried here would be to excavate other contemporaneous sites in the Riwat area to assess the degree and type of inter-assemblage variability. If, however, the site was one with a special function, it is of interest to debates over the advent of modern human beings, and a recognisably modern way of hunter-gatherer behaviour. Neanderthal behaviour is commonly presented these days as "pre-human" in the sense that they did not have the type or depth of advance planning seen in Europe in the later Upper Palaeolithic with *Homo sapiens sapiens* (see e.g. Dennell, 1983: 79-102). Site 55 is intriguing in that (assuming, as always, that the dating is correct!) it implies that this type of behaviour was being practised whilst Neanderthals were still inhabiting Europe. It remains to be seen how typical or atypical site 55 is of its period in Asia.

ii) the Middle and early Upper Pleistocene

Undoubtedly the largest gap in our present knowledge of the Pleistocene of northern Pakistan concerns the Middle and early Upper Pleistocene (i.e. between 700 ka and 45 ka). Unsurprisingly, therefore, next to nothing is known of the palaeolithic record of this period. In European terms, this time-span corresponds to most of the Lower and almost all the Middle Palaeolithic. At present, the only securely-dated material that we know of in northern Pakistan from this 600,000 year period are the three hand-axes and a few flakes from Dina and Jalalpur; and the hand-axe and disc core from the upper conglomerate at Riwat. Five pieces of stratified material from 600,000 years is an obvious invitation and plea for future fieldwork. What can be suggested as possible pointers to the future are that on historical grounds (see chapter 2), we see no reason to accept Paterson's assertion that there were two independent traditions of hand axe- and flake-users in the Punjab during this time. Instead, we would offer a more parsimonious suggestion, that there was only one Middle Pleistocene tradition in the Punjab, and this used and discarded both hand-axes and flakes. Given that four of the five artefacts known from secure contexts from this period are bifaces, and that the Acheulean is very well-documented in India, we feel it is also likely that northern Pakistan has a Middle Pleistocene Acheulean awaiting discovery. It is unfortunately too early to speculate on the age and significance of the type of Levallois cores and flakes that we have noticed on erosional surfaces in the Riwat area.

iii) A pre-Acheluean?

By far the most surprising discovery we made was that of hominid- struck pieces in the lower conglomerate horizon at Riwat that we have subsequently dated to ca. 2.0 Ma. We recognise fully that this is the most contentious part of our results, since this impinges directly upon prevailing views of human origins and Eurasian colonisation. We do not see at present how we could have reached any other conclusion over this material. We remain satisfied that R001, R014 and R88/1 were deliberately struck, and that they are wholly different from the overwhelming amount of naturally-modified clasts that we have observed in this and other conglomerate horizons. Although other pieces are more debatable as instances of hominid-flaking, the point remains that in our view - and in the view of most of these who have seen the replica or original of R001 - at least one piece was deliberately flaked. It thus has to be explained, however awkwardly. We have shown that flaking occured before incorporation into that horizon; and we think also that the context has not subsequently been reworked. There is no doubt in our minds that the lower conglomerate horizon is overlain by 70m of Upper Siwalik deposits, and we have shown that these must be between 0.9 and 2.47 Ma old on palaeomagnetic grounds. We also have no doubts that the section at Riwat is an integral part of the Soan Syncline. The dating of that folding to 1.9 Ma is the result of the work of other investigators, but we see no reason to disbelieve their conclusion.

If our conclusion that artefacts were being flaked in Pakistan ca. 2 Ma ago is rejected on the grounds that the dating is subsequently proven to be too old, then we have no cause for complaint. Where we would complain is if it is rejected simply on the grounds that is at odds with current models of human origins. Models should encourage rather than restrict further work; a discovery should not be rejected or ignored simply on grounds of inconvenience or unpopularity.

Our overall feeling after ten field seasons in northern Pakistan is one of humility towards the landscape we have been investigating. It is not a landscape (*contra* De Terra and Paterson) that yields quick and easy answers; it is fascinating and complicated, but often deceptive and unyielding. Rewards can often seem scant when set against the amount of time and effort required to obtain them. In many ways, these feelings should be evident in this study. Rather than providing a large set of answers to the problems initiated by and resulting from De Terra and Paterson, we feel we have probably ended with a longer and more thorough set of questions. There has clearly been some useful progress in recent years in unravelling the Plio- Pleistocene geology and chronology of the Upper- and post-Siwaliks. What is singularly lacking from current research on both the Siwaliks and post-Siwaliks is a climatic record that can stand independently, and be compared with what is now known from deep-sea cores and other sources. The work presented here on the loess sequence of the Potwar and Soan valley goes some way towards providing a climatic record for the Late Pleistocene, but obviously needs to be extended back in time, and supplemented by other kinds of data. Probably the greatest geological challenge now facing researchers in this area is to identify, and to gauge the effects of, the numerous climatic changes of the last 2-3 million years, especially in a region so dramatically affected by the recent uplift of the Himalayas. As stated earlier, the fascination of this region during the Pleistocene lies in the study of the interplay between climatic change and tectonic instability, and particularly in the way these two factors affected the hydrological regimes of streams and rivers, and vegetational and faunal communities. Another major problem is that the geological sequence of the Potwar contains some enormous gaps and discontinuities that were signally ignored or overlooked by De Terra when he placed his erosional unconformity in the Soan area between the Middle and Upper Siwaliks. In so doing, he gave the misleading impression that the Upper Siwalik record of this area was far more continuous than it now appears to have been. One hiatus that is particularly relevant here is between the end of the folding of the Soan Valley ca. 1.9 Ma ago, and the oldest-surviving post-Siwalik deposits that in some parts of the valley may be a million or more years younger. Palaeolithic archaeologists are inevitably entrapped in such problems. If there is a pre-Acheulean in Pakistan and India, it is clearly of great general significance, but equally clearly, it will be difficult to isolate in an area where the landscape has changed so profoundly, and where so much has been eroded. The present scarcity of evidence for whatever constitutes the palaeolithic of the Middle Pleistocene is considerably bound up with the scarcity of evidence for landscapes of this period. Although the loess landscape of the Potwar is one of the best- preserved upper pleistocene landscapes in the Old World, it also masks a hunter-gatherer landscape that is at present represented only by one excavated site, and hardly anything is yet known of what preceded or followed it. If this volume encourages others to supplement or even refute some of the conclusions that we have reached, it will have had some value.

12.1 References

Agrawal, D.P. (1982) **The Archaeology of India.** Scandinavian Institute of Asian Studies Monograph **46**. London and Malmo: Curzon Press.

Agrawal, D.P. (1988) Palaeoclimatic data from Kashmir: a synthesis and some correlations. *Proceedings of the Indian National Science Academy* **54A**: 333-342.

Agrawal, D.P. and Gupta, S.K. (1988) Climatic changes in India during the last 700,000 years *Proceedings of the second International Conference on the Palaeoenvironment of East Asia from the Mid-Tertiary* **1**: 488-499.

Aitken, M.J. (1976) Thermoluminescent age evaluation and assessment of error limits: revised system. *Archaeometry* **18**: 233-238.

Aitken, M.J. (1985) **Thermoluminescence Dating**. London and New York: Academic Press.

Aitken, M.J. and Alldred, J.C. (1972) The assessment of error limits in thermoluminescent dating. *Archaeometry* **14**: 257-267.

Akazawa, T. (1975) Preliminary notes on the Middle Palaeolithic assemblage from the Shanidar cave. *Sumer* **31**: 3-10.

Allchin, B. (1963) The Indian Stone Age sequence. *Journal of the Royal Anthropological Institute* **93 (2)**: 210-234.

Allchin, B. (1981) The palaeolithic of the Potwar Plateau, Punjab: a fresh approach. *Paléorient* **7 (1)**: 123-134.

Allchin, B. and Allchin, R. (1968) **The Birth of Indian Civilisation: India and Pakistan before 500 B.C.** London: Penguin.

Allchin, F.R. and Allchin, B. (1982) **The Rise of Civilisation in India and Pakistan.** Cambridge: Cambridge University Press.

Allchin, B., Goudie, A. and Hegde, K. (1978) **The Prehistory and Palaeogeography of the Great Indian Desert**. London and New York: Academic Press.

Anderson, R. van V. (1927) Tertiary stratigraphy and orogeny of the northern Punjab. *Bulletin of the Geological Society of America* **38**: 665-720.

Bar-Yosef, O. and Belfer-Cohen, A. (1988) The early upper palaeolithic in Levantine caves. In **The Early Upper Palaeolithic: Evidence from Europe and the Near East**, edited by J.E. Hoffecker and C.A. Wolf, 23-42. British Archaeological Reports (International Series) **437**.

Binford, L.R. (1977) Olorgesaillie deserves more than the usual book review *Journal of Anthropological Research* **33**: 493-502.

Binford, L.R. (1978a) Dimensional analysis of behavior and site structure: learning from an Eskimo hunting stand. *American Antiquity* **43**: 330-361.

Binford, L.R. (1978b) **Nunamiut Ethnoarchaeology**. London and New York: Academic Press.

Binford, L.R. (1981) **Bones: Ancient Men and Modern Myths**. London and New York: Academic Press.

Binford, L.R. (1987) Searching for camps and missing the evidence? Another look at the lower palaeolithic. In **The Pleistocene Old World**, edited by O. Soffer, 17-31. New

York and London: Plenum.

Binford, L.R. and Binford, S.R. (1966) A preliminary analysis of functional variability in the Mousterian of Levallois facies. *American Anthropologist* **68 (2)**: 238-295.

Birley, R. (1982) **The Innocent Anthropologist**. London: Penguin.

Black, D. (1925) Asia and the dispersal of primates. *Bulletin of the Geological Society of China* **4**: 133-183.

Black, D. (1926) Tertiary man in Asia: the Chou Kou Tien discovery. *Nature* **118**: 733-734.

Black, D. (1934) On the discovery, morphology, and environment of Sinanthropus pekinensis. *Philosophical Transactions of the Royal Society of London* Series B: **223**: 57-121.

Boriskovsky, P.I. (1978) Some problems of the palaeolithic of South and Southeast Asia. In **Early Palaeolithic in South and East Asia**, edited by F. Ikawa-Smith, 87- 95. Mouton: The Hague.

Bowall, L., Frean, R., McKeogh, K.J., Rendell, H.M. and Townsend, P.D. (1987) Sensitivity changes in the TL of quartz and feldspar after bleaching. *Crystal Lattice Defects and Amorphous Materials* **16**: 37-43.

Bowler, P.J. (1986) **Theories of Human Evolution: A Century of Debate 1844-1944**. Oxford: Blackwell.

Breuil, H. (1939) The Pleistocene succession in the Somme valley. *Proceedings of the Prehistoric Society* **5**: 33-38.

Breuil, H. and Koslowski, L. (1931) Étude de stratigraphie paléolithique dans le nord de la France, la B'elgique et l'Angleterre. La vallée de la Somme. *Anthropologie* **41**: 449-488; (1932) **42**: 27-47; 291-314.

Broome, R. (1925) Some notes on the Taungs skull. *Nature* **115**: 569-571.

Burbank, D.W. (1983) The chronology of intermontane-basin development in the northwestern Himalaya and the evolution of the Northwest Syntaxis. *Earth and Planetary Science Letters* **64**: 77-92.

Burbank, D.W. and Johnson, G.D. (1982) Intermontane-basin development in the past 4 Myr in the north-west Himalaya. *Nature* **298**: 432-436.

Burbank, D.W. and Johnson, G.D. (1983) The late Cenozoic chronologic and stratigraphic development of the Kashmir intermontane basin, northwestern Himalaya. *Palaeogeography, Palaeoclimatology and Palaeoecology* **43**: 205-235.

Burbank, D.W. and Raynolds, R.G.H. (1984) Sequential late Cenozoic disruption of the northern Himalayan foredeep. *Nature* **311**: 114-118.

Burbank, D.W. and Tahirkheli, R.A.K. (1985) The magnetostratigraphy, fission track dating, and stratigraphic evolution of the Peshawar intermontane basin, northern Pakistan. *Bulletin of the Geological Society of America* **96**: 539-552.

Butler, R.W.H. and Prior, D.J. (1988) Tectonic controls on the uplift of the Nanga Parbat Massif, Pakistan Himalayas. *Nature* **333**: 247-250.

Cahen, D., Keeley, L.H. and Noten, F. van (1979). Stone tools, tool- kits and human behavior in prehistory. *Current Anthropology* **20**: 661-683.

Clark, J.G.D. (1952) **Prehistoric Europe: The Economic Basis**. Cambridge:

Cambridge University Press.

Clark, J.D. and Kurashina, H. (1981) A study of the work of a modern tanner in Ethiopia and its relevance for archaeological interpretation. In **Modern Material Culture: The Archaeology of Us**, edited by R.A. Gould and M.B. Schiffer, 303-321. London and New York: Academic Press.

Colbert, E.H. (1935) Siwalik mammals in the American Museum of Natural History. *Transactions of the American Philosophical Society (N.S.)* **26**.

Cotter, G.de P. (1929) The erratics of the Punjab. *Records of the Geological Survey of India* **61**: 327-336.

Cotter, G.de P. (1933) The geology of the part of Attock District west of longitude 72° 45'E. *Memoirs of the Geological Survey of India* **55**: 63-161.

Coulson, A.L. (1938) Pleistocene glaciation in northwestern India. *Records of the Geological Survey of India* **72**: 422-39.

Crandell, D.R. (1971) Postglacial lahars from Mount Rainier volcano, Washington. *U.S. Geological Survey Professional Paper* **677**.

Dainelli, G. (1922) Studi sul glaciale. In **Spedizione italiano de Filippi nell' Himalaia, Caracorum e Turchestan Cinese (1913-1914) serie II, Risultati Geologici e Geografici 3**

Dart, R. (1925) Australopithecus africanus: the man-ape of southern Africa. *Nature* **115**: 195-199.

Davis, R.S. (1987) The Soan in Central Asia? Problems in lower palaeolithic culture history. In **Studies in the Archaeology of India and Pakistan**, edited by J. Jacobson, 1-17. Warminster: Aris and Phillips.

Dennell, R.W. (1983) **European Economic Prehistory: A New Approach**. Orlando: Academic Press.

Dennell, R.W. (1984) The importance of the Potwar Plateau, Pakistan, to studies of early man. In **South Asian Archaeology 1981**, edited by B. Allchin, 10-19. Cambridge: Cambridge University Press.

Dennell, R.W. (1986) Reply to R.K. Ganjoo "On dated lower palaeolithic artfacts from northern Pakistan". *Current Anthropology* **27**: 186.

Dennell, R.W. (1989) Reply to "Early artefacts from Pakistan? Some questions for the excavators" by M. Hemingway and D. Stapert. *Current Anthropology* **30 (3)**: 318-322.

Dennell, R.W., Rendell, H.M. and Halim, M. (1985) New perspectives on the palaeolithic of northern Pakistan. In **South Asian Archaeology 1983**, edited by J. Schotsmans and M. Taddei, 9-20. Naples: Istituto Universitario Orientale Series Minor 23.

Dennell, R.W., Rendell, H.M. and Hailwood, E. (1988a) Early tool-making in Asia: two-million-year-old artefacts in Pakistan. *Antiquity* **62**: 98-106.

Dennell, R.W., Rendell, H.M. and Hailwood, E. (1988b) Late Pliocene artefacts from northern Pakistan. *Current Anthropology* **29 (3)**: 495-498.

Derbyshire, E. (1983) Origin and characteristics of some Chinese loess at two locations in China. In **Eolian Sediments and Processes**, edited by M.E. Brookfield and T.S. Ahlbrandt, 69-90. Amsterdam: Elsevier.

De Terra, H. and Teilhard de Chardin, P. (1936) Observations on the Upper Siwalik formations and later Pleistocene deposits in India. *Proceedings of the American Philosophical Society* **76**: 791-822.

De Terra, H. and Paterson, T.T. (1939) **Studies on the Ice Age in India and Associated Human Cultures**. Washington: Carnegie Institute Publication 493.

Dodia, R., Agrawal, D.P., and Vora, A.B. (1985) New pollen data from the Kashmir bogs: a summary. In **Climate and Geology of Kashmir and Central Asia**, edited by D.P. Agrawal, S. Kusumgar and R.V. Krishnamurthy, 101-108. City: Publisher.

Dreimanis, A. and Vagners, U.J. (1971) Bimodal distribution of rock and mineral fragments in basal tills. In **Till : A Symposium**, edited by R.P. Goldthwait, 237-250. Ohio State University Press.

Dreimanis, A., Hutt, G., Raukas, A. and Whippey, P.W. (1978) Dating methods of Pleistocene deposits and their problems: 1 Thermoluminescence dating. *Geoscience Canada* **5**: 55-60.

Dubois, E. (1898) On Pithecanthropus erectus: a transitional form between man and the apes. *Scientific Transactions of the Royal Dublin Society* **6**: 1-18.

Duplessy, J.C. (1982) Glacial to interglacial contrasts in the northern Indian Ocean. *Nature* **295**: 494-498.

Dupree, L. (1972) Prehistoric research in Afghnistan (1959 - 1966). *Transactions of the American Philosophical Society* **62(4)**: 3-84.

Eberl, B. (1930) **Die Eiszeitenfolge im Nordlichen Alpenvorlande**. Augsburg: Benno Filser.

Falconer, H. and Cautley, E. (1845) **Fauna Antiqua Sivalensis**. London: Smith, Elder and Co.

Fatmi, A.N. (1973) Lithostratigraphic units of the Kohat - Potwar Province, Indus Basin, Pakistan. *Memoirs of the Geological Survey of Pakistan* **10**: 1-80.

Gamble, C. (1984) Subsistence and society in palaeolithic Europe. In **Prehistoric Europe**, edited by T. Champion, 59-88. Orlando and London: Academic Press.

Gamble, C. (1986) **The Palaeolithic Settlement of Europe**. Cambridge: Cambridge University Press.

Gill, W.D. (1952) The stratigraphy of the Siwalik Series in the northern Potwar, Punjab, Pakistan. *Quarterly Journal of the Geological Society of London* **107**: 395-421.

Goudie, A.S. (1978) Dust storms and their geomorphological significance. *Journal of Arid Environments* **1**: 291-310.

Goudie, A.S. (1983) Dust storms in space and time. *Progress in Physical Geography* **7**: 502-530.

Goudie, A.S., Rendell, H.M. and Bull, P.A. (1984) The loess of Tajik SSR. In **International Karakoram Project 1**: 399-412, edited by K.J. Miller. Cambridge: Cambridge University Press.

Graziozi, P. (1964) Prehistoric research in the northwestern Punjab. Scientific Reports of the Italian Expedition to the Karakorum (K2) and Hindu Kush 1 (part 5): 55-74. Leiden: E.J. Brill.

Gronmow, B., Meldgaard, M. and Nielsen, J.B. (1983) Aasivissuit - The great summer camp. Archaeological, ethographical and zoo-archaeological studies of a caribou-hunting site in West Greenland. *Man and Society* **5**: 5-96.

Gupta, S.P. (1979) **The Archaeology of Soviet Central Asia and the Indian Borderlands**. Delhi: B.R. Publishing Co.

Hammond, M. (1982) The expulsion of the Neanderthals from human ancestry: Marcellin Boule and the social context of scientific research. *Social Studies of Science* **12**: 1-36.

Hemingway, M. and Stapert, D. (1989) Early artefacts from Pakistan? Some questions for the excavators. *Current Anthropology* **30 (3)**: 318-322.

Hole, F. and Flannery, K.V. (1967) The prehistory of south western Iran: a preliminary report. *Proceedings of the prehistoric society* **33**: 147-206.

Holmes, J.A. (1988) Pliocene and Quaternary environmental change in Kashmir, north-west Himalayas. Unpublished D.Phil. thesis, University of Oxford.

Holmes, J.A. and Street-Perrott, F.A. (in press) The quaternary history of Kashmir, north-west Himalaya: a revision of De Terra and Paterson's sequence. *Zeitschrift für Geomorphologie N.F. Supplementband* **76 The Neogene of the Himalayas.**

Ikawa-Smith, F. (1978) Introduction: The early palaeolithic of East Asia. In **Early Palaeolithic in South and East Asia**, edited by F. Ikawa-Smith, 1-10. The Hague: Mouton.

Isaac, G.L. (1975) Middle Pleistocene stratigraphy and cultural patterns in East Africa. In **After the Australopithecines: Stratigraphy, Ecology and Cultural Change in the Middle Pleistocene**, edited by K.W. Butzer and G. Isaac, 495- 542. The Hague: Mouton.

Isaac, G.L. (1978) The food-sharing behavior of proto-human hominids. *Scientific American* **238**: 90-118.

Jenkinson, R.D.S., Rendell, H.M., Dennell, R.W., Jah, M.A. and Sutherland, S.A.S. (in press) Upper Siwalik palaeoenvironments and palaeoecology in the Pabbi Hills, northern Pakistan. In **The Neogene of the Himalayas** *Zeitschrift für Geomorphologie, Supplementband* **76**

Johnson, E. (1972) Notes on a palaeolithic site survey in Pakistan. *Asian Perspectives* **15**: 60-65.

Johnson, G.D., Zeitler, P.K., Naeser, C.W., Johnson, N.M., Summers, D.M., Frost, C.D., Opdyke, N.D. and Tahirkheli, R.A.K. (1982) Fission-track ages of Late Neogene and Quaternary volcanic sediments, Siwalik Group, northern Pakistan. *Palaeogeography, Palaeoclimatology and Palaeoecology* **37**: 63-93.

Johnson, N.M., Opdyke, N.D., Johnson, G.D., Lindsay, E.H. and Tahirkheli, R.A.K. (1982) Magnetic polarity stratigraphy and ages of Siwalik group rocks of the Potwar Plateau, Pakistan. *Palaeogeography, Palaeoclimatology and Palaeoecology* **37**: 17-42.

Jones, P. (1979) Effects of raw materials on biface manufacture. *Science* **204**: 835-836.

Keeley, L H. (1980) **Experimental Determination of Stone Tool Uses: a Microwear Analysis**. Chicago: University of Chicago Press.

Keller, H.M., Tahirkheli, R.A.K., Mirza, M.A., Johnson, G.D., Johnson, N.M. and Opdyke, N.D. (1977) Magnetic polarity stratigraphy of the Upper Siwalik deposits, Pabbi Hills, Pakistan. *Earth and Planetary Science Letters* **36**: 187-201.

Khan, M.J., Opdyke, N.D. and Shroder, J.F. (1985) Bain diamictite: lithology, age and origin. *Geological Bulletin of the University of Peshawar* **18**: 53-64.

Klein,R.G. (1987) Reconstructing how early people exploited animals: problems and prospects. In **The Evolution of Human Hunting**, edited by H. Nitecki and D.V. Nitecki, 11-45. New York and London: Plenum.

Koenigswald, G.H.R. von (1936) Early palaeolithic stone implements from Java. *Bulletin of the Raffles Museum Series B* **1: 52-60.**

Koenigswald, G.H.R. von (1956) **Meeting Prehistoric Man.** London: Thames and Hudson.

Krantz, G. (1972) Soan tool types from Ghila Khan. *Asian Perspectives* **15**: 66-82.

Kravtchenko, K.N. (1964) Soan Formation - upper unit of Siwalik group in Potwar. *Science & Industry* **2** (**3**): 230-233.

Kuenen, P.H. (1969) Origin of quartz silt. *Journal of Sedimentary Petrology* **39**: 1631-1633.

Kusumgar, S. and Agrawal, D.P. (1985) Chronology of the Kashmir loess. In **Climate and Geology of Kashmir and Central Asia**, edited by D.P. Agrawal, S. Kusumgar and R.V. Krishnamurty, 119-121. Place and Publisher.

Leakey, L.S.B. (1936) Fossil human remains from Kanam and Kanjera, Kenya Colony. *Nature* **138**: 643-645.

Leakey, L.S.B. (1951) **Olduvai Gorge**. Cambridge: Cambridge University Press.

Leakey, L.S.B. (1953) **Adam's Ancestors**. London: Methuen.

Lewin, R. (1987) **Bones of contention**. London: Penguin.

Lewis, G.E. (1934) Preliminary notice of new man-like apes from India. *American Journal of Science* Ser. **5**: 27: 161-179.

Lewis, G.E. (1937) A new Siwalik correlation. *American Journal of Science* **33**: 191-204.

Lindsay, E.H., Johnson, N.M. and Opdyke, N.D. (1980) Correlation of Siwalik faunas. In **Aspects of Vertebrate History: Essays in Honor of Edwin Harris Colbert**, edited by L.L. Jacobs, 309-319. Flagstaff: Museum of Northern Arizona.

Lydekker, R. (1878) Notes on the geology of Kashmir, Kishtwar, and Pangi. *Records of the Geological Survey of India* **11**: 30-64.

Lydekker, R. (1883) The geology of the Kashmir and Chamba territories and the British district of Khagan. *Memoirs of the Geological Survey of India* **22**: 1-344.

Manabe, S. and Hahn, D.G. (1977) Simulation of the tropical climate of an Ice Age. *Journal of Geophysical Research* **82**: 3889-3911.

Mankinen, E.A. and Dalrymple, G.B. (1979) Revised geomagnetic polarity time scale for the interval 0-5 My B.P. *Journal of Geophysical Research* **84**: 615-626.

Marks, A.E. and Reid Ferring, C. (1988) The early upper palaeolithic of the Levant. In **The Early Upper Palaeolithic: Evidence from Europe and the Near East**, edited by J.E. Hoffecker and C.A. Wolf, 43-72. British Archaeological Reports (International Series) 437.

Mason, O. (1891) Aboriginal skin-dressing: a study based on material in the U.S. National Museum. *Annual report of the Smithsonian Institution for 1889*, 553-589.

Matthew, W.D. (1914) Climate and evolution. *Annals of the New York Academy of Sciences* **24**: 171-318.

McMurty, M.G. (1980) Geological map of the central portion of the Soan syncline, Rawalpindi Distt., Pakistan. Dartmouth College, New Hampshire: Unpublished report.

Medlicott, H.B. (1864) On the geological structure and relation of the southern portion of Himalayan ranges between the river Ganges and Ravi. *Memoirs of the Geological Survey of India* **3** (**2**): 1-212.

Medlicott, H.B. (1876) Notes upon the sub-Himalayan series in the Jammu Hills. *Records of the Geological Survey of India* **9**: 49-57.

Middlemiss, C.S. (1896) The geology of Hazara and the Black Mountains. *Memoirs of the Geological Survey of India* **26**: 1-302.

Moragne, J.H. (1979) Magnetic polarity stratigraphy and timing of deformation for an Upper Siwalik sedimentary sequence, Soan Syncline, Pakistan. Dartmouth College, New Hampshire: B.A. thesis.

Morris, T.O. (1938) The Bain Boulder-bed. *Quarterly Journal of the Geological Society of London* **94: 385-421**.

Moss, A.J. and Green, P. (1975) Sand and silt grains: predetermination of their formation and properties by microfractures in quartz. *Journal of the Geological Society of Australia* **22**: 485-495.

Movius, H.L. (1944) Early man and Pleistocene stratigraphy in southern and eastern Asia. *Papers of the Peabody Museum of American Archaeology and Ethnology* **19** (**3**).

Movius, H. (1948) The lower palaeolithic cultures of southern and eastern Asia. *Transactions of the American Philosophical Society* N.S. **38** (**4**): 329-420.

Movius, H. (1978) Southern and East Asia: conclusions. In **Early Palaeolithic in South and East Asia**, edited by F. Ikawa-Smith, 351-355. The Hague: Mouton.

Murty, M.L.K. (1979) Recent research on the Upper Palaeolithic in India. *Journal of Field Archaeology* **6** (**3**): 301-320.

Nahon, D. and Trompette, R. (1982) Origin of siltstones: glacial grinding versus weathering. *Sedimentology* **29**: 25-35.

Nickling, W.G. (1983) Grain-size characteristics of sediment transported during dust storms. *Journal of Sedimentary Petrology* **53**: 1011-1024.

Norin, E. (1925) Preliminary notes on the late Quaternary glaciation of the northwest Himalaya. *Geografiska Annaler* **7**: 165-194.

Oakley, K.P. and Leakey, M.D. (1937) Report on excavations at Jaywick Sands, Essex (1934) with some observations on the Clactonian industry and on the fauna and geological significance of the Clacton Channel. *Proceedings of the Prehistoric Society* **3**: 216-260.

Ohel, M.Y. (1979) The Clactonian: an independent complex or an integral part of the acheulean? *Current Anthropology* **20**: 685-726.

Opdyke, N.D., Johnson, G.D., Johnson, N.M., Tahirkheli, R.A.K. and Mirza, M.A. (1979) Magnetic polarity stratigraphy and vertebrate palaeontology of the Upper Siwalik Sub-Group of northern Pakistan. *Palaeogeography, Palaeoclimatology and Palaeoecology* **27**: 1-34.

Pant, R.K., Krishnamurthy, R.V., Tandon, S.K. and Bisht, K. (1985) Loess lithostratigraphy of the Kashmir Basin, India. In **Climate and Geology of Kashmir and Central Asia**, edited by D.P. Agrawal, S. Kusumgar and R.V. Krishnamurthy, 123-129. Place and publisher.

Paterson, T.T. (1937) Studies in the palaeolithic succession in England: No. I. The Barnham sequence. *Proceedings of the Prehistoric Society* **3**: 87-135.

Paterson, T.T. (1945) Core, culture and complex in the Old Stone Age. *Proceedings of the Prehistoric Society* **11**: 1-19.

Paterson, T.T. and Drummond, H.J.H. (1962) *Soan, the Palaeolithic of Pakistan.* Karachi: Dept. of Archaeology, Govt. of Pakistan, Memoir **2**.

Paterson, T.T. and Fagg, B.E.B. (1940) Studies on the palaeolithic succession in England No. II. The Upper Brecklandian Acheul (Elveden). *Proceedings of the Prehistoric Society* **6**: 1-29.

Patriat, P. and Achache, J. (1984) India-Eurasia collision chronology has implications for crustal shortening and driving mechanism of plates. *Nature* **311**: 615-621.

Penck, A. and Bruckner, E. (1909) **Die Alpen im Eiszeitalter.** (3 vols). Leipzig: Tauchnitz.

Pennock, E.S., Lillie, R.J., Zaman, A.S.H. and Yousaf, M. (1989) Structural interpretation of seismic reflection data from Eastern Salt Range and Potwar Plateau, Pakistan. *Bulletin, American Association of Petroleum Geologists* **73 (7)**: 841-857.

Pilgrim, G.E. (1910) Preliminary note on a revised classification of the Tertiary freshwater deposits of the north-west Punjab. *Records of the Geological Survey of India* **40 (3)**: 185-205.

Pilgrim, G.E. (1913) The correlation of the Siwaliks with mammal horizons of Europe. *Records of the Gological Survey of India* **43 (4)**: 264-326.

Porter, S.C. (1970) Quaternary glacial record in Swat Kohistan, West Pakistan. *Bulletin of the Geological Society of America* **81**: 1421-1426.

Potts, R. (1984) Home base and early hominids. *American Scientist* **72**: 338-347.

Prell, W.L. (1984) Monsoonal climate of the Arabian Sea during the Late Quaternary: a response to changing solar radiation. In **Milankovitch and Climate Vol 1**, edited by A. Berger, J. Imbrie, J. Hays, G. Kukla and B. Saltzman, 349- 366. Dordrecht: Reidel.

Prell, W.L., Hutson, W.H., Williams, D.F.,Be, A.W.H., Geitzenauer, K. and Molfino, B. (1980) Surface circulation of the Indian Ocean during the Last Glacial maximum, approximately 18,000 yr BP. *Quaternary Research* **14**: 309-336.

Price, T.D.(1978) Mesolithic settlement systems in the Netherlands. In **The Early Postglacial Settlement of Northern Europe**, edited by P.M. Mellars, 81-113. London: Duckworth.

Pye, K. (1983) Grain surface textures and carbonate content of Late Pleistocene loess from West Germany and Poland. *Journal of Sedimentary Petrology* **53**: 973-980.

Pye, K. (1984) Loess *Progress in Physical Geography* **8**: 176-217.

Pye, K. (1987) **Aeolian Dust and Dust Deposits.** London and San Diego: Academic Press.

Pye, K. and Sperling, C.H.B. (1983) Experimental investigation of silt formation by static breakage processes: the effect of temperature, moisture and salt on quartz dune sand and granitic regolith. *Sedimentology* **30**: 49-62.

Ranov, V.A. and Davis, R.S. (1979) Toward a new outline of the Soviet Central Asian palaeolithic. *Current Anthropology* **20 (2)**: 249-270.

Raynolds, R.G.H. (1980) The Plio-Pleistocene structural and stratigraphic evolution of the Eastern Potwar Plateau, Pakistan. Dartmouth College, New Hampshire: Ph.D. thesis.

Raynolds, R.G.H. and Johnson, G.D. (1985) Rates of Neogene depositional and deformational processes, north-west Himalayan foredeep margin, Pakistan. In **The Chronology of the Geological Record**, edited by N.J. Snelling, 297- 311. Oxford: Blackwell Scientific Publications.

Reed, R. (1972) **Ancient Skins, Parchments and Leathers**. London and New York: Academic Press.

Reid Moir, J. (1932) Presidential address: The culture of Piocene Man. *Proceedings of the Prehistoric Society of East Anglia* **7**: 1-17.

Rendell, H.M. (1983) Problems of estimation of water content history of loesses. *P.A.C.T.* **9**: 523-530.

Rendell, H.M. (1984) The Pleistocene sequence in the Soan Valley, northern Pakistan. In **South Asian Archaeology 1981**, edited by B. Allchin, 3-9. Cambridge: Cambridge University Press.

Rendell, H.M. (1985) The precision of moisture content estimates in the thermoluminescence dating of loess from northern Pakistan. *Nuclear Tracks* **10**: 763-768.

Rendell, H.M. (1988) Environmental changes during the Pleistocene in the Potwar Plateau and Peshawar Basin, northern Pakistan. *Proceedings of the Indian National Science Academy* **54A**: 390-398.

Rendell, H.M., Gamble, I.J.A. and Townsend, P.D. (1983) Thermoluminescence dating of loess from the Potwar Plateau, northern Pakistan. *P.A.C.T.* **9**: 555-562.

Rendell, H.M. and Dennell, R.W. (1985) Dated lower palaeolithic artefacts from northern Pakistan. *Current Anthropology* **26**: 293.

Rendell, H.M. and Dennell, R.W. (1987) Thermoluminescence dating of an upper Pleistocene site, northern Pakistan. *Geoarchaeology* **2 (1)**: 63-67.

Rendell, H.M., Hailwood, E.A. and Dennell, R.W. (1987) Magnetic polarity stratigraphy of Upper Siwalik Sub-Group, Soan Valley, Pakistan: implications for early human occupance of Asia. *Earth and Planetary Science Letters* **85 (4)**: 488-496.

Rendell, H.M. and Townsend, P.D. (1988) TL dating of a 10 metre loess profile in Pakistan. *Quaternary Science Reviews* **7**: 251-255.

Riezebos, P.A. and Van der Waals, L. (1974) Silt-sized quartz particles: a proposed source. *Sedimentary Geology* **12**: 279-285.

Rogers, E.S. (1973) The quest for food and furs: the Mistassini Cree, 1953-1954. *National Museum of Canada Publications in Ethnology* **5**.

Said, M. and Majid, M. (1977) The Pleistocene history of the Bar Daman area, Peshawar Valley. *Journal of Science and Technology, Peshawar* **1**: 39-47.

Sankalia, H.D. (1974) **Prehistory and Protohistory of India and Pakistan**. (2nd ed.). Poona: Deccan College.

Sankalia, H.D. (1978) The early palaeolithic in India and Pakistan. In **Early Palaeolithic in South and East Asia**, edited by F. Ikawa-Smith, 97-127. The Hague: Mouton.

Seminov, S.A. (1973) **Prehistoric Technology**. Bath: Adams and Dart.

Shah, S.M.I. (1977) Stratigraphy of Pakistan. *Memoirs of the Geological Survey of Pakistan* **12**: 1-138.

Shah, S.M.I. (1984) Siwaliks: and introduction. In **Contribution to the Geology of Siwaliks of Pakistan**, edited by S.M.I. Shah and D. Pilbeam. *Memoirs of the Geological Survey of Pakistan* **11**: i-vi.

Shah, S.M.I. and Pilbeam, D. (1984) Contributions to the geology of Siwaliks of Pakistan. *Memoirs of the Geological Survey of Pakistan* **11**: 1-77.

Singh, I.B. (1982) Sedimentation pattern in the Karewa Basin, Kashmir Valley, India and its geological significance. *Journal of the Palaeontological Society of India* **27**: 71-110.

Smalley, I.J. (1966) The properties of glacial loess and the formation of loess deposits. *Journal of Sedimentary Petrology* **36**: 669-676.

Smalley, I.J. (1971) "In-situ" theories of loess formation and the significance of the calcium carbonate content of loess. *Earth-Science Reviews* **7**: 67-85.

Smalley, I.J. (1980) The formation of loess materials and loess deposits: some observations on the Tashkent loess. *Geophysichen und Geologischen Veröffentlichen der Karl Marx Universität, Leipzig* **2**: 247-257.

Smalley, I.J. and Vita-Finzi, C. (1968) The formation of fine particles in sandy deserts and the nature of 'desert' loess. *Journal of Sedimentary Petrology* **38**: 766-774.

Smalley, I.J. and Cabrera, J.G. (1970) The shape and surface texture of loess particles. *Bulletin of the Geological Society of America* **81**: 1591-1596.

Smalley, I.J. and Cabrera, J.G. (1971) The shape and surface texture of loess particles: reply. *Bulletin of the Geological Society of America* **82**: 2361-2364.

Smalley, I.J. and Krinsley, D.H. (1978 Loess deposits associated with deserts. *Catena* **5**: 53-66.

Smalley, I.J. and Smalley, V. (1983) Loess material and loess deposits: formation, distribution and consequences. In **Eolian Sediments and Processes**, edited by M.E. Brookfield and T.S. Ahlbrandt, 51-68. Amsterdam: Elsevier.

Soffer. O. (1985) **The Upper Palaeolithic of the Central Russian Plain**. London and San Diego: Academic Press

Söhnge, P.G., Visser, D.J. and Lowe, C. van Riet (1937) The geology and archaeology of the Vaal river basin. *Memoirs of the Geological Survey of South Africa, Pretoria* **35**: 1-184.

Solecki, R.S. (1963) Prehistory in Shanidar valley, northern Iraq. *Science* **139**: 179-193.

Spiess, A.E. (1979) **Reindeer and Caribou Hunters**. London and New York: Academic Press.

Stapert, D. (1976) Some natural surface modifications on flint in the Netherlands. *Palaeohistoria* **18**: 7-41.

Stiles, D. (1978) Paleolithic artifacts in Siwalik and Post-Siwalik deposits of northern Pakistan. *Papers of the Kroeber Anthropological Society* **53** & **54**: 129-148.

Theobald, W. (1877) Occurrence of erratics in the Potwar. *Records of the Geological Survey of India* **10**: 140-143.

Todd, K.R.U. (1932) Prehistoric man around Bombay. *Proceedings of the Prehistoric*

Society of East Anglia **7**: 35-42.

Toth, N. and Schick, K.D. (1986) The first million years: the archaeology of proto-human culture. In **Advances in Archaeological Method and Theory Vol. 9**, edited by M.B. Schiffer, 1-96. Orlando: Academic Press.

Tsoar, H. and Pye, K. (1987) Dust transport and the question of desert loess formation. *Sedimentology* **34**: 139-153.

Turnbull, C. (1974) **The Mountain People**. London: Picador.

Van Campo, E. (1986) Monsoon fluctuations in two 20,000-yr BP. oxygen- isotope/pollen records off southwest India. *Quaternary Research* **26**: 376-388.

Vermeersch, P.M. and Peer, P. van (1988) The early upper palaeolithic in Egypt. In **The Early Upper Palaeolithic: Evidence from Europe and the Near East**, edited by J.E. Hoffecker and C.A. Wolf, 1-22. British Archaeological Reports (International Series) 437.

Visser, C.F. and Johnson, G.D. (1978) Tectonic control of Late Pliocene molasse sedimentation in a portion of the Jhelum re-entrant, Pakistan. *Geologische Rundschau* **67**: 15-37.

Wadia, D.N. (1928) The geology of Poonch State (Kashmir) and adjacent portions of the northern Punjab. *Memoirs of the Geological Survey of India* **51**: 185-370.

Wang, C-Y, Shi, Y-L. and Zhou, W-H. (1982) Dynamic uplift of the Himalaya. *Nature* **298**: 553-556.

Warnke, D.A. (1971) The shape and surface texture of loess particles: discussion. *Bulletin of the Geological Survey of America* **82**: 2357-2360.

Wayland, E.J. (1934) Rifts, rivers, rains and early man in Uganda. *Journal of the Royal Anthropological Institute of London* **64**: 333-352.

Whalley, W.B., Marshall, J.R. and Smith, B.J. (1982) Origin of desert loess from some experimental observations. *Nature* **300**: 433-435.

Wheeler, R.E.M. (1966) **Alms For Oblivion: An Antiquary's Scrapbook**. London: Weidenfeld and Nicolson.

Williams, M.A.J. and Clarke, M.F. (1984) Late Quaternary environments in north-central India.*Nature* **308**: 633-635.

Wintle, A.G. and Huntley, D.J. (1979) Thermoluminescence dating of a deep-sea sediment core. *Nature* **279**: 710-712.

Wintle, A.G. and Huntley, D.J. (1980) Thermoluminescence dating of ocean sediments. *Canadian Journal of Earth Science* **17**: 348-360.

Wynne, A.B. (1877) Notes on the Tertiary zone and underlying rocks in the northwest Punjab. *Records of the Geological Survey of India* **10**: 107-132.

Wynne, A.B. (1879) Further notes on the geology of the upper Punjab. *Records of the Geological Survey of India* **12**:114-133.

Wynne, A.B. (1881) On the connection between traveled blocks in the upper Punjab and a supposed glacial period in upper India. *Geological Magazine* **8**: 97-99.

Yeats, R.S. and Lawrence, R.D. (1984) Tectonics of the Himalayan thrust belt in northern Pakistan. In **Marine Geology and Oceanography of Arabian Sea and Coastal Pakistan**, edited by B.U. Haq and J.D. Milliman, 177-198. New York, Van Norstrand

Reinhold.

Yeats, R.S., Khan, S.H. and Akhtar, M. (1984) Late Quaternary deformation of the Salt Range of Pakistan. *Bulletin of the Geological Society of America* **95**: 958-966.

Yeats, R.S. and Hussain, A. (1987) Timing of structural events in the Himalayan foothills of northwestern Pakistan. *Bulletin of the Geological Society of America* **99**: 161-176.

Zeitler, P.K., Johnson, N.M., Naeser, C.W. and Tahirkheli, R.A.K. (1982) Fission-track evidence for Quaternary uplift of the Nanga Parbat region, Pakistan. *Nature* **298**: 255-257.

Zeitler, P.K., Tahirkheli, R.A.K., Naeser, C.W. and Johnson, N.M. (1982) Unroofing history of a suture zone in the Himalaya of Pakistan by means of fission-track annealing ages. *Earth and Planetary Science Letters* **57**: 227-240.

Zeuner, F.E. (1946) **Dating the Past : an Introduction to Geochronology**. London: Methuen.

Zeuner, F.E. (1950) Stone age and Pleistocene chronology in Gujarat. *Deccan College Monograph Series* **6**: 1-46.

Zeuner, F.E. (1959) **The Pleistocene Period: its Climate, Chronology and Faunal Successions**. London: Hutchinson.

Zimmerman, D.W. (1971) Thermoluminescence dating using fine grains from pottery. *Archaeometry* **13**: 29-52.

Appendix A

Description and size of artefacts from site 55, Riwat.

Number	Type	Colour	Condition	Length	Width	Depth
0010	Retouched flake †	brown	fresh	51	38	8
0011	Flake		fresh			
0012	Flake		fresh			
0015	Blade †	grey	fresh	49	19	7
0017	Micro-blade †	green	fresh	27	13	3
0026	Flake		fresh			
0030	Flake		fresh			
0040	Blade fragment †	grey	fresh	28	22	9
0046	Flake		fresh			
0052	Retouched flake	brown	fresh	43	30	8
0054	Flake		fresh			
0062	Blade tip	grey	fresh	32	21	5
0065	Broken blade †	honey	abraded	48	22	15
0071	Flake †	brown-grey	fresh	104	68	20
0073	Retouched flake	grey	fresh	74	58	17
0074	Broken blade †	grey	fresh	22	22	7
0075	Flake †	grey	fresh	45	37	10
0076	Broken blade	grey	fresh	37	16	10
0077	Retouched flake	grey-green	fresh	36	32	14
0079	Flake	brown-green	weathered	22	14	5
0080	Blade tip †	white	fresh	30	27	9
0081	Flake †	grey	fresh	59	38	11
0082	Retouched flake †	grey	fresh	65	44	14
0083	Core	grey-brown	fresh	70	65	38
0084	Pebble-core	brown-grey	fresh	68	62	24
0087	Blade †	blue-brown	fresh	42	20	9
0101	Cortical flake	grey-white	weathered	26	21	10
0102	Broken blade †	brown-green	fresh	35	30	10
0103	Cortical flake	green	fresh	32	18	5
0104	Flake	grey	fresh	68	52	18
0105	Broken blade †	grey-green	fresh	37	32	13
0106	Blade fragment	green-brown	fresh	33	25	5
0107	Blade fragment	brown-green	abraded	26	17	4
0108	Blade fragment	grey-green	fresh	16	13	3
0109	Blade tip †	brown-green	fresh	36	20	9
0110	Cortical flake	white-honey	fresh	53	22	8
0112	Core	brown-grey	abraded	94	80	56
0113	Broken blade †	honey-green	fresh	36	22	10
0114	Broken flake	blue-green	fresh	30	30	10
0115	Cortical flake	brown-green	fresh	100	56	31
0116	Flake †	grey-green	fresh	70	58	20
0117	Core fragment †	honey	fresh	61	54	38
0118	Flake †	honey-green	fresh	61	38	10
0119	Cortical flake	grey-blue	fresh	60	47	18
0120	Flake †	grey-green	fresh	45	39	11
0121	Retouched cortical flake †	brown	fresh	92	70	30
0122	Cortical flake	brown-grey	fresh	45	32	11
0123	Flake	blue-grey	fresh	60	35	18
0124	Retouched cortical flake †	blue-grey	fresh	46	38	12
0125	Retouched flake †	blue-grey	fresh	57	54	11
0126	Broken blade †	grey	fresh	42	22	10

APPENDIX A (continued)

Number	Type	Colour	Condition	Length	Width	Depth
0127	Cortical flake	blue	fresh	30	25	19
0128	Broken flake	blue	fresh	37	32	13
0130	Flake †	brown-green	fresh	102	62	21
0131	Cortical flake †	brown	fresh	87	74	28
0132	Broken blade †	blue	fresh	53	32	8
0133	Blade fragment †	honey-green	fresh	49	35	12
0134	Cortical flake	brown	fresh	65	45	23
0135	Blade fragment †	blue-green	fresh	50	43	15
0136	Flake †	honey	fresh	29	28	8
0137	Retouched flake †	grey-green	fresh	41	32	10
0140	Blade tip	green	fresh	12	11	2
0141	Broken blade †	green	fresh	26	18	6
0143	Core	grey	abraded	98	90	74
0144	Core	grey-green	abraded	124	91	40
0145	Core fragment	grey	abraded	75	71	37
0146	Broken blade	white-grey	abraded	49	43	21
0147	Broken blade	honey	fresh	42	19	11
0148	Cortical flake	grey-honey	fresh	72	45	18
0149	Flake	honey	weathered	45	45	12
0150	Core fragment	white	abraded	87	58	37
0151	Core	white	weathered	88	62	41
0152	Core fragment	grey-honey	fresh	66	45	32
0153	Cortical flake	brown-green	fresh	82	60	33
0154	Core	grey-brown	fresh	93	85	64
0155	Flake †	green-grey	fresh	78	60	21
0156	Flake	grey	fresh	72	21	9
0157	Cortical flake †	brown	fresh	68	41	15
0158	Flake (5 pieces)	green	fresh	80	38	11
0159	8 flake pieces	grey	fresh	-	-	-
0160	Cortical flake	grey-brown	weathered	40	37	13
0161	Retouched flake †	grey-green	fresh	68	47	15
0162	Core	green-brown	weathered	130	65	37
0163	Blade fragment †	grey	fresh	41	39	14
0164	Flake	grey	fresh	42	28	9
0165	Core fragment	grey	fresh	58	47	35
0166	Cortical flake	honey	abraded	43	16	9
0168	Core fragment	honey	weathered	59	35	20
0169	Core fragment	honey	fresh	120	66	33
0171	Core fragment	honey	fresh	43	22	22
0172	Cortical flake	grey	fresh	70	66	31
0173	Cortical flake	grey	fresh	52	37	14
0175	Retouched blade †	grey-green	fresh	30	16	6
0177	Cortical flake	grey	fresh	74	51	12
0178	Cortical flake	honey	weathered	50	43	18
0180	Cortical flake	grey	weathered	44	31	9
0181	Core fragment †	grey	fresh	65	64	32
0182	Cortical flake	grey	fresh	51	30	9
0183	Core fragment	grey	fresh	81	74	62
0200	Biface	grey	abraded	182	100	55
0201	Cortical flake	green-brown	fresh	63	48	17
0202	Core fragment †	brown-green	fresh	85	51	50
0203	Blade fragment	grey-honey	abraded	44	35	10
0204	Core fragment	grey	fresh	61	48	30

APPENDIX A (continued)

Number	Type	Colour	Condition	Length	Width	Depth
0205	Flake	grey-brown	fresh	108	64	27
0206	Biface †	honey	abraded	115	72	43
0208	Core	honey	abraded	85	80	40
0209	Cortical flake	honey-brown	weathered	61	51	25
0210	Blade †	brown	fresh	76	44	19
0211	Cortical flake	brown-green	fresh	43	35	13
0212	Core	grey-brown	abraded	96	80	60
0214	Blade fragment †	grey-green	fresh	33	20	6
0215	Blade fragment	honey	abraded	34	27	12
0216	Notched cortical flake †	grey-blue	fresh	74	34	20
0217	Cortical flake †	grey	weathered	37	20	4
0218	Flaking debris	brown-grey	abraded	36	26	9
0219	Bladelet fragment	green	fresh	9	8	1
0220	Cortical flake	green-brown	fresh	82	52	38
0221	Cortical flake	green-brown	fresh	65	38	29
0222	Cortical flake	brown-grey	weathered	150	57	40
0223	Blade fragment	honey	fresh	24	18	10
0224	Cortical flake	grey-brown	fresh	28	19	6
0225	Flake †	brown-green	fresh	35	18	11
0226	Flaking debris	grey-brown	abraded	39	20	12
0227	Cortical flake	grey-brown	fresh	29	16	8
0228	Flaking debris	green-brown	fresh	30	12	10
0229	Cortical flake	grey	fresh	90	64	38
0230	Cortical flake	red-brown	fresh	45	21	12
0231	Cortical flake	grey	weathered	42	33	18
0232	Cortical flake	brown	fresh	37	22	20
0233a	Core	grey-brown	weathered	114	60	56
0233b	Flaking debris	grey	abraded	82	43	23
0234	Cortical flake	red-brown	weathered	22	15	8
0235	Cortical flake	grey-brown	fresh	82	30	27
0236	Cortical flake	green	weathered	19	11	5
0237	Cortical flake	brown	fresh	91	46	39
0238	Cortical flake	green-brown	fresh	29	22	9
0239	Retouched blade †	brown	fresh	78	39	15
0240	Cortical flake	grey-brown	weathered	27	20	9
0241	Flaking debris	green-brown	abraded	21	16	7
0243	Core fragment	brown	fresh	36	28	7
0244	Cortical flake	grey-brown	weathered	44	26	13
0245	Cortical flake	honey-brown	fresh	49	44	18
0246	Flake	grey	weathered	57	25	12
0247	Retouched cortical flake	grey	fresh	80	38	19
0248	Cortical flake	red-brown	fresh	69	38	14
0249	Flaking debris	grey-green	weathered	25	15	6
0250	Flake	grey-green	weathered	24	20	5
0251	Flaking debris	grey-green	fresh	37	21	12
0252	Flaking debris	honey-grey	abraded	36	24	11
0253	Cortical flake	green-brown	abraded	45	34	13
0254	Flaking debris	green-brown	abraded	27	15	7
0255	Core fragment	red-brown	weathered	105	100	59
0257	Core fragment	grey	weathered	83	52	50
0258	Retouched blade fragment †	brown-green	fresh	37	32	10
0259	Struck cobble	grey-brown	fresh	50	40	25
0260	Core	brown	fresh	105	92	80

APPENDIX A (continued)

Number	Type	Colour	Condition	Length	Width	Depth
0261	Cortical flake	grey-brown	fresh	64	52	26
0262	Core fragment	grey	fresh	60	59	39
0263	Cortical flake	honey	abraded	49	48	18
0264	Core fragment	brown	fresh	52	32	26
0265	Flaking debris	grey-brown	abraded	44	40	24
0267	Cortical flake	grey	fresh	41	27	8
0268	Blade fragment †	grey	fresh	30	25	13
0269	Cortical flake	black-brown	weathered	45	29	18
0270	Core fragment	brown	weathered	55	39	32
0271	Flake	honey	fresh	105	60	23
0272	Core fragment	honey-grey	fresh	47	42	29
0273	Core fragment	grey-pink	abraded	90	85	40
0274	Split pebble	grey	fresh	73	45	30
0275	Core fragment	grey-honey	weathered	82	58	41
0276	Cortical flake	brown	fresh	59	29	19
0277	Flake	grey	fresh	79	28	25
0278	Core	grey-honey	fresh	109	109	40
0279	Cortical flake	green	fresh	45	30	8
0281	Core	brown	fresh	77	54	44
0282	Core	grey-brown	weathered	70	70	69
0283	Cortical flake	grey-brown	fresh	67	27	24
0284	Flake	green-brown	fresh	35	23	9
0285	Cortical flake	grey	fresh	47	32	12
0286	Flaking debris	grey-brown	abraded	60	36	27
0287	Core fragment	grey-brown	fresh	100	61	20
0288	Core trimming flake	brown	weathered	80	52	30
0289	Cortical flake	grey	fresh	90	65	34
0290	Core fragment	green-brown	fresh	71	45	26
0291	Core	grey-brown	fresh	100	57	50
0292	Core	grey-brown	fresh	74	46	38
0293	Blade †	grey	fresh	45	31	11
0294	Core	white-grey	abraded	129	112	92
0295	Core	grey	fresh	126	120	84
0296	Core fragment	brown-grey	abraded	62	43	24
0297	Core	grey	fresh	89	73	42
0298	Cortical flake	brown-grey	abraded	43	32	18
0299	Broken blade †	white-grey	abraded	47	35	15
0300	Cortical flake	grey	fresh	46	24	9
0301	Core trimming flake	grey	fresh	55	51	20
0302	Core trimming flake	brown	fresh	89	57	18
0303	Retouched blade fragment †	honey	fresh	56	40	10
0304	Cortical flake	green	fresh	40	26	8
0305	Flake †	honey-green	fresh	27	18	5
0306	Retouched blade fragment	green-brown	fresh	21	20	7
0307	Blade †	brown	fresh	63	33	10
0308	Flake	grey	fresh	50	31	10
0309	Core	brown	fresh	80	74	44
0310	Cortical flake	white-grey	fresh	35	25	10
0311	Blade	green	fresh	35	15	10
0312	Flake	green	fresh	38	19	6
0313	Blade fragment †	grey	fresh	24	16	6
0314	Flake	red-green	fresh	36	29	7
0315	Cortical flake	brown-green	fresh	45	31	10

APPENDIX A (continued)

Number	Type	Colour	Condition	Length	Width	Depth
0316	Flake	grey	fresh	24	15	4
0317	Flake	grey	fresh	30	27	10
0318	Core	grey	fresh	135	81	62
0319	Flake †	green	fresh	24	22	4
0320	Flake	honey	fresh	29	15	6
0321	Flake	honey	fresh	29	24	8
0322	Cortical flake	honey	fresh	37	21	10
0323	Flake	brown	fresh	37	32	6
0324	Core fragment	honey	fresh	62	57	24
0325	Blade †	green-brown	fresh	55	26	7
0326	Cortical flake	grey	weathered	65	30	22
0327	Broken blade †	grey	fresh	40	36	11
0328	Core	grey-honey	weathered	95	66	66
0329	Core fragment	grey-brown	fresh	77	50	50
0330	Core fragment	grey	fresh	70	70	31
0331	Chipped blade †	green	fresh	53	39	14
0332	Core fragment	brown	fresh	90	67	50
0333	Flake	green-brown	fresh	38	22	11
0334	Blade †	grey	fresh	59	23	5
0335	Core	green-brown	fresh	130	95	68
0336	Blade fragment	honey	fresh	35	25	6
0337	Cortical flake	honey	fresh	41	36	16
0338	Blade †	honey	fresh	56	27	20
0339	Flake	honey	abraded	32	22	5
0340	Core	grey-brown	fresh	100	50	40
0341	Cortical flake	grey-brown	fresh	53	26	14
0342	Blade †	brown-honey	fresh	56	30	13
0343	Flaking debris	brown	abraded	45	24	8
0344	Core	green	fresh	98	88	61
0345	Cortical flake	green	fresh	31	17	10
0346	Struck cobble	brown	abraded	160	80	75
0347	Blade fragment	white-brown	weathered	27	24	10
0348	Flake	grey-brown	abraded	40	34	12
0349	Flake	brown	fresh	62	33	10
0350	Retouched flake	blue	fresh	49	29	10
0351	Retouched blade fragment †	grey-brown	fresh	34	33	14
0352	Core	grey	fresh	150	125	90
0353	Core trimming flake	grey-white	fresh	68	37	18
0354	Cortical flake	grey-honey	fresh	30	29	16
0355	Core	honey-grey	abraded	81	56	48
0356	Core	grey-brown	abraded	110	110	80
0357	Scraper	brown	fresh	46	31	14
0358	Core	grey	abraded	110	52	30
0359	Flake	honey-grey	fresh	55	41	14
0360	Blade fragment	honey-brown	fresh	25	18	6
0361	Broken blade	green-brown	fresh	54	48	16
0362	Blade fragment	blue	fresh	36	24	11
0363	Blade tip	grey	weathered	17	16	5
0364	Flake	grey-green	fresh	37	15	7
0365	Blade tip	grey	fresh	16	14	3
0366	Flaking debris	white-grey	fresh	29	21	10
0367	Core	grey-brown	fresh	95	46	42
0368	Cortical flake	grey-blue	fresh	35	23	11

Number	Type	Colour	Condition	Length	Width	Depth
0369	Blade tip	grey-green	fresh	15	9	3
0370	Scraper	green	fresh	59	33	12
0371	Flake	grey-green	fresh	28	20	10
0372	Broken blade	green	fresh	40	35	12
0373	Cortical flake	grey-green	fresh	51	40	16
0374	Scraper	green-brown	abraded	57	48	16
0375	Cortical flake	green	fresh	50	38	15
0376	Flake	green-brown	fresh	39	21	10
0377	Cortical flake	green	fresh	26	17	8
0378	Blade tip	grey-green	fresh	14	14	5
0379	Core	brown	fresh	140	100	35
0380	Core	grey	fresh	113	80	60
0381	Core	brown-green	fresh	84	65	59
0382a	Blade fragment	brown	fresh	20	19	4
0382b	Bladelet tip	grey-green	fresh	18	7	3
0383	Retouched flake †	green	fresh	45	43	16
0384	Cortical flake	grey-brown	fresh	50	35	18
0385	Core	honey-grey	fresh	50	35	18
0385	Core	honey-grey	fresh	106	75	55
0386	Core	brown-green	fresh	145	100	60
0387	Core	green-brown	fresh	128	112	81
0388	Flake	green	fresh	31	24	12
0389	Cortical flake	grey-green	fresh	95	34	25
0390	Blade fragment	green	fresh	35	23	6
0391	Blade tip	grey	fresh	20	13	4
0392	Flake	green-brown	fresh	35	30	9
0393	Blade fragment	grey-green	fresh	14	14	4
0394	Cortical flake	brown	fresh	27	26	7
0395	Blade	green-brown	fresh	88	41	11
0396	Blade tip	green	fresh	43	39	14
0397	Flake	grey-white	abraded	35	33	6
0398	Flaking debris	honey-grey	weathered	35	23	11
0399	Blade tip	white-grey	weathered	18	15	9
0400	Flake	grey-green	fresh	28	27	12
0401	Blade fragment	grey-green	fresh	21	17	5
0402	Cortical flake	white	weathered	42	28	9
0404	Blade tip	brown-green	fresh	20	18	4
0405	Blade fragment	blue	fresh	59	29	12
0406	Cortical flake	red-brown	fresh	43	27	9
0407	Blade fragment	grey-green	fresh	29	29	8
0408	Bladelet	green	fresh	62	27	12
0409	Broken blade	brown	fresh	46	22	10
0410	Flaking debris	green	abraded	17	17	3
0411	Bladelet tip	green	fresh	11	10	2
0412	Flaking debris	red-brown	abraded	20	16	3
0413	Broken blade tip	honey	fresh	26	26	14
0414	Broken blade	brown-green	fresh	47	33	10
0415	Flaking debris	grey-honey	abraded	23	20	9
0416	Cortical flake	green-brown	fresh	28	11	10
0417	Blade tip	blue-grey	fresh	27	18	6
0418	Blade tip	grey	fresh	11	9	4
0419	Blade tip	grey	fresh	20	10	4
0420	Blade fragment	green	fresh	35	21	11

APPENDIX A (continued)

Number	Type	Colour	Condition	Length	Width	Depth
0421	Broken blade	green-brown	fresh	62	41	12
0422	Cortical flake	grey	abraded	55	38	27
0423	Cortical flake	grey-white	abraded	52	27	15
0424	Blade tip	grey-green	fresh	22	20	3
0425	Blade fragment	green-brown	fresh	27	22	10
0426	Core	honey	fresh	97	95	45
0427	Blade fragment	grey-green	fresh	23	22	9
0428	Core	grey-brown	fresh	75	59	37
0429	Cortical flake	grey-white	fresh	30	21	12
0430	Blade tip	brown	fresh	17	14	5
0431	Flake	brown-green	fresh	31	24	16
0432	Cortical flake	green	fresh	44	31	23
0433	Point	brown-green	fresh	31	13	5
0434	Core	brown-black	fresh	135	78	61
0435	Blade tip	white	fresh	22	13	5
0436	Split cobble	grey-honey	abraded	50	42	22
0437	Cortical flake	green	weathered	39	25	10
0439	Cortical flake	grey-white	fresh	41	36	17
0440	Cortical flake	grey-white	abraded	62	26	17
0441	Core	brown	fresh	185	100	43
0440	Cortical flake	grey-white	abraded	62	26	17
0441	Core	brown	fresh	185	100	43
0442	Flaking debris	green-white	abraded	24	18	8
0443	Flake	grey-honey	fresh	31	22	4
0444	Blade fragment	honey	fresh	31	20	10
0446	Core	grey	fresh	42	25	20
0447	Retouched cortical flake	brown	fresh	58	42	13
0448	Blade tip	beige	fresh	16	16	4
0449	Core trimming flake	green	weathered	48	26	9
0450	Cortical flake	honey	fresh	22	13	4
0451	Flake	brown-green	fresh	62	31	6
0452	Broken blade	green	fresh	45	23	14
0453	Blade tip	green	fresh	20	18	3
0454	Cortical flake	grey-green	fresh	64	58	20
0455	Core	grey-brown	fresh	180	140	65
0456	Cortical flake	grey-green	abraded	34	27	12
0457	Core	grey	fresh	160	114	85
0458	Core	grey	fresh	89	65	40
0459	Cortical flake	green-brown	fresh	49	25	15
0460	Core	grey-red	fresh	50	42	30
0461	Core	grey	fresh	106	70	57
0462	Core	grey	fresh	82	53	45
0463	Broken blade	brown-green	fresh	48	41	8
0464	Cortical flake	grey-brown	abraded	52	35	24
0465	Broken blade	green	fresh	73	36	27
0466	Cortical flake	beige	abraded	65	54	18
0467	Core	honey-grey	fresh	106	55	44
0468	Core	honey	fresh	78	59	32
0469	Core trimming flake	brown	fresh	40	37	15
0470	Core	grey	abraded	75	40	38
0471	Core trimming flake	green	fresh	53	52	16
0472	Cortical flake	grey-white	abraded	43	36	20
0473	Cortical flake	grey-white	abraded	58	33	20

APPENDIX A (continued)

Number	Type	Colour	Condition	Length	Width	Depth
0474	Core fragment	honey	fresh	80	45	36
0475	Blade tip	grey	fresh	30	28	8
0476	Core trimming flake	green	weathered	45	33	12
0477	Blade fragment	grey-blue	fresh	23	20	5
0478	Blade fragment	grey	abraded	30	22	10
0479	Core	grey-brown	abraded	70	50	47
0481	Core	honey-grey	fresh	90	55	45
0482	Core	brown-grey	weathered	100	90	85
0483	Cortical flake	green	fresh	29	25	12
0484	Core	grey-brown	weathered	128	105	90
0486	Scraper	grey-green	fresh	42	18	10
0487	Core	brown-grey	fresh	90	63	60
0488	Core trimming flake	grey	fresh	65	45	22
0489	Core	grey-green	weathered	107	47	30
0490	Cortical flake	grey-green	fresh	67	40	28
0491	Cortical flake	green	fresh	60	40	24
0493	Cortical flake	honey	fresh	43	36	17
0494	Broken blade	grey-green	fresh	58	37	6
0495	Blade tip	green-brown	fresh	26	14	3
0496	Cortical flake	brown	fresh	21	16	5
0497	Flake	grey-green	fresh	37	34	16
0498	Blade fragment	grey-blue	fresh	33	23	9
0499	Scraper	green	fresh	39	26	13
0500	Blade fragment	green	fresh	22	19	6
0501	Core	grey-blue	fresh	114	87	75
0502	Cortical flake	green	abraded	35	28	10
0503	Cortical flake	grey-blue	fresh	23	20	11
0504	Blade fragment	green-grey	fresh	24	19	6
0505	Blade tip	green	fresh	30	22	4
0506	Retouched flake †	green-brown	fresh	54	26	12
0507	Broken blade	brown	fresh	33	18	9
0508	Retouched blade fragment †	grey	fresh	19	18	5
0509	Blade tip	grey-blue	fresh	38	27	10
0510	Broken blade	honey	fresh	40	26	14
0511	Core fragment	grey	fresh	63	52	28
0512	Cortical flake	grey-green	fresh	30	20	5
0513	Core trimming flake	grey-brown	fresh	95	60	22
0514	Broken blade	brown	fresh	46	18	11
0515	Cortical flake	grey-green	fresh	75	41	15
0516	Core	honey-grey	fresh	80	70	41
0517	Blade	grey-honey	fresh	48	25	22
0518	Blade fragment	green	fresh	30	30	12
0519	Broken blade tip	grey	weathered	16	8	2
0520	Broken blade	grey-brown	abraded	39	30	11
0521	Flaking debris	green	fresh	20	20	4
0522	Flake	grey-green	fresh	45	41	20
0523	Flaking debris	green	fresh	32	26	13
0524	Split pebble	grey	fresh	49	40	24
0525	Cortical flake	green	fresh	37	26	16
0526	Flaking debris	white-honey	fresh	41	29	15
0527	Flaking debris	grey	weathered	20	20	7
0528	Blade fragment	green	abraded	28	17	9
0529	Cortical flake	grey-brown	abraded	85	47	30

APPENDIX A (continued)

Number	Type	Colour	Condition	Length	Width	Depth
0530	Flake	honey	fresh	37	35	10
0531	Cortical flake	grey-green	fresh	67	39	20
0532	Blade tip	honey	weathered	21	18	8
0533	Hammerstone	brown	weathered	87	62	44
0534	Cortical flake	green	fresh	37	26	12
0535	Core fragment	grey-honey	abraded	54	37	22
0536	Cortical flake	brown	fresh	69	66	27
0537a	Flake	honey	fresh	31	18	12
0537b	Flake	green	fresh	31	17	9
0538	Blade fragment	grey	fresh	30	25	7
0539	Core trimming flake	green	fresh	54	49	10
0540	Flake	grey	fresh	38	31	10
0541	Cortical flake	grey-honey	fresh	29	16	8
0542	Flake	grey-honey	fresh	32	20	4
0543	Cortical flake	grey-honey	abraded	47	33	19
0544	Blade tip	green-brown	fresh	40	19	7
0545	Core fragment	grey	fresh	68	38	37
0547	Cortical flake	grey	fresh	21	19	12
0548	Broken blade tip	honey	fresh	11	8	2
0549	Cortical flake	black	fresh	38	17	13
0550	Blade tip	grey-green	fresh	29	24	10
0551	Cortical flake	honey	fresh	40	28	15
0552	Core	brown-grey	fresh	87	60	34
0553	Cortical flake	brown	fresh	95	31	25
0555	Broken blade	green	fresh	66	41	16
0556	Blade tip	grey	abraded	32	31	13
0557	Flake	honey	fresh	56	39	11
0558	Core fragment	white-grey	fresh	47	36	27
0559	Core fragment	grey-brown	fresh	25	22	15
0560	Cortical flake	grey-white	abraded	38	30	17
1000	Retouched flake	brown	fresh	120	75	35
1001	Cortical flake	grey-blue	fresh	76	64	32
1002	Cortical flake	grey-blue	fresh	44	33	13
1003	Cortical flake	green	abraded	80	80	35
1004	Cortical flake	brown	fresh	68	52	26
1005	Halved cobble	grey	abraded	75	60	50
1006	Broken blade	honey-grey	fresh	36	32	13
1007	Flake	grey	abraded	40	35	20
1008	Core	grey-honey	abraded	120	95	95
1009	Cortical flake	grey-blue	fresh	42	40	22
1010	Cortical flake	grey-honey	abraded	88	67	39
1011	Retouched flake †	honey-brown	fresh	70	37	12
1012	Cortical flake	blue	abraded	101	46	32
1013	Cortical flake	brown-green	fresh	72	60	34
1014	Core trimming flake	green-blue	fresh	135	69	25
1015	Cortical flake	brown-green	abraded	90	60	51
1016	Cortical flake	grey-blue	abraded	94	55	51
1017	Cortical flake	brown-honey	fresh	108	90	46
1018	Core trimming flake	honey	fresh	93	68	22
1019	Cortical flake	brown-honey	fresh	60	41	36
1020	Core trimming flake	green-brown	fresh	75	60	34
1021	Cortical flake	green-brown	abraded	41	36	6

APPENDIX A (continued)

Number	Type	Colour	Condition	Length	Width	Depth
1022	Cortical flake	honey	fresh	40	30	12
1023	Flake	green	fresh	56	35	18
1024	Core	brown	fresh	170	132	126
1025	Cortical flake	honey	abraded	64	60	55
1026	Flake †	green	fresh	72	55	29
1027	Cortical flake	grey	abraded	70	62	48
1028	Cortical flake	green-honey	fresh	90	35	30
1029	Cortical flake	green	fresh	58	47	33
1030	Flake	grey-brown	fresh	70	49	23
1031	Cortical flake	white-honey	fresh	57	39	25
1032	Flake	grey-brown	abraded	65	36	34
1033	Flake	honey	abraded	53	48	15
1034	Cortical flake	red-brown	fresh	73	54	21
1035	Core	brown-honey	fresh	115	70	61
1036	Cortical flake	green	abraded	54	50	23
1037	Core	honey	fresh	82	56	44
1038	Cortical flake	green	fresh	82	64	34
1039	Cortical flake	brown-grey	abraded	64	38	26
1040	Flake	honey-grey	abraded	61	59	21
1041	Flake	honey	abraded	75	43	21
1042	Core	green-brown	fresh	150	70	45
1043	Cortical flake	honey-grey	abraded	77	61	41
1044	Cortical flake	green	fresh	37	29	5
1045	Flake	honey	abraded	78	49	21
1046	Retouched flake †	brown	fresh	61	40	18
1047	Cortical flake	grey-honey	fresh	78	73	29
1048	Core	grey-blue	fresh	115	75	41
1049	Quartered cobble	honey	fresh	63	49	41
1050	Cortical flake	grey-blue	fresh	85	55	50
1051	Core	grey	fresh	125	95	80
1052	Cortical flake	grey	abraded	68	52	34
1053	Cortical flake	honey	fresh	69	51	33
1054	Cortical flake	brown-green	fresh	60	45	25
1055	Core †	brown-green	fresh	140	85	38
1056	Cortical flake	grey-blue	abraded	58	26	5
1057	Cortical flake	grey	abraded	130	58	51
1058	Core trimming flake	grey	abraded	85	50	21
1059	Cortical flake	blue	fresh	45	29	11
1060	Flake	white-honey	abraded	83	56	24
1061	Cortical flake	grey	fresh	30	24	13
1062	Cortical flake	red-brown	fresh	54	54	28
1063	Core	grey	fresh	76	72	70
1064	Core †	green-brown	fresh	115	108	48
1065	Cortical flake	grey-green	fresh	83	39	35
1066	Core	blue	fresh	86	58	43
1067	Flake	blue-grey	fresh	59	42	15
1068	Core trimming flake	grey	fresh	79	66	28
1069	Core trimming flake	honey	abraded	130	115	46
1070	Cortical flake	grey-white	fresh	68	40	20
1071	Flake	green-grey	abraded	46	24	8
1072a	Blade †	grey	abraded	43	20	16
1072b	Flake	grey	fresh	33	10	6
1073	Blade †	grey-red	fresh	45	19	12

APPENDIX A (continued)

Number	Type	Colour	Condition	Length	Width	Depth
1074	Flake †	red-brown	fresh	62	40	15
1075	Core	grey-honey	abraded	90	50	44
1076	Flake †	green-brown	fresh	24	19	5
1077	Cortical flake	grey-brown	abraded	43	38	14
1078	Flake †	white	fresh	34	26	7
1079	Cortical flake	grey	abraded	62	30	15
1080	Cortical flake †	grey	fresh	62	36	18
1081	Core fragment	grey	fresh	80	50	34
1082	Flake	red-grey	fresh	43	31	16
1083	Flake †	grey	fresh	66	38	14
1084	Retouched flake †	grey	fresh	32	31	8
1085	Core fragment	grey	abraded	58	56	29
1086	Flaking debris	honey	abraded	62	24	24
1087	Retouched flake †	grey-brown	fresh	61	32	12
1088	Retouched cortical flake †	green	abraded	70	47	22
1089	Cortical flake †	red-brown	abraded	110	100	26
1090	Core fragment	brown-green	abraded	70	60	26
1091	Cortical flake †	green-brown	abraded	63	50	22
1092	Cortical flake	grey	abraded	50	43	20
1093	Retouched blade fragment †	grey-green	fresh	31	22	10
1094	Flake	grey-green	fresh	28	25	6
1095	Cortical flake	grey-green	fresh	72	53	30
1096	Core	grey-green	abraded	101	81	76
1097	Cortical flake	green	fresh	61	33	23
1098	Core	grey-green	abraded	94	82	68
1099	Flake	green	fresh	55	42	11
1100	Cortical flake †	green	fresh	62	57	13
1101	Core	green-brown	fresh	113	82	71
1102	Blade tip †	grey-green	fresh	25	18	6
1103	Retouched cortical flake †	grey	fresh	82	68	17
1104	Cortical flake	grey	fresh	53	35	32
1105	Core fragment	honey	abraded	90	70	42
1106	Core	grey-brown	abraded	78	76	73
1107	Core fragment	grey-green	fresh	80	53	27
1108	Retouched flake †	green	fresh	56	30	12
1109	Cortical flake	honey	abraded	144	44	27
1110	Core fragment/flake †	grey-green	fresh	40	27	18
1111	Cortical flake	honey	abraded	65	65	28
1112	Cortical flake	green-brown	fresh	78	77	26
1113	Core †	brown-green	fresh	119	81	73
1114	Core fragment	green-brown	fresh	117	70	40
1115	Cortical flake	green	fresh	51	32	27
1116	Core †	brown	fresh	128	107	62
1117	Retouched blade †	brown	fresh	50	26	18
1118	Cortical flake	green-blue	abraded	109	80	36
1119	Retouched blade fragment †	green-blue	fresh	53	40	5
1120	Retouched blade fragment †	grey-green	fresh	22	18	4
1121	Core fragment †	grey-green	fresh	118	110	44
1122	Core †	grey-green	fresh	114	103	63
1123	Core	brown-green	fresh	121	99	35
1124	Retouched cortical flake †	grey-green	abraded	73	54	16
1125	Core	green-brown	fresh	100	81	47
1126	Cortical flake	green	abraded	48	44	16

APPENDIX A (continued)

Number	Type	Colour	Condition	Length	Width	Depth
1127	Cortical flake	grey-green	fresh	70	43	19
1128	Core trimming flake	grey-green	fresh	58	49	14
1129	Core fragment	grey	fresh	41	26	15
1130	Cortical flake	grey-green	abraded	70	70	21
1131	Flaking debris	grey-red	abraded	51	50	25
1132	Cortical flake †	green	fresh	102	66	26
1133	Core fragment †	green-brown	fresh	85	57	36
1134	Cortical flake †	green	fresh	76	54	18
1135	Core	grey-blue	fresh	112	76	72
1136	Flake †	green	abraded	38	30	10
1137	Broken blade	green-honey	fresh	28	23	13
1138	Core	white-grey	abraded	83	78	41
1139	Flake	brown	fresh	52	34	15
1140	Core	green-brown	abraded	113	81	70
1141	Retouched blade †	green	fresh	65	31	15
1142	Retouched blade †	brown-green	fresh	67	35	9
1143	Retouched cortical flake †	brown-green	fresh	101	78	28
1144	Retouched flake †	red-brown	abraded	130	73	22
1145	Core	green-brown	abraded	130	90	66
1146	Core fragment	brown-green	abraded	78	60	60
1147	Cortical flake †	honey	fresh	72	68	20
1148	Core fragment †	green	fresh	97	69	27
1149	Cortical flake	brown	fresh	43	39	13
1150	Flake	honey	fresh	62	36	20
1151	Cortical flake	green	fresh	34	230	12
1152	Retouched flake †	green	fresh	50	39	9
1153	Retouched blade fragment †	green	fresh	31	21	9
1154	Blade fragment †	green	fresh	21	12	5
1155	Core	green	fresh	115	60	32
1156	Cortical flake	green-brown	fresh	40	30	10
1157	Blade fragment	blue-green	fresh	35	30	10
1158	Flaking debris	grey-black	fresh	31	20	6
1159	Cortical flake †	green	fresh	68	59	31
1160	Cortical flake †	grey-brown	fresh	105	59	21
1161	Flake	grey	fresh	61	44	20
1162	Retouched blade fragment †	grey	fresh	31	17	11
1163	Flake †	grey	fresh	46	27	14
1164	Cortical flake	honey-brown	weathered	64	49	12
1165	Cortical flake †	honey-brown	fresh	53	27	17
1166	Flake †	honey-brown	fresh	80	43	17
1167	Flake †	grey	fresh	21	18	10
1168	Flaking debris	white-honey	weathered	45	28	8
1169	Core	brown	fresh	140	110	50
1170	Flaking debris	grey-green	abraded	46	40	38
1171	Flake †	brown-green	fresh	47	28	13
1172	Cortical flake †	green-brown	fresh	92	50	26
1173	Cortical flake †	white-honey	fresh	70	53	27
1174	Blade †	brown-red	fresh	50	29	14
1175	Retouched flake †	grey	fresh	22	18	3
1176	Retouched cortical flake	grey-green	fresh	47	28	15
1177	Retouched flake	honey	fresh	64	50	15
1178	Flake	grey-honey	fresh	64	29	20
1179	Retouched cortical flake †	red-green	fresh	45	40	9

APPENDIX A (continued)

Number	Type	Colour	Condition	Length	Width	Depth
1180	Retouched flake †	green-brown	fresh	50	29	15
1181	Flake †	green-brown	fresh	51	44	14
1182	Flake †	green-brown	fresh	29	27	9
1183	Cortical flake	grey-green	fresh	35	26	14
1184	Core †	honey	fresh	93	70	31
1185	Broken blade tip †	honey	fresh	17	7	5
1186	Flaking debris	grey-green	fresh	48	33	27
1187	Flake	grey-green	fresh	24	13	5
1188	Flake	grey-green	fresh	22	13	6
1189	Cortical flake †	green-brown	fresh	31	27	10
1190	Retouched broken flake †	green	fresh	38	22	14
1191	Retouched blade fragment †	grey	fresh	26	18	11
1192	Core trimming flake	brown	fresh	87	84	25
1193	Core	brown-grey	fresh	110	83	66
1194	Core	honey	abraded	100	90	55
1195	Core †	green	fresh	113	82	65
1196	Retouched blade fragment †	green	fresh	34	22	9
1197	Core	grey-brown	fresh	127	76	32
1198	Retouched blade fragment †	green	fresh	41	29	10
1199	Retouched flake	grey-honey	fresh	38	31	11
1200	Blade fragment †	grey-green	fresh	26	17	7
1201	Cortical flake	honey	abraded	82	54	22
1202	Blade fragment	brown	abraded	29	20	9
1204	Cortical flake	brown-honey	fresh	67	40	17
1205	Core	brown-green	fresh	87	75	50
1206	Cortical flake	honey	fresh	21	21	9
1207	Blade fragment †	green	fresh	21	10	10
1208	Blade fragment †	green-brown	fresh	34	20	7
1209	Flaking debris	green-brown	fresh	32	15	5
1210	Cortical flake	honey	fresh	37	21	7
1211	Flake †	honey-grey	fresh	49	38	11
1212	Retouched flake †	honey-green	fresh	45	43	14
1213	Cortical flake	red-brown	fresh	30	20	9
1214	Flaking debris	grey-black	abraded	38	28	9
1215	Cortical flake	green	fresh	35	17	10
1216	Core fragment	brown-green	fresh	50	45	36
1217	Flake †	green	fresh	29	19	9
1218	Flaking debris	brown-green	fresh	28	21	7
1219	Retouched blade fragment †	green-brown	fresh	35	25	10
1220	Biface †	brown-black	abraded	136	80	35
1221	Retouched cortical flake †	brown-green	fresh	78	50	17
1222	Flaking debris	grey-cream	abraded	46	35	16
1223	Blade fragment	green	abraded	40	26	12
1224	Blade fragment	cream	abraded	71	25	9
1225	Broken blade	grey	fresh	34	16	9
1226	Flaking debris	green	fresh	25	20	10
1227	Flaking debris	green	fresh	15	15	6
1228	Flaking debris	green	fresh	37	26	18
1229	Flake †	honey-brown	fresh	31	30	15
1230	Retouched blade †	brown	fresh	65	35	10
1231	Cortical flake	brown	fresh	36	23	10
1232	Flake	brown-green	fresh	42	24	13
1233	Retouched flake †	grey-blue	fresh	39	32	11

APPENDIX A (continued)

Number	Type	Colour	Condition	Length	Width	Depth
1234	Blade †	green-grey	fresh	55	31	13
1235	Core fragment	honey	fresh	66	48	36
1236	Core fragment	honey-brown	fresh	70	58	37
1237	Core fragment †	green	fresh	85	57	22
1238a	Flaking debris	green	fresh	13	11	6
1238b	Flaking debris	green	fresh	38	27	13
1239	Retouched cortical flake †	honey	fresh	82	56	22
1240	Core fragment	green	abraded	100	94	37
1241	Broken blade †	green	fresh	37	27	12
1242	Flaking debris	grey-black	abraded	61	48	18
1243	Core fragment	green-yellow	fresh	82	73	27
1244	Flake	green	fresh	51	20	15
1245	Core trimming flake	grey	fresh	50	39	11
1246	Flaking debris	green	fresh	16	15	5
1247	Blade †	brown	fresh	65	25	15
1248	Flaking debris	green	fresh	19	18	6
1249	Broken blade	olive	abraded	38	35	9
1250	Cortical flake	grey	fresh	33	18	9
1251	Broken blade †	grey-green	fresh	33	20	8
1252	Flake	grey	abraded	31	24	12
1253	Cortical flake	grey	fresh	44	25	19
1254	Flake †	grey-brown	fresh	66	47	16
1255	Blade †	grey-green	fresh	41	18	9
1256	Broken blade	grey-brown	fresh	40	30	10
1257	Blade fragment	red-brown	fresh	15	10	4
1258	Blade †	white-cream	abraded	75	29	17
1259	Broken blade †	grey	fresh	29	17	12
1260	Cortical flake	grey	fresh	49	40	17
1261	Blade fragment	green	fresh	22	15	5
1262	Blade fragment	green-brown	fresh	21	18	4
1263	Broken blade †	honey	abraded	50	40	13
1264	Blade fragment	green	fresh	14	11	6
1265	Blade fragment †	green-brown	fresh	33	31	11
1266	Blade fragment	green	fresh	15	11	5
1267	Cortical flake †	grey	abraded	75	60	41
1268	Cortical flake	white	abraded	50	26	16
1269a	Broken point	honey	abraded	22	15	10
1269b	Broken blade tip	grey	fresh	12	9	3
1270	Core fragment	brown	abraded	109	88	33
1271	Broken blade tip	grey	fresh	15	13	2
1272	Core	green-brown	abraded	65	61	58
1273	Cortical flake	grey	fresh	45	27	13
1274	Broken blade	grey-green	fresh	16	15	4
1275a	Blade fragment	grey-green	abraded	31	12	10
1275b	Broken blade tip	grey	fresh	15	13	6
1276	Blade fragment	grey	fresh	21	20	3
1277	Core	grey-brown	fresh	83	47	28
1278a	Retouched flake †	grey	abraded	74	51	16
1278b	Retouched blade †	grey	abraded	25	22	5
1279	Flake †	grey-honey	fresh	40	33	12
1280	Core fragment	grey-brown	fresh	63	42	30
1281	Core fragment †	blue-grey	fresh	66	55	30
1282	Core †	grey-honey	fresh	113	100	60

APPENDIX A (continued)

Number	Type	Colour	Condition	Length	Width	Depth
1283	Core †	grey-honey	fresh	100	90	48
1284	Flake	honey-pink	weathered	46	40	16
1285	Blade tip †	grey	fresh	25	19	10
1286	Cortical flake	honey	abraded	65	47	25
1287	Core	green-brown	fresh	101	97	33
1288	Blade fragment	green	abraded	31	18	9
1289	Retouched blade fragment †	grey-green	fresh	24	16	4
1290	Cortical flake	grey	fresh	29	23	6
1291	Core †	grey	fresh	112	102	48
1292	Broken blade	brown	fresh	36	22	10
1293	Flaking debris	green-brown	fresh	55	50	21
1294	Blade tip †	grey	fresh	13	12	4
1295	Blade †	green-brown	fresh	53	26	15
1296	Core †	brown	fresh	141	106	45
1297	Retouched cortical flake	green-brown	fresh	69	43	16
1298	Blade tip	green	weathered	24	20	6
1299	Flaking debris	green-brown	weathered	40	22	12
1300	Core fragment	grey-brown	abraded	73	61	44
1301	Core	grey-brown	abraded	119	72	60
1302	Cortical flake	red-brown	abraded	70	60	27
1303	Core fragment	grey-honey	fresh	67	59	32
1304	Blade †	honey-white	weathered	38	17	9
1305	Retouched blade †	brown	fresh	35	25	6
1306	Flake	grey	fresh	28	26	8
1307	Flaking debris	brown	fresh	60	54	15
1308	Flake	red-brown	fresh	50	44	13
1309	Flaking debris	grey	fresh	35	23	14
1310	Blade fragment †	honey	weathered	40	27	15
1311	Broken blade †	grey-blue	abraded	40	26	15
1312	Flake	honey-white	weathered	45	45	8
1313	Blade fragment	honey	fresh	21	19	6
1314	Flake †	grey	fresh	36	35	10
1315	Core fragment	grey-brown	abraded	61	45	17
1316	Flake	honey	abraded	33	20	12
1317	Core fragment	brown-honey	fresh	65	36	15
1318	Core	green	fresh	81	78	45
1319	Flaking debris	honey	fresh	36	22	15
1320	Cortical flake	grey	fresh	33	17	8
1321	Blade fragment †	grey	fresh	32	12	5
1322	Retouched flake/scraper †	brown-green	fresh	75	53	20
1323	Core trimming flake	green-brown	fresh	97	50	18
1324	Flake	grey-brown	fresh	88	45	20
1325	Blade fragment †	yellow-grey	fresh	15	14	4
1326	Flake	honey	abraded	51	38	5
1327	Cortical flake †	green	fresh	57	44	22
1328	Retouched core †	grey-green	fresh	64	52	26
1329	Flaking debris	grey	fresh	31	26	8
1330	Blade †	grey	abraded	50	31	13
1331	Core trimming flake †	green-brown	fresh	46	37	16
1332	Cortical flake †	green	fresh	52	46	13
1333	Retouched flake	green-brown	fresh	91	70	22
1334	Retouched flake †	brown-green	fresh	77	55	18
1335	Flake †	grey	fresh	39	32	11

APPENDIX A (continued)

Number	Type	Colour	Condition	Length	Width	Depth
1336	Cortical flake	green	abraded	43	31	20
1337	Core fragment	honey	fresh	120	56	35
1338	Broken flake †	grey	fresh	24	23	7
1339	Flake †	grey	fresh	82	47	27
1340	Flake	brown	fresh	79	68	36
1341	Cortical flake	grey-brown	fresh	41	27	11
1342	Retouched blade fragment †	brown-green	fresh	19	17	5
1343	Flaking debris	green	fresh	19	17	8
1344	Cortical flake	green-brown	abraded	35	23	12
1345	Flake	red-brown	fresh	22	20	5
1346	Retouched blade fragment †	green	fresh	23	22	8
1347	Flake	grey-green	fresh	21	18	3
1348	Flake	yellow-brown	fresh	76	72	30
1349	Flake †	grey	fresh	55	38	12
1350	Retouched blade fragment †	red-brown	fresh	25	15	6
1351	Blade fragment †	green	fresh	23	19	5
1352	Blade fragment	grey-green	fresh	20	13	4
1353	Core fragment	honey	fresh	75	70	42
1354	Blade/?burin †	green	fresh	35	18	6
1355	Flake	green	fresh	80	60	20
1356	Flake †	honey	fresh	44	33	16
1357	Retouched flake †	brown-green	fresh	45	40	13
1358	Flaking debris	grey-white	abraded	37	29	11
1359	Cortical flake	green	fresh	43	16	14
1360	Flaking debris	grey-green	fresh	21	13	3
1361	Retouched cortical flake †	grey-green	fresh	70	53	20
1362	Flaking debris	grey-blue	abraded	54	24	11
1363a	Flaking debris	green	fresh	28	25	6
1363b	Flaking debris	green	abraded	25	24	5
1364	Core fragment	grey	fresh	64	24	16
1365a	Flake/?scraper †	grey	fresh	78	53	23
1365b	Flaking debris	grey	fresh	24	11	4
1366	Flaking debris	honey	abraded	36	35	14
1367	Flake	grey-green	fresh	23	19	4
1368	Retouched flake †	grey-blue	fresh	75	45	18
1369	Retouched flake †	green-grey	fresh	75	47	18
1370	Flake †	green	fresh	72	45	18
1371	Flake	green	fresh	48	30	11
1372	Core fragment	grey-blue	fresh	45	39	11
1373	Core fragment	honey	fresh	52	25	18
1374	Flaking debris	grey-blue	fresh	31	25	12
1375	Broken blade †	grey	fresh	22	14	10
1376	Retouched blade fragment †	grey-green	fresh	28	22	6
1377	Blade tip †	grey	fresh	21	18	6
1378	Cortical flake	green-honey	fresh	50	41	20
1379	Broken blade tip	red-brown	fresh	26	19	4
1380	Flaking debris	grey-blue	abraded	28	24	13
1381	Cortical flake †	grey	fresh	64	48	15
1382	Retouched cortical flake †	grey	fresh	73	45	21
1383	Retouched blade fragment †	grey	fresh	46	40	15
1384	Core	grey-green	fresh	113	76	45
1385	Cortical flake †	green-brown	fresh	68	55	20
1386	Cortical flake	honey	abraded	42	31	9

APPENDIX A (continued)

Number	Type	Colour	Condition	Length	Width	Depth
1387	Flaking debris	honey	abraded	32	22	9
1388	Cortical flake	honey	fresh	40	38	17
1389	Broken blade †	grey	fresh	50	35	14
1390	Flaking debris	grey-green	fresh	27	18	5
1391	Cortical flake	green	fresh	33	24	11
1392	Flake	green	fresh	54	46	24
1393	Core trimming flake	green-brown	fresh	60	40	12
1394	Flake	honey-brown	fresh	53	37	10
1395a	Cortical flake	green-grey	fresh	80	40	19
1395b	Cortical flake	green	abraded	72	37	24
1395c	Cortical flake	green	abraded	52	25	18
1396	Flaking debris	green	fresh	32	16	10
1397	Flaking debris	green	abraded	36	34	12
1398	Retouched blade fragment †	grey-green	fresh	40	27	9
1399	Broken blade †	green	fresh	44	25	10
1400	Core	green	fresh	65	52	35
1401	Cortical flake	grey-green	fresh	42	32	12
1402	Core fragment †	green	fresh	73	46	40
1403	Flaking debris	brown	abraded	34	31	16
1404	Cortical flake	green	abraded	47	29	10
1405	Core fragment/flake †	brown-green	fresh	90	51	35
1406	Retouched flake †	grey	fresh	37	34	9
1407	Flake	grey	fresh	43	39	18
1408	Retouched flake †	green-brown	fresh	62	50	6
1409	Retouched flake †	grey	fresh	51	49	13
1410	Retouched blade fragment †	grey	fresh	51	27	9
1411	Cortical flake	honey	fresh	41	40	20
1412	Flaking debris	grey-blue	fresh	36	34	14
1413	Core fragment	grey	abraded	85	73	32
1414	Flaking debris	white	abraded	82	43	17
1415	Retouched blade fragment †	grey-green	fresh	50	32	10
1416	Cortical flake †	grey-brown	fresh	69	30	11
1417	Flake †	grey	fresh	30	17	8
1418	Retouched blade fragment †	grey	fresh	30	26	9
1419	Cortical flake	honey	fresh	51	36	12
1420	Blade fragment †	grey	abraded	26	18	5
1421	Core/biface	green-brown	abraded	114	70	24
1422	Core	grey	abraded	97	63	35
1423a	Blade fragment †	grey-honey	fresh	17	16	2
1423b	Blade fragment	grey-honey	fresh	16	10	3
1424a	Retouched blade fragment †	grey-green	fresh	18	15	3
1424b	Retouched blade fragment †	grey	fresh	15	12	5
1425	Core fragment †	honey-blue	fresh	190	68	37
1426	Retouched cortical flake †	green-blue	fresh	57	45	14
1427	Biface †	red-brown	abraded	120	85	40

†= see illustrations

Appendix B

Coordinates of artefacts from spring season of 1981, site 55, Riwat.

Number	X	Y	Square	Type	Number	X	Y	Square	Type
0010	1327	1088	B3 NW	flake	0011	1318	1033	B3 NW	flake
0012	1350	1057	B3 NW	flake	0015	1187	0917	C4 SW	blade
0017	1281	1099	B3 NE	blade	0026	1460	0975	A4 SE	flake
0030	1432	1075	A3 NE	flake	0040	1506	1024	A3 NW	blade
0046	1336	1003	B3 NW	flake	0052	1150	0935	C4 SW	flake
0054	1210	0964	B4 SE	flake	0062	1447	0990	A4 SE	blade
0065	1536	1035	A3 NW	blade	0071	1217	0971	B4 SE	flake
0073	1348	1006	B3 NW	flake	0074	1249	0968	B4 SE	blade
0075	1453	0998	A4 SE	flake	0076	1375	1193	B3 SW	blade
0077	1343	1311	B2 SW	flake	0079	1180	1118	C3 SW	flake
0080	1161	1056	C3 NW	blade	0081	1363	1420	B1 NW	flake
0082	1008	1379	C2 SE	flake	0083	0972	1465	D1 NW	core
0087	1250	0959	B4 SE	blade					

Coordinates of artefacts from spring season of 1982, site 55, Riwat.

Number	X	Y	Square	Type	Number	X	Y	Square	Type
0101	1518	1137	A3 SW	flake	0102	1481	1129	A3 SE	blade
0103	1475	1122	A3 SE	flake	0104	1423	1065	A3 NE	flake
0105	1421	1039	A3 NE	blade	0106	1337	1028	B3 NW	blade
0107	1356	0958	B4 SW	blade	0108	1197	0946	C4 SW	blade
0109	1348	0961	B4 SW	blade	0110	1287	0940	B4 SE	flake
0112	1547	1281	A2 NW	core	0113	1390	0923	B4 SW	blade
0114	1466	0929	A4 SE	flake	0115	1520	0879	A4 NW	flake
0116	1452	0981	A4 SE	flake	0117	1524	0969	A4 SW	core
0118	wall footing			flake	0119	1126	0909	C4 SW	flake
0120	1440	0868	A4 NE	flake	0121	wall footing			flake
0122	0926	0780	D5 SW	flake	0123	0905	0731	D5 SW	flake
0124	0915	0723	D5 SW	flake	0125	0965	0803	D4 NW	flake
0126	1128	0831	C4 NW	blade	0127	1176	0889	C4 NW	flake
0128	1039	0770	C5 SE	flake	0130	1301	0760	B5 SW	flake
0131	1330	0800	B5 SW	flake	0132	1300	0794	B5 SE	blade
0133	1344	0830	B4 NW	blade	0134	1306	0837	B4 NW	flake
0135	1542	1032	A3 NW	blade	0136	1544	1051	A3 NW	flake
0137	1516	1050	A3 NW	flake	0140	1510	1371	A2 SW	blade
0141	1453	1265	A2 NE	blade	0142	1495	1324	A2 SE	bone
0143	1491	1369	A2 SE	core	0144	1539	1346	A2 SW	core
0145	1442	1256	A2 NE	core	0146	1433	1239	A2 NE	blade
0147	1438	1240	A2 NE	blade	0148	1454	1256	A2 NE	flake
0149	1473	1259	A2 NE	flake	0150	1457	1265	A2 NE	core
0151	1461	1270	A2 NE	core	0152	1421	1248	A2 NE	core
0153	1396	1239	B2 NW	flake	0154	1389	1214	B2 NW	core
0155	1418	1297	A2 NE	flake	0156	1396	1221	B2 NW	flake
0157	1417	1287	A2 NE	flake	0158	1459	1241	A2 NE	flake
0159	1471	1235	A2 NE	flake	0160	1454	1288	A2 NE	flake
0161	1485	1358	A2 SE	flake	0162	1530	1376	A2 SW	flake
0163	A3 SW/NW		A4 SW		0164	A3 SW/NW		A4 SW	
0165	A3 SW/NW		A4 SW		0166	A3 SW/NW		A4 SW	
0168	A3 NW/SW		A4 SW		0169	A3 SW/NW		A4 SW	
0171	A3 NW/SW		A4 SW		0172	A3 SW/NW		A4 SW	
0173	A3 NW/SW		A4 SW		0175	A3 SW/NW		A4 SW	
0177	A3 NW/SW		A4 SW		0178	A3 SW/NW		A4 SW	
0180	A3 NW/SW		A4 SW		0181	A3 SW/NW		A4 SW	
0182	A3 NW/SW		A4 SW		0183	1401	1335	A2 SE	core

Missing numbers denote split and quartered pebbles/cobbles that were afterwards discarded
Note: 142 (bone) = sub-fossil, probably recent.

Appendix C

Coordinates of artefacts from autumn season of 1982, site 55, Riwat.

Number	X	Y	Square	Type	Number	X	Y	Square	Type
0200	0671	1075	E3 NE	biface	0201	0615	0871	E4 NE	flake
0202	0614	0908	E4 SE	core	0203	0650	0960	E4 SE	blade
0204	0774	0982	E4 SW	core	0205	0618	0986	E4 SE	flake
0206	0573	0811	F4 NW	biface	0207	0777	0944	E4 SW	s-cob
0208	0561	0917	F4 SW	core	0209	0586	0898	F4 NW	flake
0210	0562	0962	F4 SW	blade	0211	0583	0915	F4 SW	flake
0212	0542	0867	F4 NW	core	0214	0558	0871	F4 NW	blade
0215	0576	0910	F4 SW	blade	0216	0641	0951	E4 SE	flake
0217	0636	0948	E4 SE	flake	0218	0625	0917	E4 SE	debris
0219	0678	0941	E4 SE	blade	0220	0683	0966	E4 SE	flake
0221	0734	0977	E4 SW	flake	0222	0701	0963	E4 SW	flake
0223	0720	0969	E4 SW	blade	0224	0793	0991	E4 SW	flake
0225	0773	0971	E4 SW	flake	0226	0781	0983	E4 SW	debris
0227	0774	0964	E4 SW	flake	0228	0787	0978	E4 SW	debris
0229	0757	0967	E4 SW	flake	0230	0750	0971	E4 SW	flake
0231	0775	0957	E4 SW	flake	0232	0781	0971	E4 SW	flake
0233A	0771	0969	E4 SW	core	0233B	0771	0973	E4 SW	debris
0234	0789	0984	E4 SW	flake	0235	0791	0979	E4 SW	flake
0236	0784	0981	E4 SE	flake	0237	0775	0962	E4 SW	flake
0238	0767	0957	E4 SW	flake	0239	0677	0959	E4 SE	blade
0240	0698	0944	E4 SE	flake	0241	0701	0974	E4 SW	debris
0243	0670	0971	E4 SE	core	0244	0678	0963	E4 SE	flake
0245	0672	0968	E4 SE	flake	0246	0669	0963	E4 SE	flake
0247	0676	0971	E4 SE	flake	0248	0672	0961	E4 SE	flake
0249	0660	0955	E4 SE	debris	0250	0646	0955	E4 SE	flake
0251	0643	0935	E4 SE	debris	0252	0660	0939	E4 SE	debris
0253	0575	0926	F4 SW	flake	0254	0578	0920	F4 SW	debris
0255	0870	0875	D4 NE	core	0257	0993	1014	D3 NE	core
0258	0903	1071	D3 NW	blade	0259	0926	1115	D3 SW	s/cob
0260	0966	0970	D4 SW	core	0261	0970	0994	D4 SW	flake
0262	0974	0971	D4 SW	core	0263	0860	1040	D3 NE	flake
0264	0861	1008	D3 NE	core	0265	0835	1034	D3 NE	debris
0267	0806	1006	D3 NE	flake	0268	0862	1012	D3 NE	blade
0269	0866	1017	D3 NE	flake	0270	0879	1017	D3 NE	core
0271	0852	1000	D4 SE	flake	0272	0854	0995	D4 SE	core
0273	0924	1145	D3 SW	core	0274	1231	1295	B2 NE	s-cob
0275	1225	1279	B2 NE	core	0276	1236	1273	B2 NE	flake
0277	1253	1315	B2 SE	flake	0278	1214	1328	B2 SE	core
0279	1218	1337	B2 SE	flake	0281	1301	1328	B2 SW	core
0282	1310	1365	B2 SW	core	0283	1336	1373	B2 SW	flake
0284	1341	1379	B2 SW	flake	0285	1341	1371	B2 SW	flake
0286	1329	1370	B2 SW	debris	0287	1308	1346	B2 SW	core
0288	1291	1357	B2 SE	flake	0289	1296	1358	B2 SE	flake
0290	1320	1350	B2 SW	core	0291	1328	1344	B2 SW	core
0292	1392	1320	B2 SW	core	0293	1361	1181	B3 SW	blade
0294	1355	1184	B3 SW	core	0295	1336	1182	B3 SW	core
0296	1378	1300	B2 NW	core	0297	1374	1298	B2 NW	core
0298	1377	1297	B2 NW	flake	0299	1245	1137	B3 SE	blade
0300	1238	1131	B3 SE	flake	0301	1236	1126	B3 SE	flake
0302	1246	1147	B3 SE	flake	0303	1240	1141	B3 SE	blade
0304	1256	1137	B3 SE	flake	0305	1298	1172	B3 SE	flake

APPENDIX C (continued)

Number	X	Y	Square	Type	Number	X	Y	Square	Type
0306	1294	1174	B3 SE	blade	0307	1297	1168	B3 SE	blade
0308	1285	1165	B3 SE	flake	0309	1280	1163	B3 SE	core
0310	1302	1174	B3 SW	flake	0311	wall footing		B3	flake
0312	wall footing		B3	flake	0313	wall footing		B3	blade
0314	wall footing		B3	flake	0315	wall footing		B3	flake
0316	wall footing		B3	flake	0317	1221	1144	B3 SE	flake
0318	1222	1135	B3 SE	core	0319	1218	1141	B3 SE	flake
0320	1296	1157	B3 SE	flake	0321	1229	1135	B3 SE	flake
0322	1232	1135	B3 SE	flake	0323	1200	1141	C3 SW	flake
0324	1241	1146	B3 SE	core	0325	1119	1107	C3 SW	blade
0326	1133	1080	C3 NW	flake	0327	1149	1136	C3 SW	blade
0328	1140	1140	C3 SW	core	0329	1129	1132	C3 SW	core
0330	1128	1123	C3 SW	core	0331	1054	1066	C3 NE	blade
0332	1085	1089	C3 NE	core	0333	1096	1093	C3 NE	flake
0334	1140	1121	C3 SW	blade	0335	1145	1120	C3 SW	core
0336	1184	1121	C3 SW	blade	0337	1189	1144	C3 SW	flake
0338	1214	1097	B3 NE	blade	0339	1221	1098	B3 NE	flake
0340	1231	1107	B3 SE	core	0341	1268	1066	B3 NE	flake
0342	1263	1075	B3 NE	blade	0343	1257	1078	B3 NE	debris
0344	1241	1071	B3 NE	core	0345	1279	1093	B3 NE	flake
0346	1251	1066	B3 NE	s-cob	0347	1272	1100	B3 NE	blade
0348	1242	1075	B3 NE	flake	0350	1250	1052	B3 NE	flake
0351	0748	0806	E4 NW	blade	0352	0749	0834	E4 NW	core
0353	0743	0813	E4 NW	flake	0354	0746	0811	E4 NW	flake
0355	0704	0876	E4 NW	core	0356	0712	0877	E4 NW	core
0357	0705	0841	E4 NW	scrpr	0358	0695	0918	E4 SE	core
0359	0711	0919	E4 SW	flake	0360	1151	1019	C3 NW	blade
0362	1175	1012	C3 NW	blade	0363	1207	1088	B3 NE	blade
0364	gully fill		F6	flake	0365	1181	1058	C3 NW	blade
0366	1178	1051	C3 NW	debris	0367	1176	1045	C3 NW	core
0368	gully fill		F6	flake	0369	1199	1020	C3 NW	blade
0370	0550	0559	F6 SW	scrpr	0371	0571	0557	F6 SW	flake
0372	gully fill		F6	blade	0373	0530	0550	F6 SW	flake
0374	0523	0554	F6 SW	scrpr	0375	0538	0488	E6 NW	flake
0376	gully fill		F6	flake	0377	gully fill		F6	flake
0378	1233	1062	B3 NE	blade	0379	1239	1024	B3 NE	core
0380	1217	1037	B3 NE	core	0381	1228	1020	B3 NE	core
0382A	1292	1048	B3 NE	blade	0382B	1294	1048	B3 NE	blade
0383	gully fill		F6	flake	0384	1283	1060	B3 NE	flake
0385	gully fill		F6	core	0386	0545	0525	F6 SW	core
0387	gully fill		F6	core	0388	gully fill		F6	flake
0389	gully fill		F6	flake	0390	1285	1053	B3 NE	blade
0391	1241	1048	B3 NE	blade	0392	1276	1054	B3 NE	flake
0393	1217	1025	B3 NE	blade	0394	1265	1056	B3 NE	flake
0395	gully fill		F6	blade	0396	gully fill		F6	blade
0397	1284	1063	B3 NE	flake	0398	1278	1064	B3 NE	debris
0399	1275	1065	B3 NE	blade					
0400	gully fill		F6	flake	0401	gully fill		F6	blade
0402	1238	1072	B3 NE	flake	0404	1220	1028	B3 NE	blade
0405	1228	1044	B3 NE	blade	0406	1229	1036	B3 NE	flake
0407	1231	1029	B3 NE	blade	0408	1240	1040	B3 NE	blade
0409	1381	1172	B3 SW	blade	0410	1401	1135	A3 SE	debris
0411	1388	1198	B3 SW	blade	0412	1359	1178	B3 SW	debris
0413	1359	1188	B3 SW	blade	0414	1362	1198	B3 SW	blade

APPENDIX C (continued)

Number	X	Y	Square	Type	Number	X	Y	Square	Type
0415	1337	1185	B3 SW	debris	0416	1303	1163	B3 SW	flake
0417	1326	1199	B3 SW	blade	0418	1399	1184	B3 SW	blade
0419	1399	1188	B3 SW	blade	0420	1373	1229	B2 NW	blade
0421	0750	0966	E4 SW	blade	0422	0752	0959	E4 SW	flake
0423	0724	0947	E4 SW	flake	0424	0795	0982	E4 SW	blade
0425	0703	0952	E4 SW	blade	0426	0744	0958	E4 SW	core
0427	0752	0965	E4 SW	blade	0428	0767	0955	E4 SW	core
0429	0714	0958	E4 SW	flake	0430	0775	0960	E4 SW	blade
0431	0749	0915	E4 SW	flake	0432	0775	0904	E4 SW	flake
0433	0770	0903	E4 SW	point	0434	0770	0879	E4 NW	core
0435	0767	0846	E4 NW	blade	0436	0651	0944	E4 SE	s-cob
0437	0663	0943	E4 SE	flake	0439	0703	0819	E4 NW	flake
0440	0620	0927	E4 SE	flake	0441	0632	0944	E4 SE	core
0442	0685	0890	E4 NE	debris	0443	surface		E5	flake
0444	0876	0779	D5 SE	blade	0446	0623	0830	E4 NE	core
0447	1390	1129	B3 SW	flake	0448	1362	1145	B3 SW	blade
0449	1344	1150	B3 SW	flake	0450	1345	1139	B3 SW	flake
0451	0880	0862	D4 NE	flake	0452	0805	0818	D4 NE	blade
0453	0850	0945	D4 SE	blade	0454	0786	0851	E4 NW	flake
0455	0729	0939	E4 SW	core	0456	0753	0818	E4 NW	flake
0457	0740	0876	E4 NW	core	0458	0787	0916	E4 SW	core
0459	0773	0914	E4 SW	flake	0460	0647	0856	E4 NE	core
0461	0642	0872	E4 NE	core	0462	0771	0974	E4 SW	core
0463	0630	0910	E4 SE	blade	0464	0616	0912	E4 SE	flake
0465	0719	0846	E4 NW	blade	0466	0713	0929	E4 SW	flake
0467	0714	0879	E4 NW	core	0468	0732	0911	E4 SW	core
0469	0799	0862	E4 NW	flake	0470	0791	0871	E4 NW	core
0471	0772	0961	E4 SW	flake	0472	0757	0965	E4 SW	flake
0473	surface		E4	flake	0474	0743	0908	E4 SW	core
0475	0841	0918	D4 SE	blade	0476	surface		E4	flake
0477	0767	0935	E4 SW	blade	0478	0764	0926	E4 SW	blade
0479	0722	0930	E4 SW	core	0481	0983	0928	D4 SW	core
0482	0722	0912	E4 SW	core	0483	0945	0896	D4 NW	flake
0484	0739	0922	E4 SW	core	0486	0739	0850	E4 NW	scrpr
0487	0795	0935	E4 SW	core	0488	0776	0939	E4 SW	flake
0489	0638	0992	E4 SE	core	0490	0659	0831	E4 NE	flake
0491	0655	0818	E4 NE	flake	0493	0678	0695	E5 NE	flake
0494	0657	0735	E5 SE	blade	0495	0841	0770	D5 SE	blade
0496	0824	0859	D4 NE	flake	0497	0854	0951	D4 SE	flake
0498	0678	0688	E5 NE	blade	0499	0682	0695	E5 NE	scrpr
0500	0821	0802	D4 NE	blade	0501	0745	0882	E4 NW	core
0502	0773	0781	E5 SW	flake	0503	0760	0791	E5 SW	flake
0504	0755	0782	E5 SW	blade	0505	0749	0777	E5 SW	blade
0506	0741	0768	E5 SW	flake	0507	0731	0783	E5 SW	blade
0508	0673	0704	E5 SE	blade	0509	0649	0716	E5 SE	blade
0510	0627	0713	E5 SE	blade	0511	0618	0714	E5 SE	core
0512	0670	0719	E5 SE	flake	0513	0678	0717	E5 SE	flake
0514	0691	0728	E5 SE	blade	0515	0691	0747	E5 SE	flake
0516	0718	0776	E5 SW	core	0517	0571	0655	F5 NW	blade
0518	0560	0588	F6 SW	blade	0519	0502	0626	F5 NW	blade
0520	1108	1205	C2 NW	blade	0521	1191	1083	C3 NW	debris
0522	1279	1246	B2 NE	flake	0523	1184	1074	C3 NW	debris
0524	1167	1062	C3 SW	s-cob	0525	1194	1085	C3 NW	flake
0526	1185	0093	C3 NW	debris	0527	1182	1082	C3 NW	debris

APPENDIX C (continued)

Number	X	Y	Square	Type	Number	X	Y	Square	Type
0528	1192	1070	C3 NW	blade	0529	1296	1152	B3 SE	flake
0530	1171	1070	C3 NW	flake	0531	1154	1050	C3 NW	flake
0532	1018	1024	C3 NE	blade	0533	1053	1018	C3 NE	hammer
0534	1067	0948	C4 SE	flake	0535	1044	0950	C4 SE	core
0536	1293	1142	B3 SE	flake	0527A	1243	1115	B3 SE	flake
0537B	1245	1116	B3 SE	flake	0538	1145	1027	C3 NW	blade
0539	1203	1096	B3 NE	flake	0540	1208	1136	B3 SE	flake
0541	1168	1169	C3 SW	flake	0542	1297	1258	B2 NE	flake
0543	1187	1123	C3 SW	flake	0544	1136	1129	C3 SW	blade
0545	1188	1156	C3 SW	core	0547	1268	1129	B3 SE	flake
0548	1012	1011	C3 NE	blade	0549	1077	1042	C3 NE	flake
0550	1108	1109	C3 SW	blade	0551	1091	1083	C3 NE	flake
0552	1155	1101	C3 SW	core	0553	1440	1217	A2 NE	flake
0555	1402	1203	A2 NE	blade	0556	1401	1133	A3 SE	blade
0557	1327	1124	B3 SW	flake	0558	1356	1131	B3 SW	core
0559	niche		B3	core	0560	0708	0862	E4 NW	flake
1000	0586	0177	F8 SW	flake	1001	0490	0020	F8 NE	flake
1002	0478	0031	F8 NE	flake	1003	0560	0030	F8 NW	flake
1004	0496	0069	F8 NE	flake	1005	0513	0094	F8 NW	s-cob
1006	0521	0136	F8 SW	blade	1007	0577	0082	F8 NW	flake
1008	0511	0145	F8 SW	core	1009	0493	0176	F8 SE	flake
1010	0511	0156	F8 SW	flake	1011	0489	0204	F7 NE	flake
1012	0540	0206	F7 NW	flake	1013	0606	0216	E7 NE	flake
1014	0478	0262	F7 NE	flake	1015	0507	0222	F7 NW	flake
1016	0532	0221	F7 NW	flake	1017	0576	0245	F7 NW	flake
1018	0632	0065	E8 NE	flake	1019	0727	0054	E8 NW	flake
1020	0739	0052	E8 NW	flake	1021	0729	0044	E8 NW	flake
1022	0724	0130	E8 SW	flake	1023	0719	0132	E8 SW	flake
1024	0739	0169	E8 SW	core	1025	0780	0196	E8 SW	flake
1026	0741	0204	E7 NW	flake	1027	0748	0151	E8 SW	core
1028	0725	0113	E8 SW	flake	1029	0638	0180	E8 SE	flake
1030	0646	0180	E8 SE	flake	1031	0641	0165	E8 SE	flake
1032	0642	0140	E8 SE	flake	1033	0643	0093	E8 NE	flake
1034	0658	0107	E8 SE	flake	1035	0611	0066	E8 NE	core
1036	0566	0146	F8 SW	flake	1037	0588	0092	F8 NW	core
1038	0704	0209	E7 NW	flake	1039	0654	0203	E7 NE	flake
1040	0618	0233	E7 NE	flake	1041	0609	0213	E7 NE	flake
1042	0650	0250	E7 NE	core	1043	0603	0275	E7 NE	flake
1044	0607	0269	E7 NE	flake	1045	0648	0274	E7 NE	flake
1046	0776	0259	E7 NW	flake	1047	0768	0270	E7 NW	flake
1048	0717	0252	E7 NW	core	1049	0716	0240	E7 NW	qcob
1050	0657	0241	E7 NE	core	1051	0672	0253	E7 NE	core
1052	0724	0288	E7 NW	flake	1053	0745	0276	E7 NW	flake
1054	0721	0307	E7 SW	flake	1055	0681	0396	E7 SE	core
1056	0683	0403	E6 NE	flake	1057	0640	0396	E7 SE	flake
1058	0622	0368	E7 SE	flake	1059	0618	0341	E7 SE	flake
1060	0622	0303	E7 SE	flake	1061	0614	0302	E7 SE	flake
1062	0618	0305	E7 SE	flake	1063	0639	0352	E7 SE	core
1064	0731	0336	E7 SW	core	1065	0626	0300	E7 NE	flake
1066	0659	0267	E7 NE	core	1067	0710	0350	E7 SW	flake
1068	0582	0320	F7 SW	flake	1069	0556	0317	F7 SW	flake
1070	0588	0366	F7 SW	flake	1071	1543	0418	A6 NW	flake
1072A	1541	0481	A6 NW	blade	1072B	1541	0481	A6 NW	flake
1073	1544	0520	A6 SW	blade	1074	1559	0563	A6 SW	flake

APPENDIX C (continued)

Number	X	Y	Square	Type	Number	X	Y	Square	Type
1075	1542	0537	A6 SW	core	1076	1505	0447	A6 NW	flake
1077	1467	0541	A6 SE	flake	1078	1532	0536	A6 SW	flake
1079	1350	0426	B6 NW	flake	1080	1350	0461	B6 NW	flake
1081	1347	0476	B6 SW	core	1082	1328	0497	B6 NW	flake
1083	1375	0604	B6 NW	flake	1084	1388	0593	B6 SW	flake
1085	1374	0574	B6 SW	core	1086	1370	0572	B6 SW	debris
1087	1202	0441	B6 NE	flake	1088	1193	0450	C6 NW	flake
1089	1209	0453	B6 NE	flake	1090	1272	0584	B6 SE	core
1091	1251	0597	B6 SE	flake	1092	1270	0582	B6 SE	flake
1093	1312	0404	B6 NW	blade	1094	1310	0412	B6 NW	flake
1095	1303	0428	B6 NW	flake	1096	1351	0451	B6 NW	core
1097	1392	0420	B6 NW	flake	1098	1378	0413	B6 NW	core
1099	1211	0457	B6 NE	flake	1100	1197	0429	C6 NW	flake
1101	1201	0439	B6 NE	core	1102	1206	0496	B6 NE	blade
1103	1233	0479	B6 NE	flake	1104	1230	0484	B6 NE	flake
1105	1254	0503	B6 SE	core	1106	1270	0497	B6 NE	core
1107	1223	0516	B6 SE	core	1108	1224	0572	B6 SE	flake
1109	1222	0578	B6 SE	flake	1110	1296	0559	B6 SE	flake
1111	1341	0591	B6 SW	flake	1112	1047	0480	C6 NE	flake
1113	1051	0473	C6 NE	core	1114	1035	0465	C6 NE	core
1115	1042	0459	C6 NE	flake	1116	1089	0452	C6 NE	core
1117	1084	0424	C6 NE	blade	1118	1177	0432	C6 NW	flake
1119	1170	0406	C6 NW	blade	1120	1192	0404	C6 NW	blade
1121	1182	0573	C6 SW	core	1122	1148	0525	C6 SW	core
1123	1140	0509	C6 SW	core	1124	1164	0485	C6 NW	flake
1125	1151	0596	C6 SW	core	1126	1025	0534	C6 SE	flake
1127	1174	0716	C5 SW	flake	1128	1171	0706	C5 SW	flake
1129	1171	0723	C5 SW	core	1130	1160	0720	C5 SW	flake
1131	1196	0714	C5 SW	debris	1132	1183	0736	C5 SW	flake
1133	1151	0646	C5 NW	core	1134	1158	0613	C5 NW	flake
1135	1189	0610	C5 NW	core	1136	1136	0635	C5 NW	flake
1137	1128	0658	C5 NW	blade	1138	1130	0638	C5 NW	core
1139	1121	0717	C5 SW	flake	1140	1140	0715	C5 SW	core
1141	1078	0727	C5 SE	blade	1142	1079	0698	C5 NE	blade
1143	1081	0691	C5 NE	flake	1144	1007	0720	C5 SE	flake
1145	1031	0701	C5 SE	core	1146	1025	0672	C5 NE	core
1147	1008	0640	C5 NE	flake	1148	1015	0616	C5 NE	core
1149	1008	0699	C5 NE	flake	1150	1006	0711	C5 NE	flake
1151	0992	0657	D5 NW	flake	1152	1194	0779	D5 NW	flake
1153	1098	0766	C5 SE	blade	1154	1121	0758	C5 SW	blade
1155	1092	0746	C5 SE	core	1156	1069	0737	C5 SE	flake
1157	1030	0740	C5 SE	blade	1158	1015	0792	C5 SE	debris
1159	1014	0796	C5 SE	flake	1160	1258	0790	B5 SE	flake
1161	1273	0798	B5 SE	flake	1162	1292	0789	B5 SE	blade
1163	1297	0769	B5 SE	flake	1164	1347	0771	B5 SW	flake
1165	1338	0776	B5 SW	flake	1166	1331	0781	B5 SW	flake
1167	1325	0752	B5 SW	flake	1168	1295	0740	B5 SW	debris
1169	1307	0744	B5 SW	core	1170	1304	0759	B5 SW	debris
1171	1292	0748	B5 SE	flake	1172	1264	0777	B5 SE	flake
1173	1219	0693	B5 NE	flake	1174	1210	0667	B5 NE	blade
1175	1244	0648	B5 NE	flake	1176	1231	0638	B5 NE	flake
1177	1261	0647	B5 NE	flake	1178	1265	0646	B5 NE	flake
1179	1282	0611	B5 NE	flake	1180	1314	0654	B5 NW	flake
1181	1302	0622	B5 NW	flake	1182	1292	0626	B5 NE	flake

APPENDIX C (continued)

Number	X	Y	Square	Type	Number	X	Y	Square	Type
1183	1373	0716	B5 SW	flake	1184	1394	0734	B5 SW	core
1185	1383	0730	B5 SW	blade	1186	1398	0644	B5 NW	debris
1187	1325	0608	B5 NW	flake	1188	1190	0442	C6 NW	flake
1189	1238	0412	B6 NE	flake	1190	1290	0451	B6 NE	flake
1191	1348	0424	B6 NW	blade	1192	1397	0955	B4 SW	flake
1193	1393	0975	B4 SW	core	1194	1395	0938	B4 SW	core
1195	1347	0936	B4 SW	core	1196	1365	0930	B4 SW	blade
1197	1369	0924	B4 SW	core	1198	1380	0912	B4 SW	blade
1199	1375	0962	B4 SW	flake	1200	1375	0956	B4 SW	blade
1201	1377	0976	B4 SW	flake	1202	1364	0982	B4 SW	blade
1204	1344	0952	B4 SW	flake	1205	1347	0998	B4 SW	core
1206	1356	0990	B4 SW	flake	1207	1303	1004	B3 NW	blade
1208	1322	0968	B4 SW	blade	1209	1317	0960	B4 SW	debris
1210	1325	0965	B4 SW	flake	1211	1273	0966	B4 SE	flake
1212	1274	0969	B4 SE	flake	1213	1270	0960	B4 SE	flake
1214	1274	0959	B4 SE	flake	1215	1271	0955	B4 SE	flake
1216	1270	0990	B4 SE	core	1217	1252	1009	B3 NE	flake
1218	1224	0993	B4 SE	debris	1219	1209	0975	B4 SE	blade
1220	1223	0955	B4 SE	biface	1221	1213	0948	B4 SE	flake
1222	1256	0959	B4 SE	debris	1223	1246	0959	B4 SE	blade
1224	1201	0941	B4 SE	blade	1225	1214	0936	B4 SE	blade
1226	1201	0887	B4 NE	flake	1227	1204	0883	B4 NE	flake
1228	1211	0892	B4 NE	core	1229	1240	0913	B4 SE	flake
1230	1248	0922	B4 SE	blade	1232	1206	0917	B4 SE	flake
1233	1249	0878	B4 NE	flake	1234	1261	0885	B4 NE	blade
1235	1313	0930	B4 SW	core	1236	1368	0836	B4 SE	core
1237	1398	0839	B4 NW	core	1238A	1326	0907	B4 SW	flake
1238B	1328	0907	B4 SW	flake	1239	1358	0886	B4 NW	flake
1240	1350	0894	B4 NW	core	1241	1293	0892	B4 NE	blade
1242	1284	0871	B4 NE	flake	1243	1224	0844	B4 NE	core
1244	1205	0836	B4 NE	flake	1245	1203	0863	B4 NE	flake
1246	1383	0832	B4 NW	debris	1247	1383	1057	B3 NW	blade
1248	1369	1064	B3 NW	debris	1249	1269	1000	B4 SE	blade
1250	1390	1038	B3 NW	flake	1251	1452	1072	A3 NE	blade
1252	1433	1103	A3 SE	flake	1253	1435	1150	A3 SE	flake
1254	1424	1146	A3 SE	flake	1255	1408	0992	A4 SE	blade
1256	1463	0998	A4 SE	blade	1257	1424	1045	A3 NE	blade
1258	1426	1100	A3 NE	blade	1259	1404	1107	A3 SE	blade
1260	1446	1086	A3 NE	flake	1261	1427	1119	A3 SE	blade
1262	1462	1138	A3 SE	blade	1263	1446	1054	A3 NE	blade
1264	1442	1020	A3 NE	blade	1265	1469	0994	A4 SE	blade
1266	1447	1146	A3 SE	blade	1267	1472	1009	A3 NE	flake
1268	1412	0964	A4 SE	flake	1269A	1428	0976	A4 SE	point
1269B	1430	0974	A4 SE	blade	1270	1451	0955	A4 SE	core
1271	1437	0962	A4 SE	blade	1272	1467	0963	A4 SE	core
1273	1430	0893	A4 NE	flake	1274	1452	0896	A4 NE	blade
1275A	1438	0922	A4 SE	blade	1275B	1438	0920	A4 SE	blade
1276	1444	0889	A4 NE	blade	1277	1445	0863	A4 NE	core
1278A	1429	0862	A4 NE	flake	1278B	1431	0860	A4 NE	blade
1279	1452	0874	A4 NE	flake	1280	1480	0858	A4 NE	core
1281	1446	0819	A4 NE	core	1282	1166	0818	C4 NW	core
1283	1143	0817	C4 NW	core	1284	1193	0839	C4 NW	flake
1285	1182	0853	C4 NW	blade	1286	1107	0847	C4 NW	flake
1287	1094	0831	C4 NE	core	1288	1013	0825	C4 NE	blade

APPENDIX C (continued)

Number	X	Y	Square	Type	Number	X	Y	Square	Type
1289	0992	0861	D4 NW	blade	1290	0990	0867	D4 NW	flake
1291	0999	0869	D4 NW	core	1292	1060	0849	C4 NE	blade
1293	1023	0883	C4 NE	debris	1294	1017	0885	C4 NE	blade
1295	1014	0905	C4 SE	blade	1296	1036	0874	C4 NE	core
1297	1038	0081	C4 NE	flake	1298	1016	0902	C4 SE	blade
1299	1060	0920	C4 SE	debris	1300	1056	0924	C4 SE	core
1301	1060	0923	C4 NE	core	1302	1064	0910	C4 NE	flake
1303	1084	0926	C4 SE	core	1304	1081	0900	C4 NE	blade
1305	1114	0905	C4 SW	blade	1306	1121	0898	C4 NW	flake
1307	1119	0952	C4 SW	flake	1308	1156	0932	C4 SW	flake
1309	1125	0970	C4 SW	debris	1310	1145	0962	C4 SW	blade
1311	1133	0907	C4 SW	blade	1312	1169	0898	C4 NW	flake
1313	1182	0897	C4 NW	blade	1314	1167	0868	C4 NW	flake
1315	0914	0798	D5 SW	core	1316	0937	0840	D4 NW	flake
1317	0985	0797	D5 SW	core	1318	0938	0800	D5 SW	core
1320	0654	0621	E5 NE	flake	1321	0689	0649	E5 NE	blade
1322	0687	0686	E5 NE	scrpr	1323	0687	0641	E5 NE	flake
1324	0685	0645	E5 NE	flake	1325	0711	0647	E5 NW	blade
1326	0734	0640	E5 NW	flake	1327	0807	0736	D5 SE	flake
1328	0799	0735	E5 SW	core	1329	0777	0729	E5 SW	flake
1330	0777	0720	E5 SW	blade	1331	O805	0727	D5 SE	flake
1332	0801	0722	D5 SE	flake	1333	0812	0694	D5 NE	flake
1334	0790	0666	E5 NW	flake	1335	0777	0646	E5 NW	flake
1336	0773	0639	E5 NW	flake	1337	0766	0592	E6 SW	core
1338	1427	0739	A5 SE	blade	1339	1462	0706	A5 SE	flake
1340	1423	0704	A5 SE	flake	1341	1433	0694	A5 NE	flake
1342	1463	0700	A5 NE	blade	1343	1437	0654	A5 NE	debris
1344	1441	0633	A5 NE	flake	1345	1532	0761	A5 SW	flake
1346	1521	0707	A5 SW	blade	1347	1554	0664	A5 NW	flake
1348	1444	0593	A6 SE	flake	1349	1445	0440	A6 NE	flake
1350	1503	0478	A6 NW	blade	1351	1540	0438	A6 NW	blade
1352	1549	0431	A6 NW	blade	1353	0982	0727	D5 SW	core
1354	0978	0683	D5 NW	burin	1355	0976	0656	D5 NW	flake
1356	0965	0655	D5 NW	flake	1357	0969	0660	D5 NW	flake
1358	0905	0779	D5 SW	debris	1359	0907	0745	D5 SW	flake
1360	0911	0676	D5 NW	flake	1361	0903	0652	D5 NW	scrpr
1362	0882	0636	D5 NE	flake	1363A	0872	0757	D5 SE	flake
1363B	0874	0760	D5 SE	debris	1364	0847	0755	D5 SE	core
1365A	0898	0709	D5 SE	scrpr	1365B	0895	0707	D5 SE	flake
1366	0865	0702	D5 SE	flake	1367	0812	0683	D5 NE	flake
1368	0840	0708	D5 NE	flake	1369	0867	0646	D5 SE	flake
1370	0857	0636	D5 NE	flake	1371	0738	0562	E6 SW	flake
1372	0729	0548	E6 SW	core	1373	0758	0551	E6 SW	core
1374	0758	0531	E6 SW	debris	1375	0685	0572	E6 SE	blade
1376	0609	0569	E6 SE	blade	1377	0699	0550	E6 SE	blade
1378	0780	0509	E6 SW	flake	1379	0771	0504	E6 SW	blade
1380	0780	0501	E6 SW	debris	1381	0764	0519	E6 SW	flake
1382	0707	0481	E6 NW	flake	1383	0719	0424	E6 NW	blade
1384	0628	0431	E6 NE	core	1385	0670	0462	E6 NE	flake
1386	0653	0464	E6 NE	flake	1387	0669	0484	E6 NE	flake
1388	0702	0483	E6 NW	flake	1389	0697	0511	E6 SE	blade
1390	0605	0446	E6 NE	flake	1391	0960	0536	D6 SW	flake
1392	0943	0591	D6 SW	flake	1393	0930	0595	D6 SW	flake
1394	0928	0599	D6 SW	flake	1395A	0939	0525	D6 SW	flake

APPENDIX C (continued)

Number	X	Y	Square	Type	Number	X	Y	Square	Type
1395B	0941	0523	D6 SW	flake	1395C	0944	0522	D6 SW	flake
1396	0854	0590	D6 SE	flake	1397	0861	0560	D6 SE	flake
1398	0809	0523	D6 SE	blade	1399	0870	0500	D6 SE	blade
1400	0910	0481	D6 SE	flake					
1401	0910	0481	D6 NW	core	1402	0988	0452	D6 NW	core
1403	0484	0410	F6 NE	flake	1404	0521	0366	F7 SW	flake
1405	0760	0328	E7 SW	core	1406	0594	0529	F6 SW	flake
1407	0577	0500	F6 NW	flake	1408	0500	0496	F6 NE	flake
1409	0530	0483	F6 NW	flake	1410	0570	0538	F6 SW	blade
1411	0510	0400	F7 SW	flake	1412	0250	0210	G7 NE	debris
1413	0320	0210	G7 NW	core	1414	0260	0080	G8 NW	debris
1415	0295	0499	G6 NE	blade	1416	0249	0650	G5 NE	flake
1417	0265	0705	G5 SE	flake	1418	0165	0690	H5 HW	blade
1419	0159	0670	H5 NW	flake	1420	0210	0565	G6 SE	blade
1421	0204	0479	G6 NE	core	1422	0181	0444	H6 NW	core
1423A	0138	0242	H7 NW	blade	1423B	0140	0241	H7 NW	blade
1424A	0110	0202	H7 NW	blade	1424B	0112	0201	H7 NW	blade
1425	1093	0413	C6 NE	core	1426	1068	0507	C6 SE	flake
1427	gully fill		H7	biface					

notes: scrpr = scraper
debris = flaking debris
s-cob = split cobble
qcob = quartered cobble

Appendix D

Coordinates and types of unlifted cobbles, cores and flakes from site 55, Riwat.

Key to types:
01: complete cobble
02: cobble with one flake detached
03: cobble with two flakes detached
04: cobble with three flakes detached
05: cobble with four flakes detached
06: cobble with five flakes detached

20: cortical flake struck once
21: cortical flake with one flake detached
22: cortical flake with two flakes detached
23: cortical flake with three flakes detached

30: cobble split in half
40: quartered cobble

The suffix A denotes that a cobble has been truncated by a transverse blow to its end, or that a cortical flake was detached from an already truncated core.

Number	Type	X	Y	Square	Number	Type	X	Y	Square
0001	01	1462	1122	A3 SE	0002	01	1430	1122	A3 SE
0003	01	1433	1120	A3 SE	0004	01	1438	1117	A3 SE
0005	01	1447	1102	A3 SE	0006	20	1442	1121	A3 SE
0007	30	1444	1113	A3 SE	0008	40A	1466	1128	A3 SE
0009	01	1439	1098	A3 NE	0010	01	1445	1095	A3 NE
0011	01	1435	1085	A3 NE	0012	01	1441	1088	A3 NE
0013	01	1435	1052	A3 NE	0014	01	1451	1055	A3 NE
0015	01	1460	1041	A3 NE	0016	01	1438	1041	A3 NE
0017	01	1434	1036	A3 NE	0018	01	1433	1031	A3 NE
0019	01	1425	1030	A3 NE	0020	01	1415	1024	A3 NE
0021	01	1427	1003	A3 NE	0022	01	1448	1013	A3 NE
0023	01	1449	1006	A3 NE	0024	01	1446	1003	A3 NE
0025	02	1406	1030	A3 NE	0026	04	1417	1080	A3 NE
0027	05	1417	1057	A3 NE	0028	05	1411	1004	A3 NE
0029	08	1454	1011	A3 NE	0030	20	1406	1036	A3 NE
0031	20	1446	1045	A3 NE	0032	20	1439	1053	A3 NE
0033	23	1424	1059	A3 NE	0034	30	1425	1093	A3 NE
0035	30A	1411	1096	A3 NE	0036	30A	1416	1091	A3 NE
0037	31A	1440	1060	A3 NE	0038	05	1401	1035	A3 NE
0039	01	1449	0998	A4 SE	0040	01	1459	0992	A4 SE
0041	01	1451	0987	A4 SE	0042	01	1432	0986	A4 SE
0043	01	1401	0979	A4 SE	0044	01	1437	0977	A4 SE
0045	01	1432	0975	A4 SE	0046	01	1414	0969	A4 SE
0047	01	1408	0960	A4 SE	0048	01	1411	0956	A4 SE
0049	01	1463	0965	A4 SE	0050	01	1434	0962	A4 SE
0051	01	1462	0957	A4 SE	0052	01	1457	0950	A4 SE
0053	01	1447	0951	A4 SE	0054	01	1452	0945	A4 SE
0055	01	1403	0933	A4 SE	0056	01	1413	0929	A4 SE
0057	01	1408	0920	A4 SE	0058	01	1449	0935	A4 SE
0059	01	1454	0934	A4 SE	0060	01	1458	0932	A4 SE
0061	01	1462	0929	A4 SE	0062	01	1452	0920	A4 SE

APPENDIX D (continued)

Number	Type	X	Y	Square	Number	Type	X	Y	Square
0063	01	1460	0910	A4 SE	0064	01	1455	0905	A4 SE
0065	02	1402	0970	A4 SE	0066	02	1439	0965	A4 SE
0067	02	1441	0959	A4 SE	0068	03	1455	0979	A4 SE
0069	03	1416	0995	A4 SE	0070	04	1444	0946	A4 SE
0071	04A	1401	0989	A4 SE	0072	05A	1418	0953	A4 SE
0073	20	1419	0963	A4 SE	0074	30	1444	0962	A4 SE
0075	30	1413	0949	A4 SE	0076	30	1415	0938	A4 SE
0077	30	1463	0948	A4 SE	0078	30	1441	0910	A4 SE
0079	30	1448	0908	A4 SE	0080	30	1405	0903	A4 SE
0081	30A	1447	0964	A4 SE	0082	30A	1466	0966	A4 SE
0083	30A	1445	0982	A4 SE	0084	04	1433	0955	A4 SE
0085	40	1449	0973	A4 SE	0086	40	1426	0979	A4 SE
0087	40	1402	0975	A4 SE	0088	40A	1411	0940	A4 SE
0089	40A	1461	0979	A4 SE					
0090	01	1468	0899	A4 NE	0091	01	1459	0897	A4 NE
0092	01	1447	0894	A4 NE	0093	01	1425	0895	A4 NE
0094	01	1412	0894	A4 NE	0095	01	1431	0889	A4 NE
0096	01	1401	0889	A4 NE	0097	01	1455	0886	A4 NE
0098	01	1412	0878	A4 NE	0099	01	1427	0854	A4 NE
0100	01	1472	0849	A4 NE	0101	01	1408	0843	A4 NE
0102	01	1401	0835	A4 NE	0103	01	1412	0835	A4 NE
0104	01	1469	0821	A4 NE	0105	01	1455	0822	A4 NE
0106	01	1466	0813	A4 NE	0107	01	1458	0816	A4 NE
0108	01	1451	0815	A4 NE	0109	01	1434	0813	A4 NE
0110	01	1419	0807	A4 NE	0111	01	1474	0803	A4 NE
0112	03	1407	0883	A4 NE	0113	03	1409	0806	A4 NE
0114	04	1470	0833	A4 NE	0115	05	1448	0866	A4 NE
0116	20	1440	0893	A4 NE	0117	20	1420	0894	A4 NE
0118	20	1403	0837	A4 NE	0119	20	1441	0817	A4 NE
0120	21	1439	0871	A4 NE	0121	21A	1403	0895	A4 NE
0122	22	1427	0869	A4 NE	0123	22	1416	0857	A4 NE
0124	22	1443	0835	A4 NE	0125	30	1428	0899	A4 NE
0126	30	1443	0860	A4 NE	0127	30	1455	0853	A4 NE
0128	30	1403	0855	A4 NE	0129	30	1419	0832	A4 NE
0130	30	1428	0813	A4 NE	0131	30	1401	0805	A4 NE
0132	30	1443	0802	A4 NE	0133	40	1442	0869	A4 NE
0134	40	1401	0818	A4 NE					
0135	01	1405	0785	A5 SE	0136	01	1465	0768	A5 SE
0137	01	1441	0771	A5 SE	0138	01	1431	0765	A5 SE
0139	01	1451	0759	A5 SE	0140	01	1452	0746	A5 SE
0141	01	1462	0741	A5 SE	0142	01	1450	0737	A5 SE
0143	01	1433	0740	A5 SE	0144	01	1419	0738	A5 SE
0145	01	1427	0729	A5 SE	0146	01	1436	0725	A5 SE
0147	01	1454	0710	A5 SE	0148	01	1419	0714	A5 SE
0149	01	1407	0714	A5 SE	0150	01	1471	0701	A5 SE
0151	02	1435	0792	A5 SE	0152	02	1465	0779	A5 SE
0153	02	1412	0767	A5 SE	0154	02	1424	0763	A5 SE
0155	02	1456	0724	A5 SE	0156	02	1462	0713	A5 SE
0157	03	1442	0749	A5 SE	0158	20	1435	0758	A5 SE
0159	20A	1444	0775	A5 SE	0160	30A	1403	0705	A5 SE
0161	40	1432	0777	A5 SE					
0162	01	1429	0699	A5 NE	0163	01	1463	0695	A5 NE
0164	01	1443	0688	A5 NE	0165	01	1450	0680	A5 NE
0166	01	1429	0679	A5 NE	0167	01	1439	0674	A5 NE
0168	01	1466	0670	A5 NE	0169	01	1412	0668	A5 NE
0170	01	1406	0665	A5 NE	0171	01	1462	0658	A5 NE
0172	01	1451	0660	A5 NE	0173	01	1445	0655	A5 NE

APPENDIX D (continued)

Number	Type	X	Y	Square	Number	Type	X	Y	Square
0174	01	1452	0644	A5 NE	0175	01	1433	0638	A5 NE
0176	01	1475	0626	A5 NE	0177	01	1447	0621	A5 NE
0178	01	1424	0619	A5 NE	0179	01	1404	0608	A5 NE
0180	01	1430	0603	A5 NE	0181	02	1434	0682	A5 NE
0182	03	1419	0660	A5 NE	0183	04	1458	0614	A5 NE
0184	20	1412	0693	A5 NE	0185	20	1458	0603	A5 NE
0186	20A	1441	0663	A5 NE	0187	30	1452	0693	A5 NE
0188	30	1420	0678	A5 NE	0189	30	1430	0667	A5 NE
0190	30A	1457	0675	A5 NE	0191	40	1402	0697	A5 NE
0192	40	1454	0657	A5 NE	0193	40A	1403	0692	A5 NE
0194	21	1433	0601	A5 NE					
0195	01	1417	0597	A6 SE	0196	01	1401	0589	A6 SE
0197	01	1425	0575	A6 SE	0198	01	1438	0563	A6 SE
0199	01	1410	0562	A6 SE	0200	01	1449	0554	A6 SE
0201	01	1401	0554	A6 SE	0202	01	1423	0546	A6 SE
0203	01	1479	0529	A6 SE	0204	01	1449	0523	A6 SE
0205	01	1425	0526	A6 SE	0206	01	1412	0525	A6 SE
0207	01	1429	0507	A6 SE	0208	02	1431	0588	A6 SE
0209	04	1477	0590	A6 SE	0210	04	1472	0583	A6 SE
0211	04	1478	0564	A6 SE	0212	20	1406	0590	A6 SE
0213	20A	1407	0553	A6 SE	0214	23	1472	0519	A6 SE
0215	30	1445	0585	A6 SE	0216	30	1407	0511	A6 SE
0217	40	1463	0590	A6 SE					
0218	01	1424	0495	A6 NE	0219	01	1413	0492	A6 NE
0220	01	1432	0490	A6 NE	0221	01	1444	0486	A6 NE
0222	01	1445	0475	A6 NE	0223	01	1469	0471	A6 NE
0224	01	1447	0452	A6 NE	0225	01	1469	0447	A6 NE
0226	01	1464	0442	A6 NE	0227	01	1445	0453	A6 NE
0228	01	1425	0453	A6 NE	0229	01	1425	0440	A6 NE
0230	01	1419	0436	A6 NE	0231	01	1405	0431	A6 NE
0232	01	1423	0427	A6 NE	0233	01	1465	0424	A6 NE
0234	01	1435	0413	A6 NE	0235	20	1474	0499	A6 NE
0236	30	1457	0432	A6 NE	0237	30	1414	0428	A6 NE
0238	30A	1459	0454	A6 NE	0239	40	1412	0485	A6 NE
0240	01	1386	1108	B3 NW	0241	01	1354	1097	B3 NW
0242	01	1385	1091	B3 NW	0243	01	1341	1091	B3 NW
0244	01	1390	1084	B3 NW	0245	01	1387	1070	B3 NW
0246	01	1344	1074	B3 NW	0247	01	1328	1077	B3 NW
0248	01	1312	1078	B3 NW	0249	01	1307	1079	B3 NW
0250	01	1305	1062	B3 NW	0251	01	1310	1059	B3 NW
0252	01	1308	1055	B3 NW	0253	01	1317	1054	B3 NW
0254	01	1384	1036	B3 NW	0255	01	1374	1031	B3 NW
0256	01	1350	1011	B3 NW	0257	01	1331	1012	B3 NW
0258	01	1309	1001	B3 NW	0259	02	1398	1022	B3 NW
0260	02	1344	1016	B3 NW	0261	03	1322	1003	B3 NW
0262	04	1367	1017	B3 NW	0263	04A	1339	1079	B3 NW
0264	05A	1397	1040	B3 NW	0265	20	1393	1085	B3 NW
0266	20	1328	1019	B3 NW	0267	21	1392	1091	B3 NW
0268	30	1392	1078	B3 NW	0269	30	1373	1075	B3 NW
0270	30	1385	1048	B3 NW	0271	30	1387	1018	B3 NW
0272	30	1340	1001	B3 NW	0273	31A	1398	1065	B3 NW
0274	40	1358	1085	B3 NW	0275	40	1388	1089	B3 NW
0276	40	1355	1075	B3 NW	0277	40	1394	1072	B3 NW
0278	40	1321	1022	B3 NW	0279	40A	1306	1010	B3 NW
0280	01	1391	0997	B4 SW	0281	01	1378	0990	B4 SW
0282	01	1365	0990	B4 SW	0283	01	1370	0985	B4 SW

APPENDIX D (continued)

Number	Type	X	Y	Square	Number	Type	X	Y	Square
0284	01	1320	0991	B4 SW	0285	01	1354	0976	B4 SW
0286	01	1341	0972	B4 SW	0287	01	1333	0970	B4 SW
0288	01	1337	0966	B4 SW	0289	01	1317	0966	B4 SW
0290	01	1394	0968	B4 SW	0291	01	1363	0964	B4 SW
0292	01	1351	0959	B4 SW	0293	01	1335	0960	B4 SW
0294	01	1385	0940	B4 SW	0295	01	1308	0944	B4 SW
0296	01	1313	0933	B4 SW	0297	01	1322	0926	B4 SW
0298	01	1316	0919	B4 SW	0299	01	1309	0913	B4 SW
0300	01	1310	0905	B4 SW	0301	01	1332	0901	B4 SW
0302	01	1386	0931	B4 SW	0303	01	1387	0927	B4 SW
0304	01	1398	0920	B4 SW	0305	01	1392	0912	B4 SW
0306	01	1386	0901	B4 SW	0307	02	1314	0961	B4 SW
0308	02	1313	0950	B4 SW	0309	02	1379	0944	B4 SW
0310	02	1391	0906	B4 SW	0311	03	1374	0999	B4 SW
0312	03	1386	0993	B4 SW	0313	03	1394	0986	B4 SW
0314	03A	1303	0960	B4 SW	0315	04	1398	0959	B4 SW
0316	04	1389	0959	B4 SW	0317	04	1327	0943	B4 SW
0318	04	1318	0945	B4 SW	0319	04	1362	0933	B4 SW
0320	04A	1309	0931	B4 SW	0321	05	1325	0985	B4 SW
0322	20	1387	0988	B4 SW	0323	20	1326	0980	B4 SW
0324	20	1332	0961	B4 SW	0325	20	1378	0913	B4 SW
0326	20	1362	0901	B4 SW	0327	20A	1332	0984	B4 SW
0328	22	1306	0982	B4 SW	0329	30	1331	0989	B4 SW
0330	30	1369	0977	B4 SW	0331	30	1355	0951	B4 SW
0332	30	1387	0946	B4 SW	0333	30	1348	0921	B4 SW
0334	31A	1327	0919	B4 SW	0335	40	1358	0992	B4 SW
0336	40	1377	0971	B4 SW	0337	40	1370	0931	B4 SW
0338	40A	1364	0960	B4 SW	0339	31	1336	0985	B4 SW
0340	01	1393	0899	B4 NW	0341	01	1343	0888	B4 NW
0342	01	1326	0889	B4 NW	0343	01	1336	0878	B4 NW
0344	01	1365	0868	B4 NW	0345	01	1354	0866	B4 NW
0346	01	1335	0866	B4 NW	0347	01	1327	0866	B4 NW
0348	01	1346	0853	B4 NW	0349	01	1397	0860	B4 NW
0350	01	1385	0853	B4 NW	0351	01	1378	0843	B4 NW
0352	01	1366	0855	B4 NW	0353	01	1398	0802	B4 NW
0354	01	1392	0821	B4 NW	0355	01	1384	0820	B4 NW
0356	01	1360	0826	B4 NW	0357	01	1359	0818	B4 NW
0358	01	1354	0810	B4 NW	0359	01	1342	0827	B4 NW
0360	01	1331	0801	B4 NW	0361	02	1391	0876	B4 NW
0362	02	1337	0862	B4 NW	0363	03	1392	0879	B4 NW
0364	03	1345	0878	B4 NW	0365	03	1309	0878	B4 NW
0366	03	1314	0850	B4 NW	0367	03	1311	0815	B4 NW
0368	03A	1309	0823	B4 NW	0369	04	1323	0897	B4 NW
0370	04	1334	0885	B4 NW	0371	04	1313	0875	B4 NW
0372	05A	1381	0876	B4 NW	0373	05A	1341	0813	B4 NW
0374	20	1370	0896	B4 NW	0375	20	1385	0885	B4 NW
0376	20	1350	0819	B4 NW	0377	20	1361	0811	B4 NW
0378	20	1359	0893	B4 NW	0379	21	1377	0899	B4 NW
0380	21	1319	0888	B4 NW	0381	30	1365	0894	B4 NW
0382	30	1350	0883	B4 NW	0383	30	1392	0867	B4 NW
0384	30	1352	0862	B4 NW	0385	30	1332	0852	B4 NW
0386	30	1371	0818	B4 NW	0387	31	1376	0890	B4 NW
0388	31	1346	0860	B4 NW	0389	31	1314	0857	B4 NW
0390	40	1327	0859	B4 NW	0391	40	1323	0854	B4 NW
0392	40	1326	0812	B4 NW	0393	40A	1393	0843	B4 NW
0394	01	1326	0797	B5 SW	0395	01	1359	0790	B5 SW
0396	01	1394	0772	B5 SW	0397	01	1322	0775	B5 SW
0398	01	1348	0759	B5 SW	0399	01	1332	0754	B5 SW

APPENDIX D (continued)

Number	Type	X	Y	Square	Number	Type	X	Y	Square
0400	01	1328	0746	B5 SW	0401	01	1389	0743	B5 SW
0402	01	1391	0736	B5 SW	0403	01	1371	0737	B5 SW
0404	01	1358	0738	B5 SW	0405	01	1335	0735	B5 SW
0406	01	1335	0726	B5 SW	0407	01	1325	0713	B5 SW
0408	01	1312	0708	B5 SW	0409	01	1396	0719	B5 SW
0410	01	1397	0713	B5 SW	0411	01	1368	0712	B5 SW
0412	01	1367	0706	B5 SW	0413	01	1312	0782	B5 SW
0414	02A	1387	0790	B5 SW	0415	03	1305	0789	B5 SW
0416	03A	1379	0744	B5 SW	0417	04	1348	0737	B5 SW
0418	05A	1392	0724	B5 SW	0419	06	1375	0773	B5 SW
0420	20	1305	0735	B5 SW	0421	30	1337	0793	B5 SW
0422	30	1397	0763	B5 SW	0423	30	1355	0730	B5 SW
0424	30	1325	0727	B5 SW	0425	30	1320	0735	B5 SW
0426	30	1315	0731	B5 SW	0427	30	1368	0716	B5 SW
0428	30	1353	0714	B5 SW	0429	30	1343	0709	B5 SW
0430	40	1398	0753	B5 SW	0431	40	1381	0703	B5 SW
0432	40A	1385	0776	B5 SW					
0433	01	1355	0699	B5 NW	0434	01	1348	0699	B5 NW
0435	01	1341	0695	B5 NW	0436	01	1334	0696	B5 NW
0437	01	1321	0695	B5 NW	0438	01	1301	0697	B5 NW
0439	01	1375	0686	B5 NW	0440	01	1398	0677	B5 NW
0441	01	1381	0667	B5 NW	0442	01	1380	0671	B5 NW
0443	01	1360	0667	B5 NW	0444	01	1370	0677	B5 NW
0445	01	1354	0682	B5 NW	0446	01	1387	0655	B5 NW
0447	01	1367	0652	B5 NW	0448	01	1374	0647	B5 NW
0449	01	1370	0648	B5 NW	0450	01	1379	0639	B5 NW
0451	01	1389	0625	B5 NW	0452	01	1304	0669	B5 NW
0453	01	1320	0669	B5 NW	0454	01	1327	0662	B5 NW
0455	01	1301	0660	B5 NW	0456	01	1313	0646	B5 NW
0457	O1	1341	0643	B5 NW	0458	01	1343	0632	B5 NW
0459	01	1323	0631	B5 NW	0460	01	1359	0625	B5 NW
0461	01	1396	0606	B5 NW	0462	01	1379	0601	B5 NW
0463	01	1351	0604	B5 NW	0464	01	1308	0608	B5 NW
0465	02	1393	0611	B5 NW	0466	02	1364	0607	B5 NW
0467	02	1356	0608	B5 NW	0468	02A	1375	0655	B5 NW
0469	03	1372	0664	B5 NW	0470	03	1355	0654	B5 NW
0471	03	1326	0603	B5 NW	0472	04	1327	0691	B5 NW
0473	04	1311	0676	B5 NW	0474	04	1397	0664	B5 NW
0475	04	1360	0679	B5 NW	0476	05	1305	0683	B5 NW
0477	05A	1319	0632	B5 NW	0478	20	1354	0633	B5 NW
0479	20	1328	0621	B5 NW	0480	20	1398	0614	B5 NW
0481	20	1372	0609	B5 NW	0482	20A	1365	0676	B5 NW
0483	21	1381	0605	B5 NW	0484	22	1387	0664	B5 NW
0485	22	1362	0638	B5 NW	0486	30	1318	0688	B5 NW
0487	30	1333	0652	B5 NW	0488	30	1371	0630	B5 NW
0489	30	1385	0644	B5 NW	0490	30	1372	0670	B5 NW
0491	30A	1366	0695	B5 NW	0492	40	1380	0680	B5 NW
0493	40	1385	0633	B5 NW	0494	40A	1394	0691	B5 NW
0495	40A	1356	0637	B5 NW	0496	40A	1359	0617	B5 NW
0497	01	1374	0599	B6 SW	0498	01	1338	0580	B6 SW
0499	01	1374	0574	B6 SW	0500	01	1340	0571	B6 SW
0501	01	1328	0568	B6 SW	0502	01	1334	0560	B6 SW
0503	01	1305	0559	B6 SW	0504	01	1346	0555	B6 SW
0505	01	1392	0543	B6 SW	0506	01	1351	0545	B6 SW
0507	01	1386	0532	B6 SW	0508	01	1363	0534	B6 SW
0509	01	1369	0526	B6 SW	0510	01	1356	0517	B6 SW
0511	01	1342	0516	B6 SW	0512	01	1361	0512	B6 SW
0513	01	1307	0513	B6 SW	0514	01	1303	0508	B6 SW

APPENDIX D (continued)

Number	Type	X	Y	Square	Number	Type	X	Y	Square
0515	01	1365	0503	B6 SW	0516	01	1335	0501	B6 SW
0517	01	1301	0501	B6 SW	0518	01	1329	0548	B6 SW
0519	03	1339	0543	B6 SW	0520	03A	1365	0519	B6 SW
0521	03A	1360	0548	B6 SW	0522	04	1356	0592	B6 SW
0523	04A	1386	0569	B6 SW	0524	05	1367	0591	B6 SW
0525	05A	1335	0509	B6 SW	0526	20	1315	0532	B6 SW
0527	20	1385	0515	B6 SW	0528	21	1328	0589	B6 SW
0529	21	1383	0589	B6 SW	0530	21	1384	0539	B6 SW
0531	30	1316	0589	B6 SW	0532	30	1344	0567	B6 SW
0533	30	1315	0547	B6 SW	0534	30	1323	0547	B6 SW
0535	30	1354	0509	B6 SW	0536	40	1301	0587	B6 SW
0537	40	1398	0562	B6 SW	0538	40	1360	0540	B6 SW
0539	40	1327	0542	B6 SW					
0540	01	1375	0494	B6 NW	0541	01	1377	0485	B6 NW
0542	01	1362	0493	B6 NW	0543	01	1354	0488	B6 NW
0544	01	1347	0487	B6 NW	0545	01	1390	0432	B6 NW
0546	01	1390	0446	B6 NW	0547	01	1372	0454	B6 NW
0548	01	1340	0452	B6 NW	0549	01	1337	0463	B6 NW
0550	01	1329	0453	B6 NW	0551	01	1322	0459	B6 NW
0552	01	1336	0409	B6 NW	0553	01	1324	0410	B6 NW
0554	04A	1361	0448	B6 NW	0555	20	1320	0496	B6 NW
0556	20	1365	0463	B6 NW					
0557	01	1261	1044	B3 NE	0558	01	1275	1028	B3 NE
0559	01	1294	1034	B3 NE	0560	30	1264	1001	B3 NE
0561	40A	1260	1003	B3 NE					
0562	01	1201	0995	B4 SE	0563	01	1227	0997	B4 SE
0564	01	1222	0990	B4 SE	0565	01	1215	0989	B4 SE
0566	01	1208	0986	B4 SE	0567	01	1235	0988	B4 SE
0568	01	1233	0983	B4 SE	0569	01	1219	0982	B4 SE
0570	01	1223	0975	B4 SE	0571	01	1256	0994	B4 SE
0572	01	1259	0990	B4 SE	0573	01	1261	0982	B4 SE
0574	01	1288	0981	B4 SE	0575	01	1283	0977	B4 SE
0576	01	1292	0962	B4 SE	0577	01	1265	0960	B4 SE
0578	01	1280	0954	B4 SE	0579	01	1263	0950	B4 SE
0580	01	1204	0947	B4 SE	0581	01	1289	0937	B4 SE
0582	01	1255	0936	B4 SE	0583	01	1211	0940	B4 SE
0584	01	1209	0934	B4 SE	0585	01	1299	0906	B4 SE
0586	01	1269	0909	B4 SE	0587	01	1244	0901	B4 SE
0588	01	1248	0909	B4 SE	0589	01	1261	0901	B4 SE
0590	01	1240	0908	B4 SE	0591	01	1219	0901	B4 SE
0592	01	1213	0902	B4 SE	0593	01	1203	0911	B4 SE
0594	02	1289	0996	B4 SE	0595	02	1266	0984	B4 SE
0596	02	1273	0978	B4 SE	0597	02A	1236	0950	B4 SE
0598	03	1240	0998	B4 SE	0599	03	1205	0961	B4 SE
0600	03	1220	0944	B4 SE	0601	04	1210	0992	B4 SE
0602	04	1246	0984	B4 SE	0603	05	1282	0983	B4 SE
0604	06A	1237	0903	B4 SE	0605	20	1278	0988	B4 SE
0606	20	1264	0963	B4 SE	0607	20	1278	0948	B4 SE
0608	20	1208	0908	B4 SE	0609	21	1216	0968	B4 SE
0610	22	1259	0911	B4 SE	0611	30	1204	0992	B4 SE
0612	30	1241	0991	B4 SE	0613	30	1290	0975	B4 SE
0614	30	1290	0969	B4 SE	0615	30	1264	0968	B4 SE
0616	30	1213	0973	B4 SE	0617	30	1261	0941	B4 SE
0618	30	1295	0922	B4 SE	0619	30	1252	0927	B4 SE
0620	30A	1250	0987	B4 SE	0621	31	1267	0938	B4 SE
0622	31	1259	0921	B4 SE	0623	40	1207	0997	B4 SE
0624	40	1230	0972	B4 SE	0625	40	1283	0958	B4 SE

APPENDIX D (continued)

Number	Type	X	Y	Square	Number	Type	X	Y	Square
0626	40	1287	0958	B4 SE	0627	40	1228	0901	B4 SE
0628	01	1202	0883	B4 NE	0629	01	1214	0868	B4 NE
0630	01	1221	0863	B4 NE	0631	01	1224	0859	B4 NE
0632	01	1229	0866	B4 NE	0633	01	1249	0864	B4 NE
0634	01	1258	0870	B4 NE	0635	01	1260	0879	B4 NE
0636	01	1253	0886	B4 NE	0637	01	1290	0887	B4 NE
0638	01	1298	0880	B4 NE	0639	01	1285	0862	B4 NE
0640	01	1293	0852	B4 NE	0641	01	1273	0842	B4 NE
0642	01	1254	0838	B4 NE	0643	01	1261	0829	B4 NE
0644	01	1247	0831	B4 NE	0645	01	1236	0823	B4 NE
0646	01	1256	0806	B4 NE	0647	02	1202	0890	B4 NE
0648	02	1257	0858	B4 NE	0649	02	1220	0830	B4 NE
0650	03	1287	0855	B4 NE	0651	03	1221	0807	B4 NE
0652	03A	1211	0806	B4 NE	0653	04	1267	0894	B4 NE
0654	04	1243	0819	B4 NE	0655	04	1263	0814	B4 NE
0656	05A	1266	0873	B4 NE	0657	20	1272	0897	B4 NE
0658	20	1285	0896	B4 NE	0659	20	1291	0879	B4 NE
0660	20	1234	0803	B4 NE	0661	21	1281	0894	B4 NE
0662	22	1235	0867	B4 NE	0663	30	1225	0891	B4 NE
0664	30	1299	0871	B4 NE	0665	30	1253	0801	B4 NE
0666	30A	1205	0855	B4 NE	0667	31A	1231	0863	B4 NE
0668	40	1224	0883	B4 NE	0669	40	1295	0867	B4 NE
0670	40	1280	0890	B4 NE					
0671	01	1278	0786	B5 SE	0672	01	1249	0789	B5 SE
0673	01	1211	0781	B5 SE	0674	01	1234	0775	B5 SE
0675	01	1258	0763	B5 SE	0676	01	1239	0758	B5 SE
0677	01	1226	0760	B5 SE	0678	01	1241	0745	B5 SE
0679	01	1231	0742	B5 SE	0680	01	1235	0740	B5 SE
0681	01	1270	0736	B5 SE	0682	01	1258	0734	B5 SE
0683	01	1220	0733	B5 SE	0684	01	1220	0728	B5 SE
0685	01	1271	0719	B5 SE	0686	01	1225	0718	B5 SE
0687	01	1297	0708	B5 SE	0688	03A	1285	0767	B5 SE
0689	04	1299	0778	B5 SE	0690	04	1297	0726	B5 SE
0691	04A	1227	0790	B5 SE	0692	04A	1270	0743	B5 SE
0693	04A	1253	0717	B5 SE	0694	05A	1275	0770	B5 SE
0695	06	1232	0735	B5 SE	0696	06	1232	0729	B5 SE
0697	06	1238	0789	B5 SE	0698	20	1219	0777	B5 SE
0699	20	1275	0777	B5 SE	0700	21	1278	0715	B5 SE
0701	22	1225	0732	B5 SE	0702	30	1269	0772	B5 SE
0703	30	1247	0754	B5 SE	0704	30	1283	0731	B5 SE
0705	30	1234	0719	B5 SE	0706	30	1263	0707	B5 SE
0707	30	1259	0702	B5 SE	0708	40	1205	0796	B5 SE
0709	40	1293	0780	B5 SE	0710	40	1247	0770	B5 SE
0711	40	1246	0763	B5 SE	0712	40	1277	0703	B5 SE
0713	40	1268	0767	B5 SE	0714	40A	1223	0725	B5 SE
0715	01	1291	0695	B5 NE	0716	01	1287	0687	B5 NE
0717	01	1255	0691	B5 NE	0718	01	1233	0690	B5 NE
0719	01	1232	0683	B5 NE	0720	01	1221	0677	B5 NE
0721	01	1258	0675	B5 NE	0722	01	1259	0669	B5 NE
0723	01	1249	0668	B5 NE	0724	01	1245	0670	B5 NE
0725	01	1258	0655	B5 NE	0726	01	1248	0650	B5 NE
0727	01	1241	0645	B5 NE	0728	01	1226	0628	B5 NE
0729	01	1296	0615	B5 NE	0730	01	1277	0611	B5 NE
0731	01	1296	0601	B5 NE	0732	03	1218	0666	B5 NE
0733	03	1234	0666	B5 NE	0734	03	1276	0665	B5 NE
0735	03	1262	0640	B5 NE	0736	04A	1271	0642	B5 NE
0737	06	1234	0602	B5 NE	0738	20	1280	0648	B5 NE

APPENDIX D (continued)

Number	Type	X	Y	Square	Number	Type	X	Y	Square
0739	30	1266	0697	B5 NE	0740	30	1204	0695	B5 NE
0741	30	1299	0682	B5 NE	0742	30	1248	0681	B5 NE
0743	30	1216	0683	B5 NE	0744	30	1247	0663	B5 NE
0745	30	1278	0659	B5 NE	0746	30	1284	0653	B5 NE
0747	30	1284	0653	B5 NE	0748	30	1201	0657	B5 NE
0749	30	1280	0635	B5 NE	0750	30	1288	0614	B5 NE
0751	40	1247	0686	B5 NE	0752	40	1258	0650	B5 NE
0753	40	1217	0645	B5 NE	0754	40	1226	0640	B5 NE
0755	01	1248	0589	B6 SE	0756	01	1295	0581	B6 SE
0757	01	1284	0578	B6 SE	0758	01	1272	0555	B6 SE
0759	01	1253	0559	B6 SE	0760	01	1293	0529	B6 SE
0761	01	1275	0538	B6 SE	0762	01	1273	0543	B6 SE
0763	01	1261	0545	B6 SE	0764	01	1239	0546	B6 SE
0765	01	1255	0519	B6 SE	0766	01	1206	0520	B6 SE
0767	01	1258	0511	B6 SE	0768	01	1240	0511	B6 SE
0769	01	1236	0501	B6 SE	0770	02	1294	0590	B6 SE
0771	02	1237	0559	B6 SE	0772	02	1226	0545	B6 SE
0773	02A	1216	0525	B6 SE	0774	03A	1299	0546	B6 SE
0775	03A	1282	0550	B6 SE	0776	04A	1244	0574	B6 SE
0777	04A	1243	0539	B6 SE	0778	30	1209	0595	B6 SE
0779	30	1239	0587	B6 SE	0780	30	1258	0575	B6 SE
0781	30	1284	0529	B6 SE					
0782	01	1268	0481	B6 NE	0783	01	1273	0472	B6 NE
0784	01	1226	0440	B6 NE	0785	01	1213	0432	B6 NE
0786	01	1259	0420	B6 NE	0787	02	1294	0496	B6 NE
0788	04A	1280	0424	B6 NE	0789	05A	1223	0492	B6 NE
0790	20	1220	0492	B6 NE	0791	20	1201	0488	B6 NE
0792	20	1234	0453	B6 NE	0793	20	1232	0460	B6 NE
0794	20	1227	0460	B6 NE	0795	21A	1296	0420	B6 NE
0796	30	1299	0445	B6 NE	0797	40	1222	0456	B6 NE
0798	01	1195	0993	C4 SW	0799	01	1191	0995	C4 SW
0800	01	1183	0999	C4 SW	0801	01	1178	0994	C4 SW
0802	01	1173	0990	C4 SW	0803	01	1176	0983	C4 SW
0804	01	1173	0985	C4 SW	0805	01	1167	0981	C4 SW
0806	01	1165	0985	C4 SW	0807	01	1156	0983	C4 SW
0808	01	1152	0993	C4 SW	0809	01	1199	0974	C4 SW
0810	01	1176	0975	C4 SW	0811	01	1145	0973	C4 SW
0812	01	1188	0950	C4 SW	0813	01	1180	0943	C4 SW
0814	01	1176	0946	C4 SW	0815	01	1166	0946	C4 SW
0816	01	1174	0954	C4 SW	0817	01	1148	0947	C4 SW
0818	01	1124	0936	C4 SW	0819	01	1123	0945	C4 SW
0820	01	1115	0950	C4 SW	0821	01	1192	0910	C4 SW
0822	01	1187	0902	C4 SW	0823	01	1182	0903	C4 SW
0824	01	1175	0917	C4 SW	0825	01	1174	0926	C4 SW
0826	01	1153	0911	C4 SW	0827	01	1135	0921	C4 SW
0828	01	1127	0914	C4 SW	0829	01	1126	0919	C4 SW
0830	01	1111	0922	C4 SW	0831	01	1113	0913	C4 SW
0832	03	1102	0931	C4 SW	0833	03A	1129	0943	C4 SW
0834	05	1186	0995	C4 SW	0835	20	1135	0950	C4 SW
0836	20	1140	0932	C4 SW	0837	20	1188	0925	C4 SW
0838	22	1157	0975	C4 SW	0839	22	1159	0917	C4 SW
0840	30	1140	0948	C4 SW	0841	30	1152	0943	C4 SW
0842	30	1178	0936	C4 SW	0843	30	1103	0915	C4 SW
0844	31	1101	0954	C4 SW	0845	40	1166	0960	C4 SW
0846	40	1156	0945	C4 SW					
0847	01	1195	0897	C4 NW	0848	01	1191	0898	C4 NW

APPENDIX D (continued)

Number	Type	X	Y	Square	Number	Type	X	Y	Square
0849	01	1192	0892	C4 NW	0850	01	1159	0893	C4 NW
0851	01	1118	0896	C4 NW	0852	01	1112	0893	C4 NW
0853	01	1117	0891	C4 NW	0854	01	1103	0885	C4 NW
0855	01	1178	0878	C4 NW	0856	01	1127	0879	C4 NW
0857	01	1133	0872	C4 NW	0858	01	1190	0870	C4 NW
0859	01	1177	0867	C4 NW	0860	01	1192	0864	C4 NW
0861	01	1187	0862	C4 NW	0862	01	1195	0859	C4 NW
0863	01	1116	0855	C4 NW	0864	01	1106	0852	C4 NW
0865	01	1136	0853	C4 NW	0866	01	1147	0847	C4 NW
0867	01	1144	0840	C4 NW	0868	01	1121	0842	C4 NW
0869	01	1168	0838	C4 NW	0870	01	1101	0833	C4 NW
0871	01	1107	0827	C4 NW	0872	01	1126	0826	C4 NW
0873	01	1183	0817	C4 NW	0874	01	1106	0811	C4 NW
0875	01	1141	0808	C4 NW	0876	01	1157	0807	C4 NW
0877	01	1167	0893	C4 NW	0878	01	1196	0851	C4 NW
0879	02	1196	0833	C4 NW	0880	02	1130	0828	C4 NW
0881	02	1112	0823	C4 NW	0882	03	1157	0821	C4 NW
0883	03	1162	0865	C4 NW	0884	03	1191	0838	C4 NW
0885	03	1133	0821	C4 NW	0886	03A	1174	0815	C4 NW
0887	04	1176	0835	C4 NW	0888	04	1136	0840	C4 NW
0889	04	1111	0801	C4 NW	0890	04A	1185	0899	C4 NW
0891	04A	1148	0898	C4 NW	0892	05	1167	0855	C4 NW
0893	05	1177	0853	C4 NW	0894	05	1192	0847	C4 NW
0895	05	1141	0822	C4 NW	0896	05A	1135	0897	C4 NW
0897	20	1189	0899	C4 NW	0898	20	1124	0869	C4 NW
0899	20	1105	0859	C4 NW	0900	20	1190	0849	C4 NW
0901	20	1155	0849	C4 NW	0902	21	1125	0860	C4 NW
0903	21	1110	0820	C4 NW	0904	22	1104	0895	C4 NW
0905	22	1142	0890	C4 NW	0906	22	1112	0849	C4 NW
0907	22	1187	0841	C4 SW	0908	23	1103	0803	C4 NW
0909	30	1155	0880	C4 SW	0910	30	1141	0863	C4 NW
0911	30	1186	0850	C4 SW	0912	30	1171	0846	C4 NW
0913	30	1113	0844	C4 NW	0914	30	1173	0837	C4 NW
0915	30	1188	0834	C4 NW	0916	30	1178	0831	C4 NW
0917	30	1172	0832	C4 NW	0918	30	1175	0824	C4 NW
0919	30	1131	0824	C4 NW	0920	30	1117	0826	C4 NW
0921	30	1114	0818	C4 NW	0922	30	1130	0801	C4 NW
0923	30A	1190	0819	C4 NW	0924	40	1163	0875	C4 NW
0925	40	1111	0871	C4 NW	0926	40	1189	0856	C4 NW
0927	40	1150	0857	C4 NW	0928	40	1106	0847	C4 NW
0929	40	1166	0843	C4 NW	0930	40	1133	0836	C4 NW
0931	40	1180	0811	C4 NW					
0932	01	1176	0793	C5 SW	0933	01	1153	0793	C5 SW
0934	01	1138	0794	C5 SW	0935	01	1106	0790	C5 SW
0936	01	1104	0775	C5 SW	0937	01	1153	0765	C5 SW
0938	01	1185	0765	C5 SW	0939	01	1103	0760	C5 SW
0940	01	1172	0739	C5 SW	0941	01	1155	0734	C5 SW
0942	01	1139	0737	C5 SW	0943	01	1189	0729	C5 SW
0944	01	1181	0722	C5 SW	0945	01	1165	0726	C5 SW
0946	01	1152	0722	C5 SW	0947	01	1145	0719	C5 SW
0948	01	1137	0722	C5 SW	0949	01	1131	0717	C5 SW
0950	01	1131	0724	C5 SW	0951	01	1122	0725	C5 SW
0952	01	1104	0718	C5 SW	0953	01	1189	0715	C5 SW
0954	01	1178	0713	C5 SW	0955	01	1165	0701	C5 SW
0956	01	1149	0705	C5 SW	0957	01	1133	0702	C5 SW
0958	02	1124	0736	C5 SW	0959	02	1169	0710	C5 SW
0960	02	1171	0771	C5 SW	0961	03	1137	0780	C5 SW
0962	03	1150	0779	C5 SW	0963	03	1149	0749	C5 SW
0964	03A	1135	0774	C5 SW	0965	03A	1176	0748	C5 SW

APPENDIX D (continued)

Number	Type	X	Y	Square	Number	Type	X	Y	Square
0966	03A	1113	0752	C5 SW	0967	04	1193	0769	C5 SW
0968	04A	1164	0751	C5 SW	0969	05	1142	0788	C5 SW
0970	20	1183	0795	C5 SW	0971	20	1145	0760	C5 SW
0972	20	1187	0743	C5 SW	0973	21	1188	0751	C5 SW
0974	21	1108	0730	C5 SW	0975	30	1166	0798	C5 SW
0976	30	1118	0788	C5 SW	0977	30	1133	0779	C5 SW
0978	30	1115	0774	C5 SW	0979	30	1198	0772	C5 SW
0980	30	1141	0768	C5 SW	0981	30	1111	0744	C5 SW
0982	30	1197	0737	C5 SW	0983	30	1113	0724	C5 SW
0984	30	1169	0718	C5 SW	0985	30	1191	0708	C5 SW
0986	30A	1122	0781	C5 SW	0987	30A	1111	0765	C5 SW
0988	40	1191	0789	C5 SW	0989	40	1157	0790	C5 SW
0990	40	1133	0712	C5 SW	0991	40A	1127	0749	C5 SW
0992	01	1185	0698	C5 NW	0993	01	1175	0699	C5 NW
0994	01	1138	0698	C5 NW	0995	01	1154	0685	C5 NW
0996	01	1133	0683	C5 NW	0997	01	1168	0675	C5 NW
0998	01	1134	0676	C5 NW	0999	01	1159	0667	C5 NW
1000	01	1138	0666	C5 NW	1001	01	1158	0660	C5 NW
1002	01	1165	0655	C5 NW	1003	01	1144	0643	C5 NW
1004	01	1111	0639	C5 NW	1005	01	1149	0634	C5 NW
1006	01	1143	0626	C5 NW	1007	01	1196	0618	C5 NW
1008	01	1133	0615	C5 NW	1009	01	1109	0611	C5 NW
1010	01	1171	0606	C5 NW	1011	01	1186	0601	C5 NW
1012	02	1104	0686	C5 NW	1013	03	1175	0646	C5 NW
1014	03	1167	0640	C5 NW	1015	04	1113	0664	C5 NW
1016	04	1146	0661	C5 NW	1017	04A	1179	0687	C5 NW
1018	04A	1145	0620	C5 NW	1019	05	1111	0673	C5 NW
1020	05	1115	0626	C5 NW	1021	05A	1139	0685	C5 NW
1022	05A	1112	0647	C5 NW	1023	06	1125	0686	C5 NW
1024	20	1178	0690	C5 NW	1025	20	1165	0647	C5 NW
1026	20	1187	0640	C5 NW	1027	30	1168	0690	C5 NW
1028	30	1142	0679	C5 NW	1029	30	1181	0672	C5 NW
1030	30	1169	0665	C5 NW	1031	30	1178	0661	C5 NW
1032	30	1120	0651	C5 NW	1033	30	1114	0605	C5 NW
1034	30A	1107	0656	C5 NW	1035	40	1172	0620	C5 NW
1036	40	1128	0647	C5 NW	1037	40A	1101	0669	C5 NW
1038	O1	1198	0585	C6 SW	1039	01	1187	0592	C6 SW
1040	01	1189	0577	C6 SW	1041	01	1194	0564	C6 SW
1042	01	1166	0591	C6 SW	1043	01	1121	0599	C6 SW
1044	01	1120	0584	C6 SW	1045	01	1162	0542	C6 SW
1046	01	1115	0540	C6 SW	1047	01	1137	0523	C6 SW
1048	01	1125	0509	C6 SW	1049	01	1198	0503	C6 SW
1050	02	1159	0555	C6 SW	1051	02	1153	0534	C6 SW
1052	02	1131	0501	C6 SW	1053	03	1179	0572	C6 SW
1054	03	1199	0517	C6 SW	1055	03	1124	0571	C6 SW
1056	03	1132	0557	C6 SW	1057	03	1121	0557	C6 SW
1058	03	1105	0556	C6 SW	1059	04	1173	0565	C6 SW
1060	04A	1112	0548	C6 SW	1061	04A	1135	0548	C6 SW
1062	04A	1195	0549	C6 SW	1063	04A	1162	0521	C6 SW
1064	04A	1188	0512	C6 SW	1065	05	1174	0549	C6 SW
1066	05	1195	0528	C6 SW	1067	05A	1136	0538	C6 SW
1068	20	1172	0537	C6 SW	1069	21A	1153	0513	C6 SW
1070	30	1172	0576	C6 SW	1071	30	1161	0576	C6 SW
1072	30	1167	0559	C6 SW	1073	30	1195	0537	C6 SW
1074	40	1113	0518	C6 SW	1075	40	1167	0510	C6 SW
1076	01	1149	0496	C6 NW	1077	01	1164	0477	C6 NW
1078	01	1145	0480	C6 NW	1079	01	1131	0482	C6 NW

APPENDIX D (continued)

Number	Type	X	Y	Square	Number	Type	X	Y	Square
1080	01	1145	0461	C6 NW	1081	01	1129	0466	C6 NW
1082	01	1131	0455	C6 NW	1083	01	1109	0457	C6 NW
1084	01	1125	0437	C6 NW	1085	01	1146	0433	C6 NW
1086	01	1139	0428	C6 NW	1087	02	1150	0452	C6 NW
1088	02	1168	0443	C6 NW	1089	02	1187	0420	C6 NW
1090	03	1156	0430	C6 NW	1091	04A	1148	0424	C6 NW
1092	30	1102	0473	C6 NW	1093	40	1112	0472	C6 NW
1094	01	1087	0936	C4 SE	1095	01	1075	0932	C4 SE
1096	01	1094	0901	C4 SE	1097	01	1077	0908	C4 SE
1098	01	1070	0904	C4 SE	1099	01	1069	0920	C4 SE
1100	01	1060	0917	C4 SE	1101	01	1053	0923	C4 SE
1102	01	1053	0917	C4 SE	1103	01	1045	0917	C4 SE
1104	01	1050	0905	C4 SE	1105	01	1044	0902	C4 SE
1106	01	1013	0911	C4 SE	1107	01	1020	0902	C4 SE
1108	04	1079	0942	C4 SE	1109	05A	1022	0913	C4 SE
1110	30	1092	0939	C4 SE	1111	30	1082	0931	C4 SE
1112	30	1067	0927	C4 SE	1113	40	1098	0941	C4 SE
1114	40A	1010	0901	C4 SE					
1115	01	1054	0899	C4 NE	1116	01	1028	0885	C4 NE
1117	01	1056	0891	C4 NE	1118	01	1075	0895	C4 NE
1119	01	1087	0892	C4 NE	1120	01	1077	0880	C4 NE
1121	01	1068	0871	C4 NE	1122	01	1033	0868	C4 NE
1123	01	1081	0859	C4 NE	1124	01	1078	0846	C4 NE
1125	01	1075	0840	C4 NE	1126	01	1089	0816	C4 NE
1127	01	1095	0809	C4 NE	1128	01	1017	0848	C4 NE
1129	01	1020	0835	C4 NE	1130	01	1012	0839	C4 NE
1131	01	1001	0827	C4 NE	1132	01	1018	0825	C4 NE
1133	01	1010	0804	C4 NE	1134	01	1041	0825	C4 NE
1135	01	1035	0810	C4 NE	1136	01	1040	0808	C4 NE
1137	01	1057	0804	C4 NE	1138	02	1029	0899	C4 NE
1139	02	1067	0880	C4 NE	1140	02	1071	0858	C4 NE
1141	02	1090	0842	C4 NE	1142	02	1099	0813	C4 NE
1143	02	1047	0813	C4 NE	1144	03	1073	0888	C4 NE
1145	03	1061	0874	C4 NE	1146	03	1086	0834	C4 NE
1147	03	1071	0825	C4 NE	1148	03	1001	0809	C4 NE
1149	04	1022	0862	C4 NE	1150	05	1070	0867	C4 NE
1151	05	1060	0838	C4 NE	1152	05	1005	0806	C4 NE
1153	05A	1099	0848	C4 NE	1154	20	1049	0854	C4 NE
1155	20	1075	0851	C4 NE	1156	20	1068	0828	C4 NE
1157	20	1029	0811	C4 NE	1158	20	1035	0801	C4 NE
1159	21	1065	0848	C4 NE	1160	21	1072	0842	C4 NE
1161	21	1082	0818	C4 NE	1162	21	1092	0801	C4 NE
1163	22	1028	0889	C4 NE	1164	22	1060	0869	C4 NE
1165	22	1034	0841	C4 NE	1166	30	1040	0887	C4 NE
1167	30	1009	0834	C4 NE	1168	30A	1066	0802	C4 NE
1169	30A	1042	0801	C4 NE	1170	40	1095	0856	C4 NE
1171	40	1053	0846	C4 NE	1172	40A	1027	0868	C4 NE
1173	02	1087	0805	C4 NE					
1174	01	1071	0799	C5 SE	1175	01	1053	0792	C5 SE
1176	01	1099	0784	C5 SE	1176	01	1083	0776	C5 SE
1178	01	1084	0786	C5 SE	1179	01	1051	0777	C5 SE
1180	01	1051	0782	C5 SE	1181	01	1059	0783	C5 SE
1182	01	1045	0786	C5 SE	1183	01	1053	0792	C5 SE
1184	01	1022	0784	C5 SE	1185	01	1014	0786	C5 SE
1186	01	1013	0769	C5 SE	1187	01	1013	0760	C5 SE
1188	01	1050	0757	C5 SE	1189	01	1037	0739	C5 SE
1190	01	1025	0739	C5 SE	1191	01	1031	0731	C5 SE

APPENDIX D (continued)

Number	Type	X	Y	Square	Number	Type	X	Y	Square
1192	01	1099	0752	C5 SE	1193	01	1073	0742	C5 SE
1194	01	1081	0738	C5 SE	1195	01	1061	0733	C5 SE
1196	01	1098	0728	C5 SE	1197	01	1092	0720	C5 SE
1198	01	1094	0709	C5 SE	1199	01	1068	0709	C5 SE
1200	01	1062	0712	C5 SE	1201	01	1054	0715	C5 SE
1202	01	1023	0722	C5 SE	1203	01	1019	0710	C5 SE
1204	02	1056	0772	C5 SE	1205	02	1035	0718	C5 SE
1206	03	1044	0781	C5 SE	1207	03	1042	0775	C5 SE
1208	03	1055	0743	C5 SE	1209	03	1044	0742	C5 SE
1210	03	1019	0720	C5 SE	1211	03A	1099	0743	C5 SE
1212	03A	1018	0737	C5 SE	1213	04	1043	0792	C5 SE
1214	04	1038	0788	C5 SE	1215	04	1075	0731	C5 SE
1216	04	1040	0709	C5 SE	1217	04A	1070	0780	C5 SE
1218	05	1030	0789	C5 SE	1219	05	1006	0771	C5 SE
1220	06	1035	0776	C5 SE	1221	20	1088	0786	C5 SE
1222	20	1045	0723	C5 SE	1223	21	1014	0775	C5 SE
1224	21	1087	0780	C5 SE	1225	21	1066	0767	C5 SE
1226	21	1080	0742	C5 SE	1227	21	1053	0733	C5 SE
1228	21	1076	0720	C5 SE	1229	23	1025	0775	C5 SE
1230	24	1091	0761	C5 SE	1231	30	1076	0793	C5 SE
1232	30	1068	0790	C5 SE	1233	30	1016	0781	C5 SE
1234	30	1077	0765	C5 SE	1235	30	1068	0759	C5 SE
1236	30	1007	0743	C5 SE	1237	30A	1027	0796	C5 SE
1238	30A	1089	0795	C5 SE	1239	30A	1096	0772	C5 SE
1240	30A	1094	0757	C5 SE	1241	30A	1076	0735	C5 SE
1242	40	1090	0751	C5 SE	1243	40	1078	0703	C5 SE
1244	40A	1046	0730	C5 SE	1245	40A	1009	0734	C5 SE
1246	40A	1045	0748	C5 SE	1247	40A	1081	0756	C5 SE
1248	40A	1033	0780	C5 SE	1249	40A	1061	0795	C5 SE
1250	01	1045	0691	C5 NE	1251	01	1083	0684	C5 NE
1252	01	1035	0680	C5 NE	1253	01	1050	0676	C5 NE
1254	01	1028	0676	C5 NE	1255	01	1045	0670	C5 NE
1256	01	1075	0663	C5 NE	1257	01	1047	0658	C5 NE
1258	01	1001	0654	C5 NE	1259	01	1094	0649	C5 NE
1260	01	1024	0640	C5 NE	1261	01	1061	0632	C5 NE
1262	01	1067	0630	C5 NE	1263	01	1086	0614	C5 NE
1264	02	1029	0690	C5 NE	1265	03	1010	0665	C5 NE
1266	03A	1063	0685	C5 NE	1267	04	1034	0647	C5 NE
1268	04A	1008	0694	C5 NE	1269	04A	1079	0678	C5 NE
1270	04A	1037	0640	C5 NE	1271	05	1070	0644	C5 NE
1272	05	1001	0637	C5 NE	1273	06	1096	0628	C5 NE
1274	06A	1070	0655	C5 NE	1275	20	1013	0634	C5 NE
1276	20	1044	0626	C5 NE	1277	20	1023	0607	C5 NE
1278	30	1070	0698	C5 NE	1279	30	1008	0685	C5 NE
1280	30	1024	0680	C5 NE	1281	30	1068	0679	C5 NE
1282	30	1041	0672	C5 NE	1283	30	1042	0664	C5 NE
1284	30	1031	0661	C5 NE	1285	30	1073	0670	C5 NE
1286	30	1093	0660	C5 NE	1287	30	1081	0652	C5 NE
1288	30	1037	0631	C5 NE	1289	30	1023	0629	C5 NE
1290	40	1095	0697	C5 NE	1291	40	1085	0623	C5 NE
1292	40	1052	0619	C5 NE	1293	40	1069	0601	C5 NE
1294	01	1057	0598	C6 SE	1295	01	1007	0588	C6 SE
1296	01	1072	0574	C6 SE	1297	01	1058	0564	C6 SE
1298	01	1032	0577	C6 SE	1299	01	1013	0574	C6 SE
1300	01	1022	0558	C6 SE	1301	01	1097	0552	C6 SE
1302	01	1003	0536	C6 SE	1303	01	1074	0528	C6 SE
1304	01	1073	0515	C6 SE	1305	01	1065	0520	C6 SE
1306	01	1043	0520	C6 SE	1307	01	1044	0509	C6 SE
1308	01	1015	0503	C6 SE	1309	02	1026	0586	C6 SE

APPENDIX D (continued)

Number	Type	X	Y	Square	Number	Type	X	Y	Square
1310	02	1088	0536	C6 SE	1311	02	1030	0529	C6 SE
1312	02	1050	0514	C6 SE	1313	03	1097	0598	C6 SE
1314	03	1078	0538	C6 SE	1315	03	1018	0538	C6 SE
1316	04	1078	0589	C6 SE	1317	04A	1054	0576	C6 SE
1318	04A	1054	0538	C6 SE	1319	04A	1078	0522	C6 SE
1320	05A	1042	0581	C6 SE	1321	06A	1088	0515	C6 SE
1322	21A	1070	0565	C6 SE	1323	21A	1088	0551	C6 SE
1324	30	1044	0590	C6 SE	1325	40	1057	0570	C6 SE
1326	02	1099	0525						
1327	01	1087	0499	C6 NE	1328	01	1059	0492	C6 NE
1329	01	1066	0483	C6 NE	1330	01	1056	0482	C6 NE
1331	01	1030	0479	C6 NE	1332	01	1097	0479	C6 NE
1333	01	1075	0470	C6 NE	1334	01	1070	0443	C6 NE
1335	01	1083	0439	C6 NE	1336	01	1068	0432	C6 NE
1337	01	1076	0425	C6 NE	1338	01	1084	0418	C6 NE
1339	01	1042	0430	C6 NE	1340	01	1003	0421	C6 NE
1341	02	1043	0491	C6 NE	1342	02	1081	0482	C6 NE
1343	03	1012	0462	C6 NE	1344	03A	1035	0441	C6 NE
1345	04A	1030	0496	C6 NE	1346	20	1028	0432	C6 NE
1347	40	1021	0445	C6 NE	1348	40	1048	0407	C6 NE
1349	01	0978	0885	D4 NW	1350	01	0988	0884	D4 NW
1351	01	0976	0881	D4 NW	1352	01	0981	0874	D4 NW
1353	01	0999	0869	D4 NW	1354	01	0958	0858	D4 NW
1355	01	0975	0850	D4 NW	1356	01	0948	0860	D4 NW
1357	01	0941	0853	D4 NW	1358	01	0930	0840	D4 NW
1359	01	0975	0821	D4 NW	1360	01	0967	0826	D4 NW
1361	01	0942	0825	D4 NW	1362	01	0929	0817	D4 NW
1363	01	0918	0810	D4 NW	1364	01	0990	0810	D4 NW
1365	01	0999	0804	D4 NW	1366	01	0971	0801	D4 NW
1367	02	0979	0835	D4 NW	1368	02	0935	0829	D4 NW
1369	03	0991	0896	D4 NW	1370	04	0991	0872	D4 NW
1371	04	0937	0830	D4 NW	1372	04	0981	0819	D4 NW
1373	05	0981	0890	D4 NW	1374	05	0980	0841	D4 NW
1375	20	0972	0868	D4 NW	1376	20	0999	0858	D4 NW
1377	22	0962	0829	D4 NW	1378	23	0973	0876	D4 NW
1379	23	0973	0805	D4 NW	1380	30	0980	0863	D4 NW
1381	30	0996	0851	D4 NW	1382	30	0987	0814	D4 NW
1383	30	0911	0805	D4 NW	1384	40	0998	0889	D4 NW
1385	40	0996	0842	D4 NW	1386	40	0986	0818	D4 NW
1387	01	0789	0741	E5 SW	1388	01	0778	0739	E5 SW
1389	01	0770	0720	E5 SW	1390	01	0796	0712	E5 SW
1391	01	0750	0717	E5 SW	1392	01	0729	0723	E5 SW
1393	01	0737	0707	E5 SW	1394	04	0798	0742	E5 SW
1395	04	0762	0735	E5 SW	1396	20	0752	0729	E5 SW
1397	30	0773	0734	E5 SW	1398	30	0761	0727	E5 SW
1399	30	0741	0716	E5 SW	1400	30	0709	0713	E5 SW
1401	30	0719	0701	E5 SW	1402	30A	0722	0712	E5 SW
1403	30A	0727	0705	E5 SW					
					1404	01	0723	0691	E5 NW
1405	01	0760	0690	E5 NW	1406	01	0736	0686	E5 NW
1407	01	0759	0684	E5 NW	1408	01	0726	0686	E5 NW
1409	01	0784	0741	E5 SW	1410	01	0770	0673	E5 NW
1411	01	9718	0670	E5 NW	1412	01	0708	0653	E5 NW
1413	01	0757	0650	E5 NW	1414	01	0729	0645	E5 NW
1415	01	0765	0635	E5 NW	1416	01	0716	0638	E5 NW
1417	01	0724	0634	E5 NW	1418	01	0707	0633	E5 NW
1419	01	0765	0612	E5 NW	1420	01	0758	0601	E5 NW
1421	01	0738	0607	E5 NW	1422	02	0748	0695	E5 NW

APPENDIX D (continued)

Number	Type	X	Y	Square	Number	Type	X	Y	Square
1423	02	0764	0663	E5 NW	1424	02	0768	0653	E5 NW
1425	02	0751	0608	E5 NW	1426	03	0721	0617	E5 NW
1427	04	0776	0658	E5 NW	1428	04	0785	0654	E5 NW
1429	04A	0723	0677	E5 NW	1430	04A	0741	0670	E5 NW
1431	04A	0799	0636	E5 NW	1432	04A	0789	0630	E5 NW
1433	05A	0746	0621	E5 NW	1434	20	0739	0678	E5 NW
1435	20	0732	0659	E5 NW	1436	20	0733	0648	E5 NW
1437	20	0741	0616	E5 NW	1438	20	0762	0603	E5 NW
1439	21	0710	0631	E5 NW	1440	22	0764	0694	E5 NW
1441	22	0740	0653	E5 NW	1442	22	0795	0647	E5 NW
1443	22	0721	0647	E5 NW	1444	30	0781	0668	E5 NW
1445	30	0786	0662	E5 NW	1446	30	0715	0646	E5 NW
1447	30A	0762	0647	E5 NW	1448	31	0726	0638	E5 NW
1449	40	0713	0688	E5 NW	1450	40	0704	0633	E5 NW
1451	01	0681	0677	E5 NE	1452	01	0671	0670	E5 NE
1453	01	0672	0661	E5 NE	1454	01	0684	0656	E5 NE
1455	01	0671	0650	E5 NE	1456	01	0656	0647	E5 NE
1457	01	0676	0644	E5 NE	1458	01	0667	0638	E5 NE
1459	01	0681	0632	E5 NE	1460	01	0693	0628	E5 NE
1461	01	0641	0631	E5 NE	1462	01	0691	0612	E5 NE
1463	01	0664	0614	E5 NE	1464	01	0642	0616	E5 NE
1465	01	0670	0603	E5 NE	1466	02	0695	0657	E5 NE
1467	02	0697	0633	E5 NE	1468	02	0632	0614	E5 NE
1469	03	0691	0675	E5 NE	1470	03A	0653	0656	E5 NE
1471	04A	0651	0607	E5 NE	1472	22	0686	0668	E5 NE
1473	22	0666	0629	E5 NE	1474	22	0675	0623	E5 NE
1475	30	0685	0683	E5 NE	1476	30	0683	0665	E5 NE
1477	30	0665	0619	E5 NE	1478	30A	0645	0605	E5 NE
1479	40	0659	0640	E5 NE	1480	40	0647	0639	E5 NE
1481	40	0625	0601	E5 NE					

Appendix E

Spot Heights of Site 55, Riwat.

(See Figure for location of points)

Number	Height (metres)	Number	Height (metres)	Number	Height (metres)	Number	Height (metres)
01	96.230	02	96.220	03	96.290	04	96.290
05	96.300	06	96.355	07	96.375	08	96.440
09	96.465	10	96.480	11	96.550	12	96.600
13	96.735	14	96.695	15	96.640	16	96.625
17	96.538	18	96.675	19	96.703	20	96.765
21	96.770	22	96.790	23	96.795	24	96.840
25	96.755	26	96.835	27	96.913	28	96.955
29	96.923	30	96.945	31	96.970	32	97.095
33	97.135	34	97.190	35	97.200	36	97.220
37	97.130	38	97.253	39	97.260	40	97.345
41	97.443	42	97.460	43	97.230	44	97.575
45	97.485	46	97.545	47	97.655	48	97.740
49	97.955	50	97.943	51	97.835	52	97.945
53	97.430	54	98.225	55	98.198	56	98.080
57	98.045	58	97.705	59	97.715	60	97.725
61	97.885	62	98.420	63	98.585	64	98.600
65	98.565	66	98.315	67	98.250	68	97.300
69	98.055	70	97.705	71	96.410	72	96.415
73	96.510	74	96.500	75	96.850	76	96.813
77	97.530	78	97.505	79	97.480	80	97.215

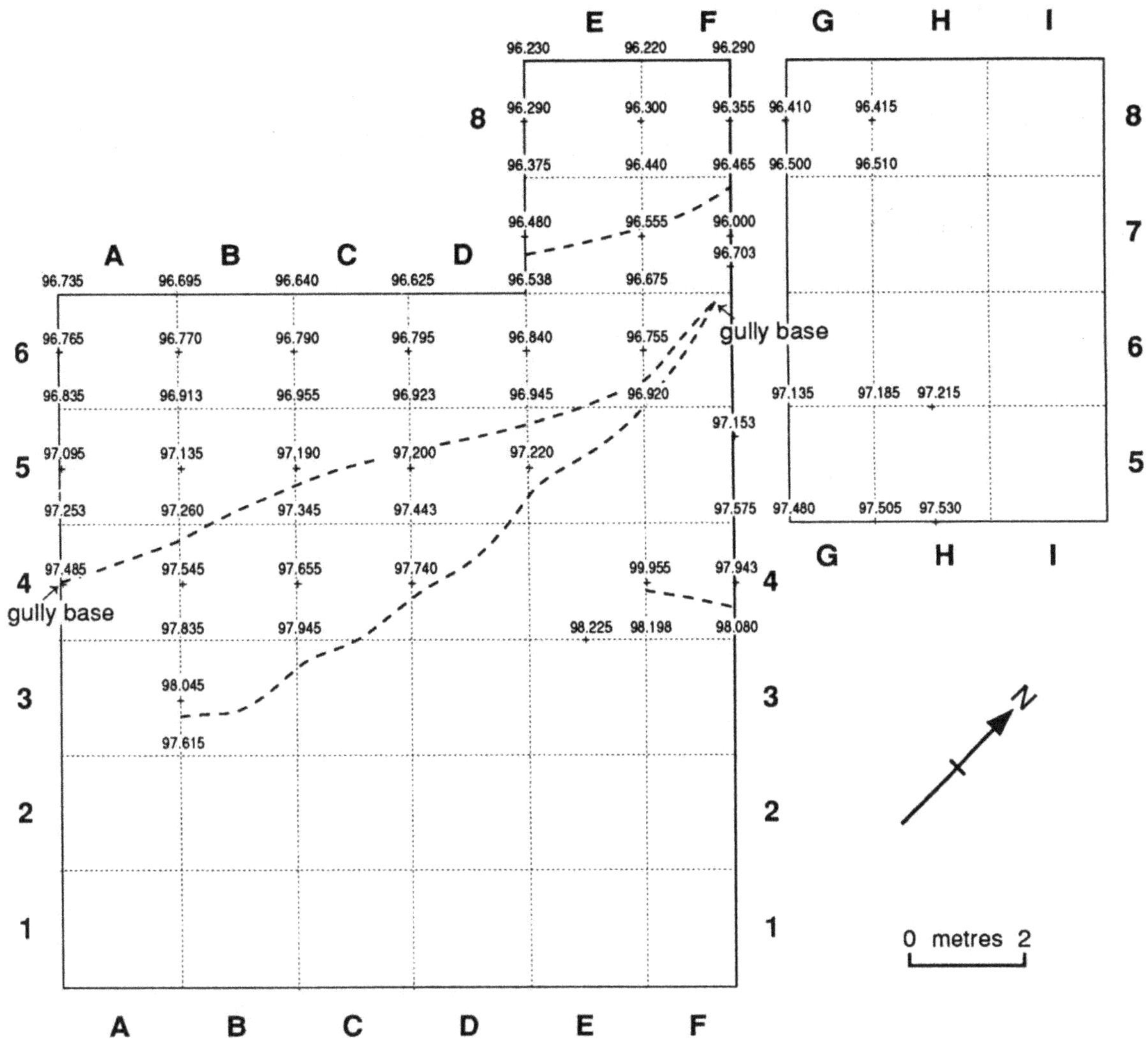

Figure E.1: Location of spot-heights site 55, Riwat

Appendix F

List of conjoining pieces from site 55, Riwat

ARTEFACTS AND CONJOINING PIECES

	ARTEFACT	CONJOINING PIECES				
1	0117 (core fragment)	0136 (flake)	1078 (flake)			
2	0120 (flake)	0137 (retouched flake)	0163 (blade fragment)			
3	0121 (retouched cortical flake)	1195 (core)	1278A (retouched flake)			
4	0124 (retouched cortical flake)	125 (retouched flake)	1368 (retouchedflake)			
5	0141 (blade fragment)	508 (retouched blade fragment)	1321 (blade fragment)	1383 (retouched blade fragment)	1406 (retouched flake)	1410 (retouched blade fragment)
6	0147 (blade fragment)	1356 (flake)				
7	0169 (core flake)	0171 (core fragment)				
8	0202 (core fragment)	211 (cortical flake)	0225 (cortical flake)			
9	0313 (blade fragment)	0334 (blade)				
10	0376 (flake)	0506 (retouched flake)	1297 (retouched cortical flake)			
11	0383 (retouched flake)	1409 (retouched flake)				
12	1024 (core)	1322 (retouched flake/ ?scraper)	1405 (core fragment)			
13	1093 (retouched blade fragment)	1343 (retouched cortical flake)				
14	1099 (flake)	1119 (retouched blade fragment)	1121 (core fragment)	1134 (cortical flake)		
15	1103 (retouched cortical flake)	1365A (flake/ ?scraper)				
16	1113 (core)	1148 (core fragment)	1392 (flake)			
17	1160 (cortical flake)	1177 (retouched flake)	1178 (flake)	1282 (core)		
18	1164 (cortical flake)	1166 (flake)				
19	1184 (core)	1185 (broken blade)				
20	1221 (retouched cortical flake)	1305 (retouched blade)				
21	1290 (cortical flake)	1291 (core)	1412 (flake spall)			
22	1338 (blade fragment)	1417 (flake)				
23	1345 (flake)	1350 (blade fragment)				
24	1400 (core)	1402 (core fragment)				

Appendix G

Colour variation in conjoining sets

CONJOINING PIECES					
0117 (honey)	0136 (honey)	1078 (white)			
0129 (grey-green)	0137 (grey-green)	0163 (grey)			
0121 (brown)	1195 (green)	1278A (grey)			
0124 (blue-grey)	0125 (blue-grey)	1368 (grey-blue)			
0141 (green)	0508 (grey)	1321 (grey)	1383 (grey)	1406 (grey)	1410 (grey)
0147 (honey)	1356 (honey)				
0169 (honey)	0171 (honey)				
0202 (brown-green)	0211 (brown-green)	0225 (brown-green)			
0313 (grey)	0334 (grey)				
0376 (green-brown)	0506 (green-brown)	1297 (green-brown)			
0383 (grey-brown)	1409 (grey)				
1024 (brown)	1322 (brown-green)	1405 (brown-green)			
1093 (grey-green)	1343 (green)				
1099 (green)	1119 (green-blue)	1121 (grey-green)	1134 (green)		
1103 (grey)	1365A (grey)				
1113 (brown-green)	1148 (green)	1392 (green)			
1160 (grey-brown)	1177 (honey)	1178 (grey-honey)	1282 (grey-honey)		
1164 (honey-brown)	1166 (honey-brown)				
1184 (honey)	1185 (honey)				
1221 (brown-green)	1305 (brown)				
1290 (grey)	1291 (grey)	1412 (grey-blue)			
1338 (grey)	1417 (grey)				
1345 (red-brown)	1350 (red-brown)				
1400 (green)	1402 (green)				

Appendix H

The photoframe used at Site 55, Riwat

R.W.Dennell, J. Payne, E. Moth

(Dept. of Archaeology and Prehistory, University of Sheffield, Sheffield S10 2TN).

Photogrammetric techniques have been around for a long time as a way of providing a permanent and unbiased record of an archaeological site. However, they have been used only intermittently because of their cost. In most instances, a photomosaic is made by using a camera placed on high scaffold tower; from cameras slung between towers; from balloons; or from a durable frame covering the entire site. If the final plan is made by stereo-plotting, the equipment and skill needed are highly expensive and well beyond most archaeological budgets (see Fussell 1982 for a useful review). On the other hand, recording a large surface photographically is far faster than recording the usual way with drawing frame, plumb bob and steel tape. It is also especially advantageous when - as at site 55 - skilled labour was at a premium. Finally, it provides an unbiased and permanent archive of an excavation.

The "Quadrupod" photographic frame that was used at site 55 was designed so that a mosaic of black and white photographs could be made of a large, sloping, stone-covered surface, thus enabling a site plan to be drawn later. The main emphases were on cost and simplicity since "*generally speaking, it is only relative accuracy and not true metric accuracy that is required*" (Anderson 1982: 203) when digging most sites.

Its construction was tightly constrained by a number of factors. First, it had to be light and easily portable. In more precise terms, it had to be carried as a bundle of rods and tubes, none of which could be more than 2m long in order to fit into a London Underground train carriage in the morning rush hour without causing undue obstruction or inconvenience to other passengers. Once at Heathrow Airport, it had to be sufficiently small and light that it could be sent as normal baggage, and without incurring excess baggage charges. However, as it was used on the excavation for several weeks, it had to be robust as well as light. It also had to be easy to assemble, use and dismantle. Finally, there had to be sufficient fine adjustment of the frame and camera mount so that the camera was always level.

Because of the limitations restricting its size (especially the length of the legs) and weight, a compromise had to be accepted over its accuracy. This comes down in a large part to the choice of lens. It would have been better to have used stronger and therefore more stable materials to build a taller frame, thus thus placing the camera at a greater height above the ground. This in turn would have enabled each 2m grid square to be covered by four shots taken with a 50 mm lens. As the minimum height of the camera above the ground surface was only 2 m, it was necessary to use a 35mm lens. This is not wholly satisfactory, as a wide angle lens distorts the shape of objects on the edge of each photograph, which then have to be corrected when drawings are made from them. In most cases, however, this was not difficult to do, and in others, a 2m square was covered by 6 shots taken by a 50 mm lens.

There was little difficulty in making the Quadrupod once materials available for its construction were found. The only specialised equipment required was a lathe

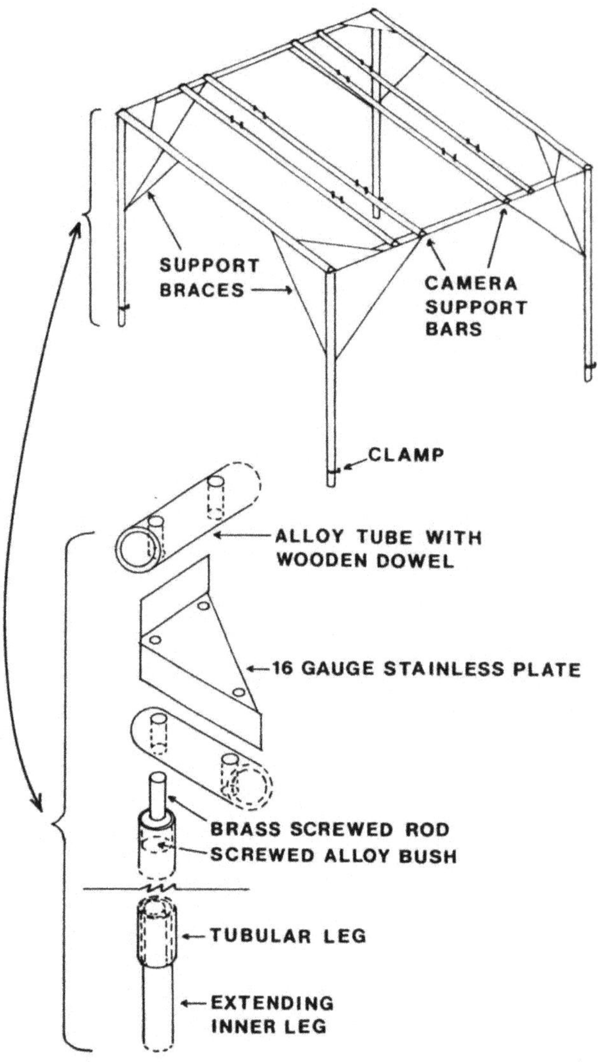

Figure H.1: Plan of quadrupod photographic frame

to turn the bushes for the top of the legs, and a jig had to be made from scrap material to ensure the holes in the top section were in line and squared up with all the others. The materials used were mostly re-cycled waste. The main frame poles and corner braces were constructed from a frame for a discarded caravan awning. The corner plates were from an off-cut of 16 guage stainless sheet steel. The extension poles for the feet were lengths of old tent poles. The only new parts were the cycle push-clamps (quick-release types) that clamped the legs when levelling the frame, and the nuts and bolts. The quadrupod frame is illustrated in Fig. H.1. The camera mount was improvised from an off-cut of plywood, with a small wooden cradle lined with strips of foam rubber for the camera (Fig. H.2). The mount was attached to the support bars by coach bolts that could be adjusted to level the camera in the same way as with more expensive versions used to level hi-fi systems. The total cost (1982 prices) was under 50 pounds sterling.

When used on the site, the leg that was positioned on the highest corner of a

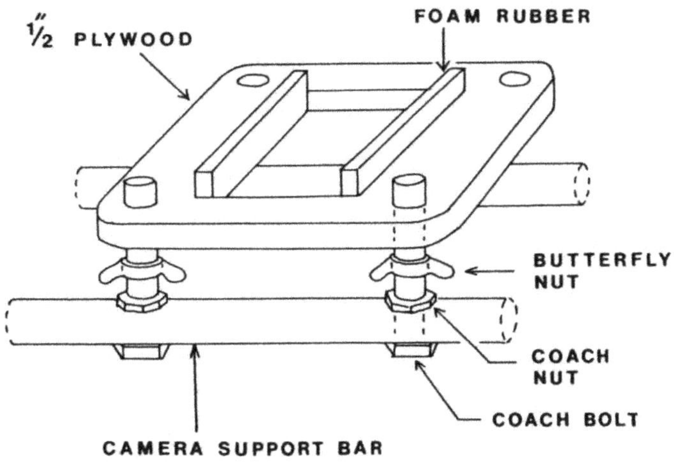

Figure H.2: Camera mount of quadrupod

particular grid square was always kept two metres long. Those further downslope were then extended until the camera support bars were horizontal. A spirit level was used to check that the legs were vertical and the cross bars horizontal. Finer levelling was then carried out on the camera mount by adjusting the legs that attached the mount to the frame. A small spirit level on the back of the camera ensured that the camera itself was level. Once level, the lens was focussed from underneath. As small a camera stop as possible was used to give maximum depth of field, thereby ensuring that everything was in focus. Shutter exposure was estimated by a hand held exposure metre, held pointing to the ground and near the camera. Once set, the shutter was released with an extension cable, triggered by a hand-held air-bulb. A letter-board of the type easily obtainable in most commercial stationers was used to record the site number, date and grid square for each shot.

It was found best to photograph several grid squares in each session so that a complete film could be developed. Photographs were developed and printed either locally in Rawalpindi, or by the Archaeology Department's own photographer. Each print was enlarged to 7" by 5". These could be overlain with tracing paper so that each stone could be drawn. The location of artefacts, the flaking of each stone, and any other features could then be recorded with a simple coding system. Afterwards, the photographs could be cut and joined (see Fig. 8.9 for an example), and these were used as the basis of the final site plan.

Overall, the Quadrupod amply satisfied its design requirements. The recording of 300 sq m of stone surface did not interrupt the course of the excavation to anything like the same extent as had the site been grided out and drawn in the

Figure H.3: Quadrupod photo-frame in use

usual way. It was also far quicker: every stone within an area of 4 sq m could be recorded in an hour or so. Finally, it paid for itself, since the alternative would have been to employ an additional draughts-person on the site for eight weeks. It is attractive to a wide range of excavations, as the basic capital outlay is less than that for a couple of wheel- barrows, or (in British terms) a few days subsistence costs for a semi-skilled digger, or the cost of hiring a photographic tower for a week-end. The Quadrupod's advantages of size, cost, simplicity and weight more than compensated for the slight loss of accuracy caused by the need to rely on a 35 mm lens. Fig. H.3 shows the quadrpod in use.

References:

Andersen, R.C. (1982) Photogrammetry: pros and cons for archaeology. *World Archaeology* **14 (2)**: 200-205.

Fussell, A. (1982) Terrestial photogrammetry in archaeology. *World Archaeology* **14 (2)**: 157-172.

Appendix I

Riwat Site Plan, Figures I.1 - I.36

Fig. I.0 Key to site plan, Riwat Site 55

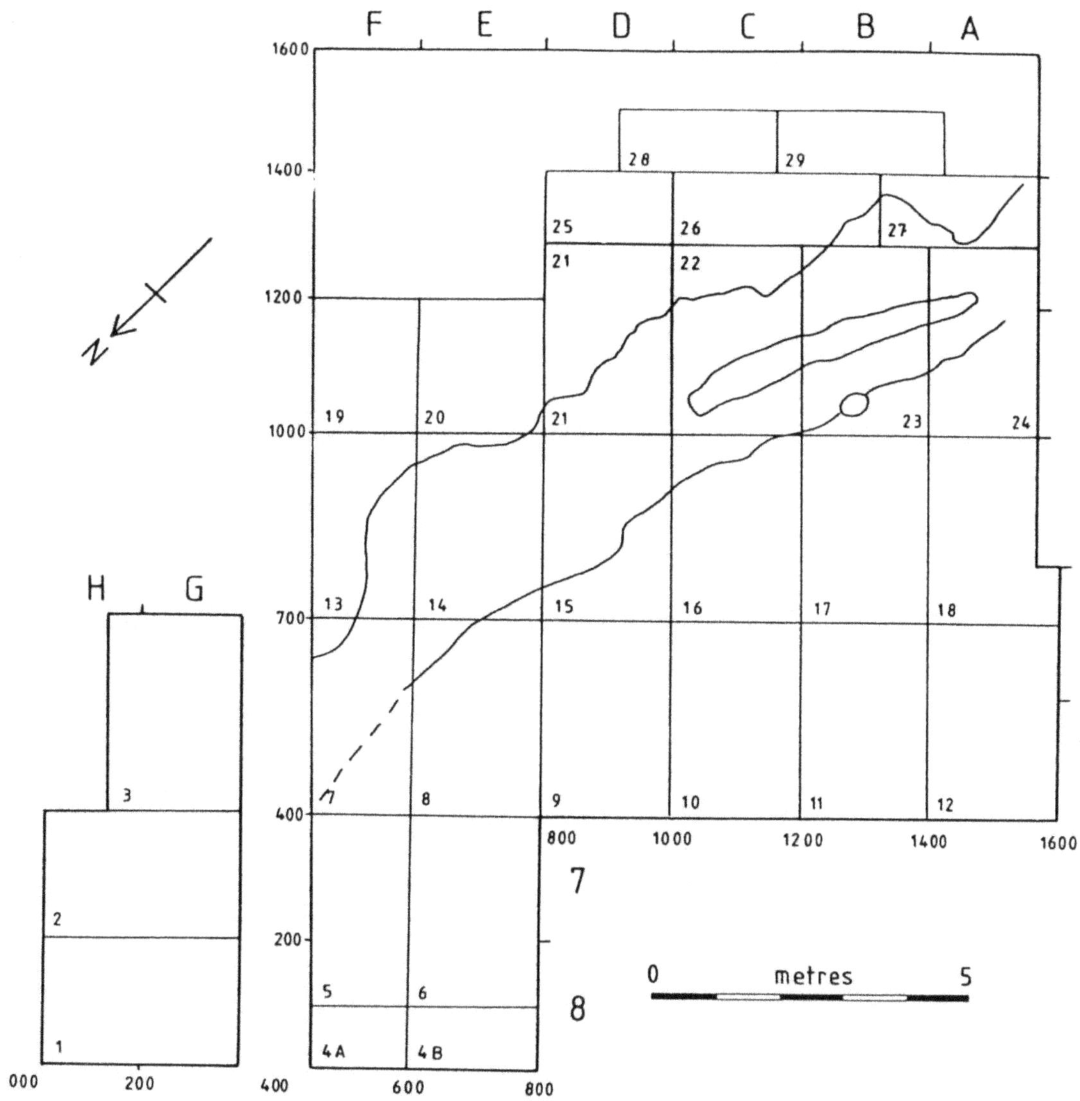

Fig. I.1 Site 55, Riwat: Squares G-H/8.

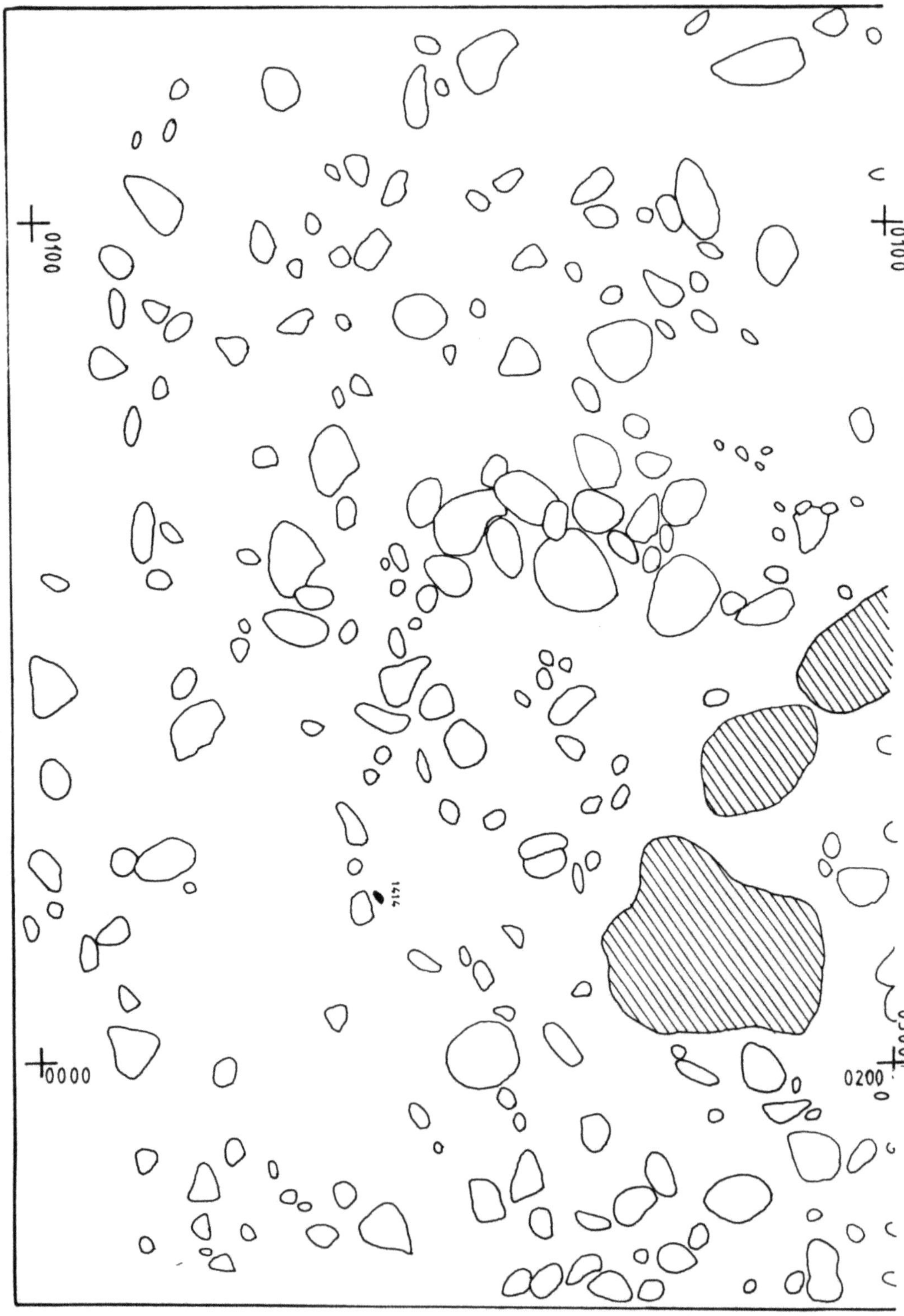

Fig. I.2 Site 55, Riwat: Squares G-H/7.

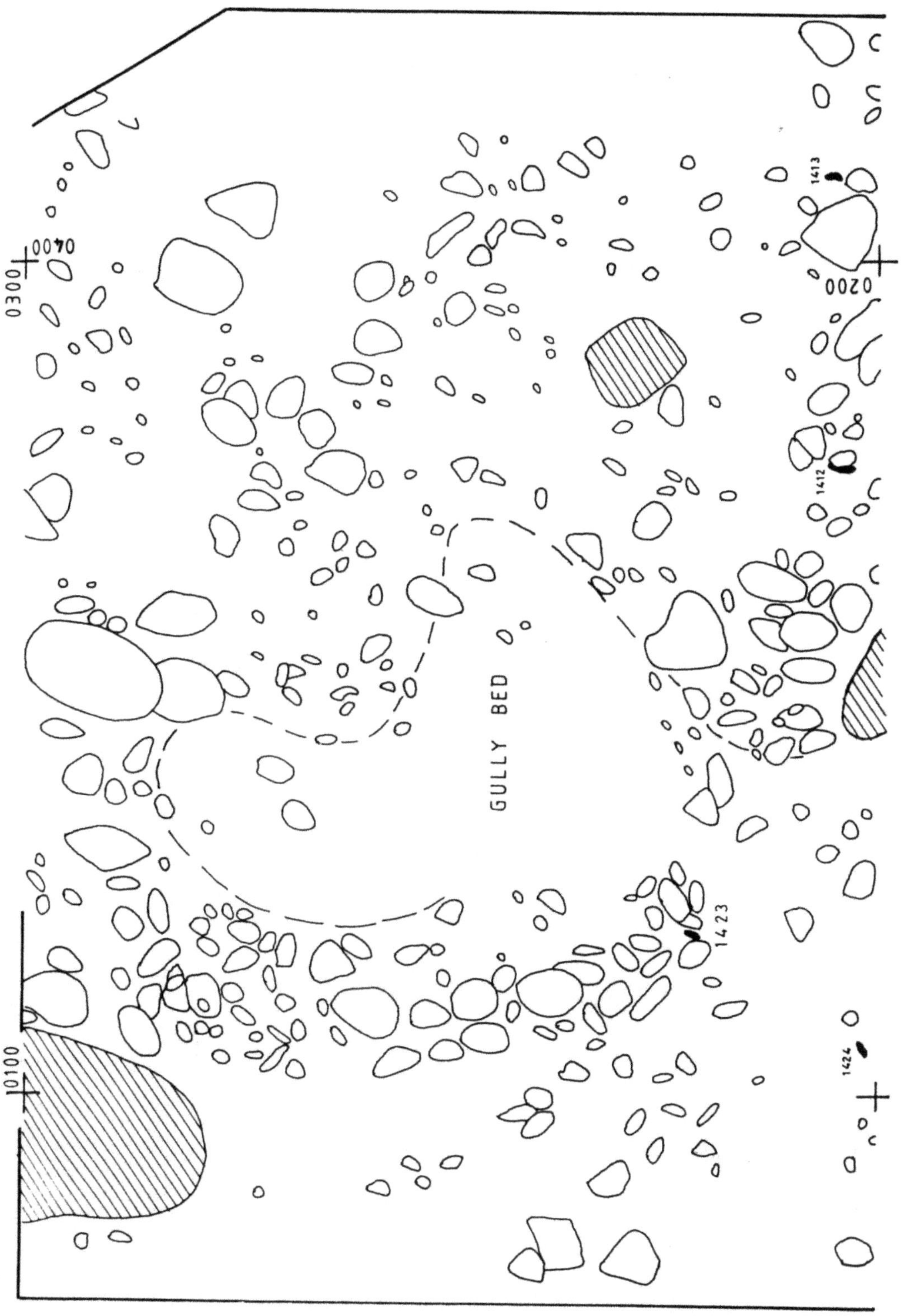

Fig. I.3 Site 55, Riwat: Squares G/5-6.

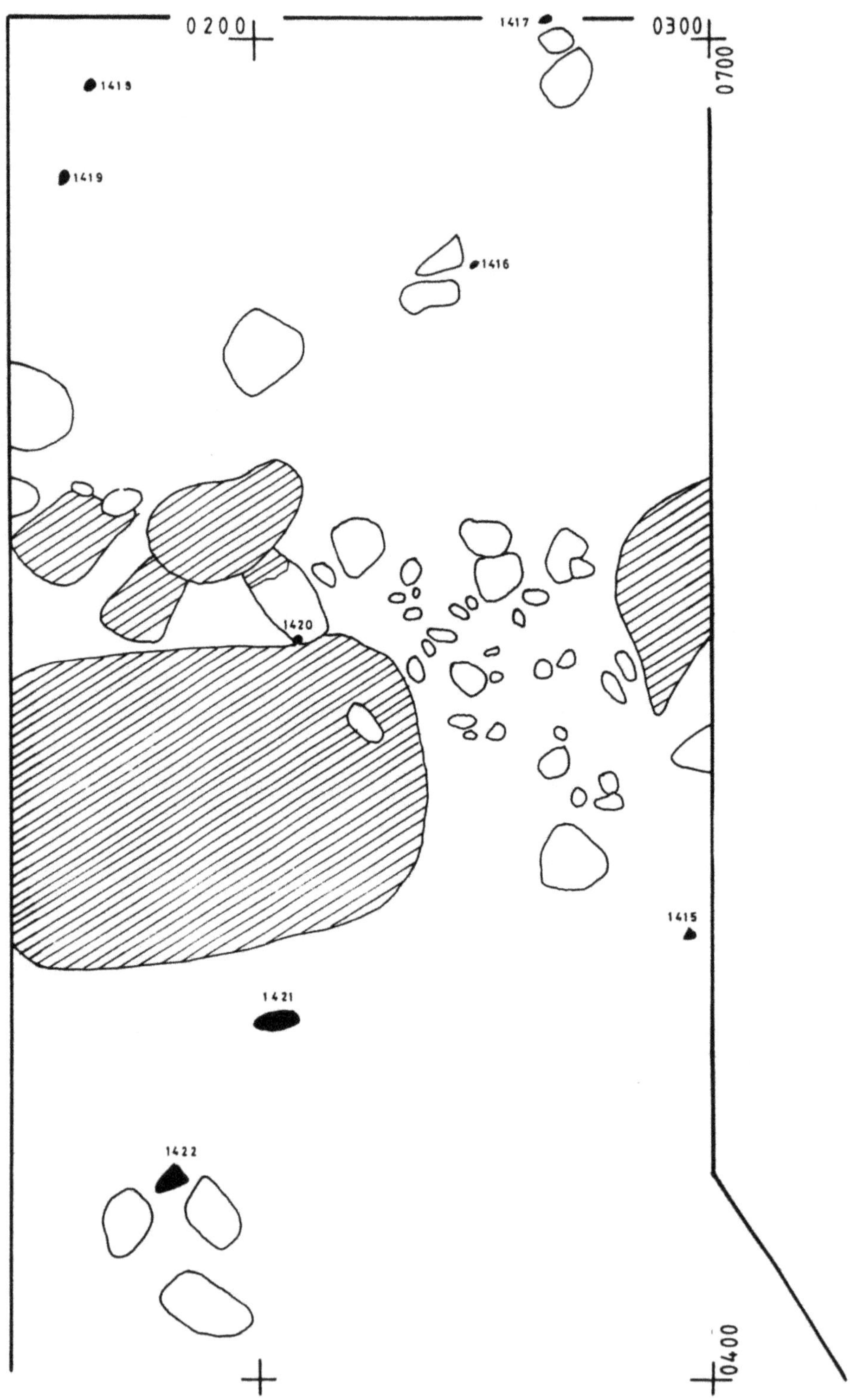

Fig. I.4 Site 55, Riwat: Squares F8 (4A) and E8 (4B).

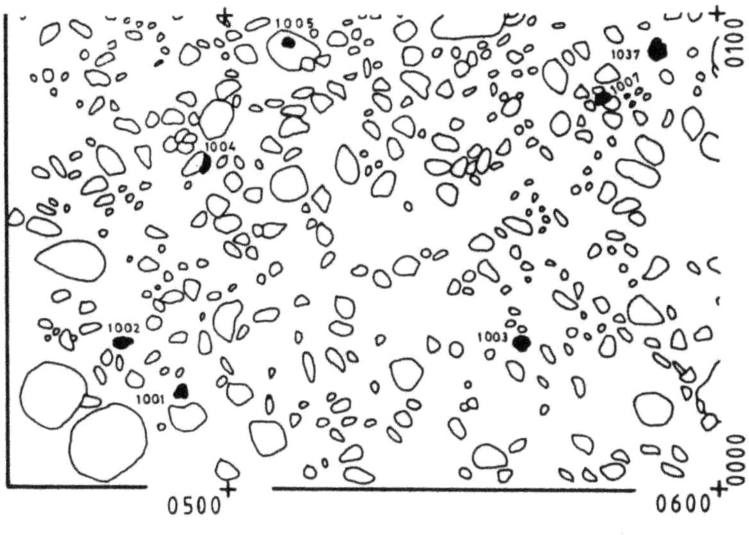

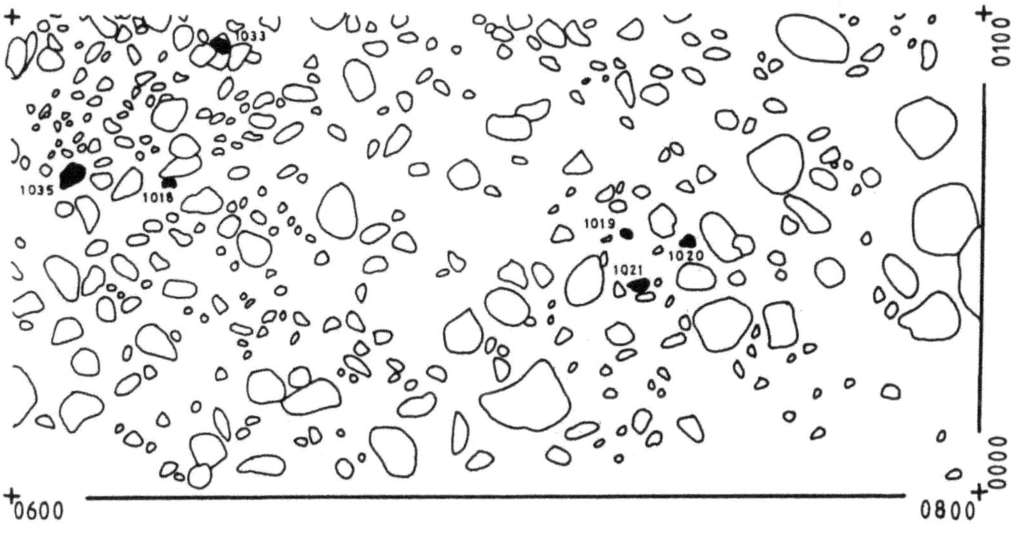

Fig. I.5 Site 55, Riwat: Squares F7-8.

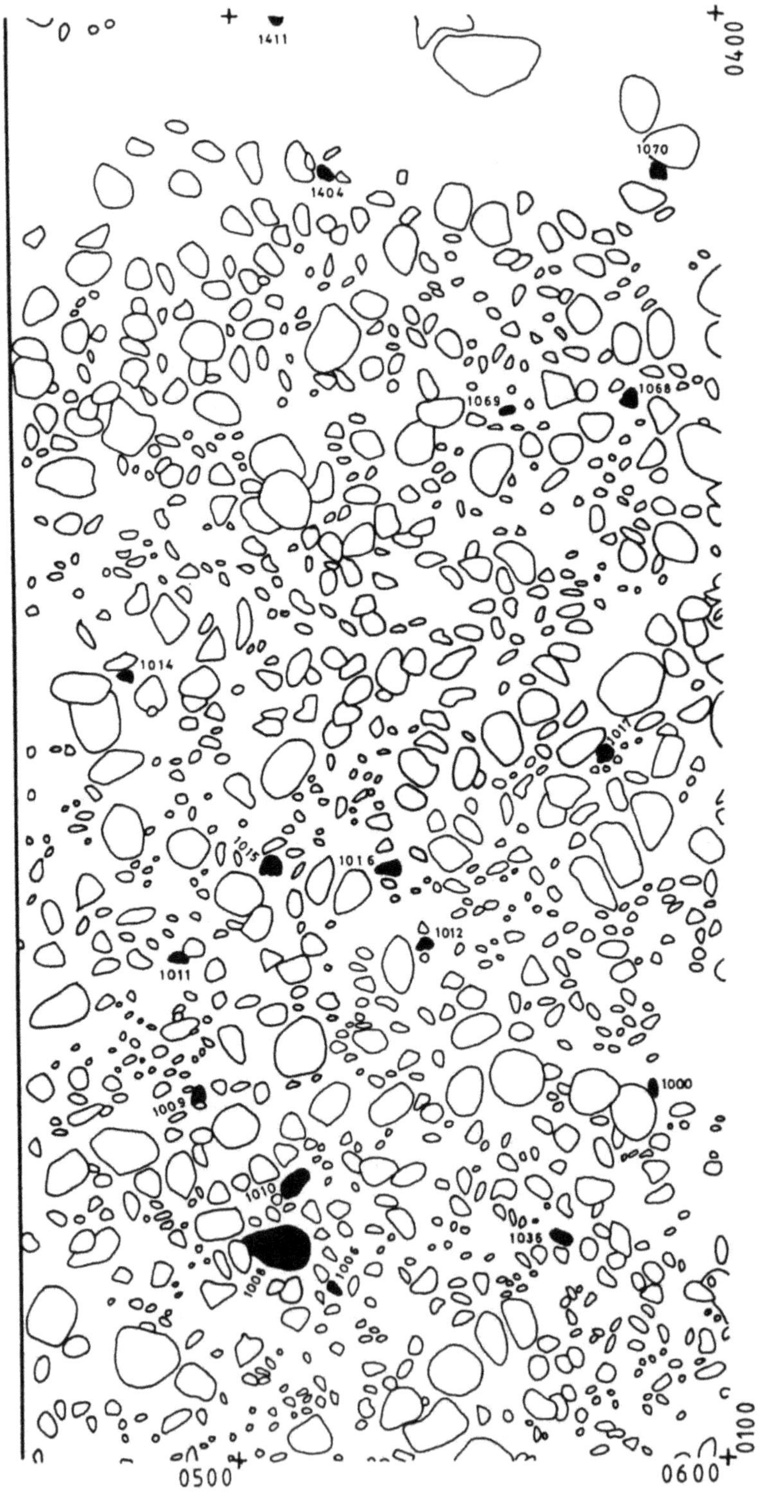

Fig. I.6 Site 55, Riwat: Squares E7-8.

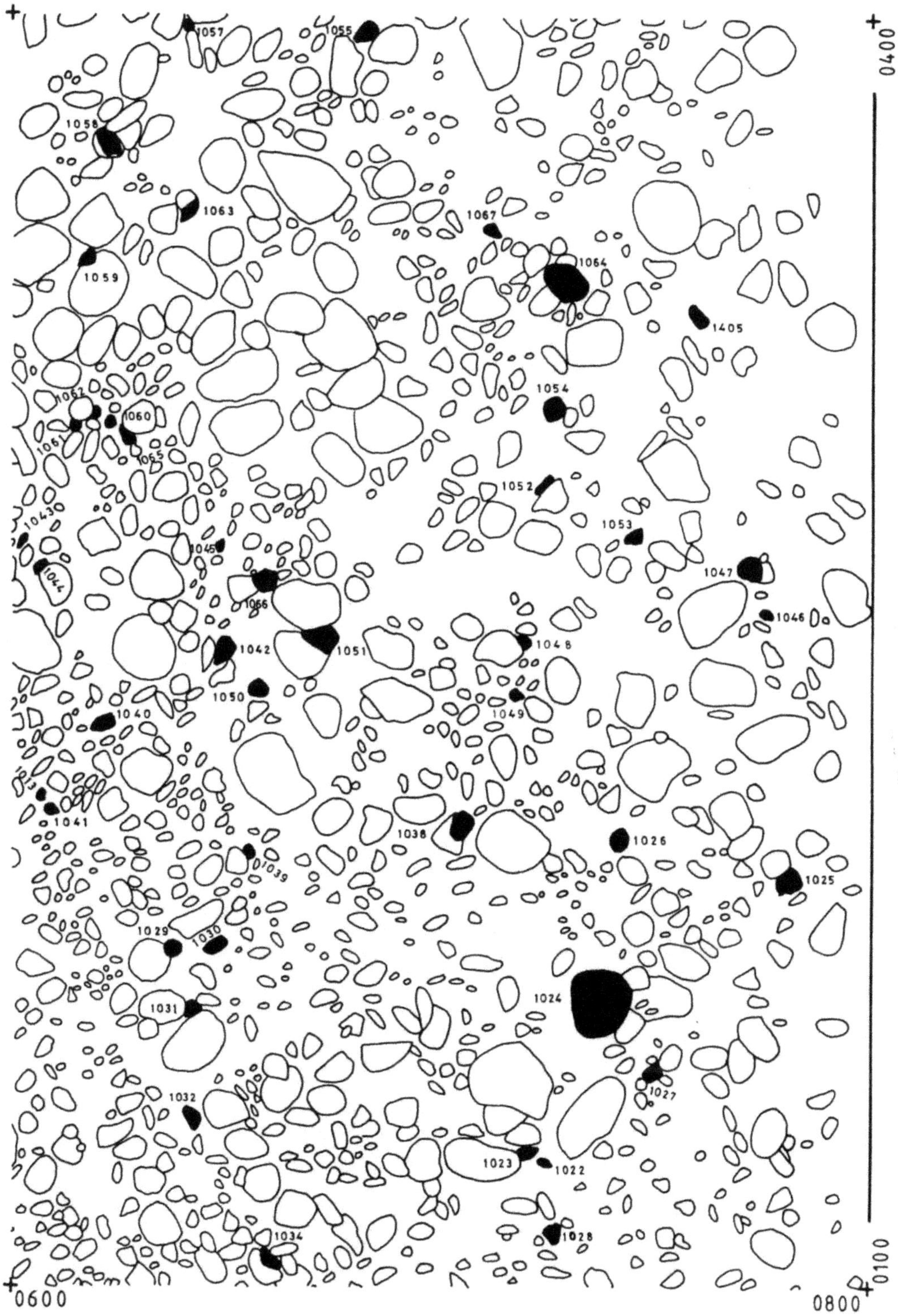

Fig. I.7 Site 55, Riwat: Squares F5-6.

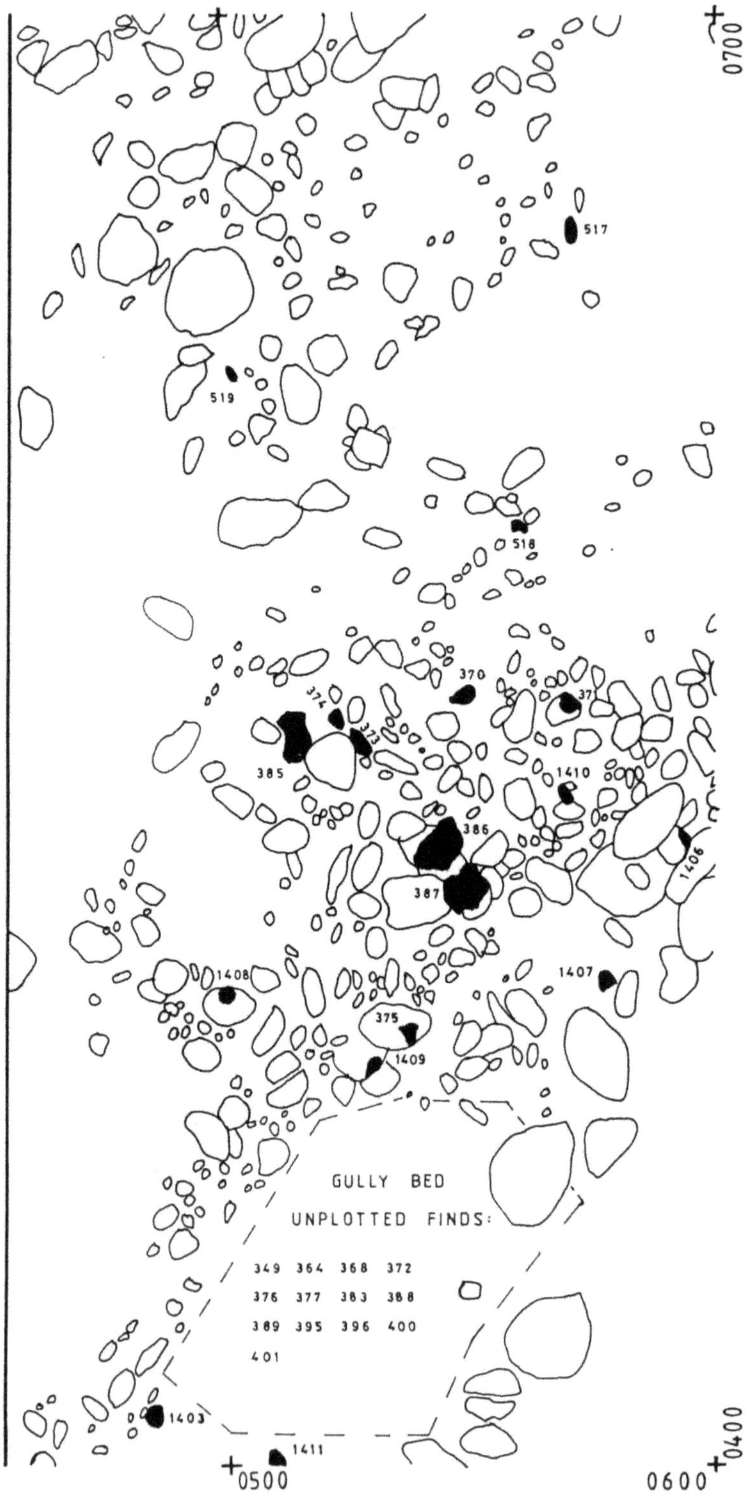

Fig. I.8 Site 55, Riwat: Squares E5-6.

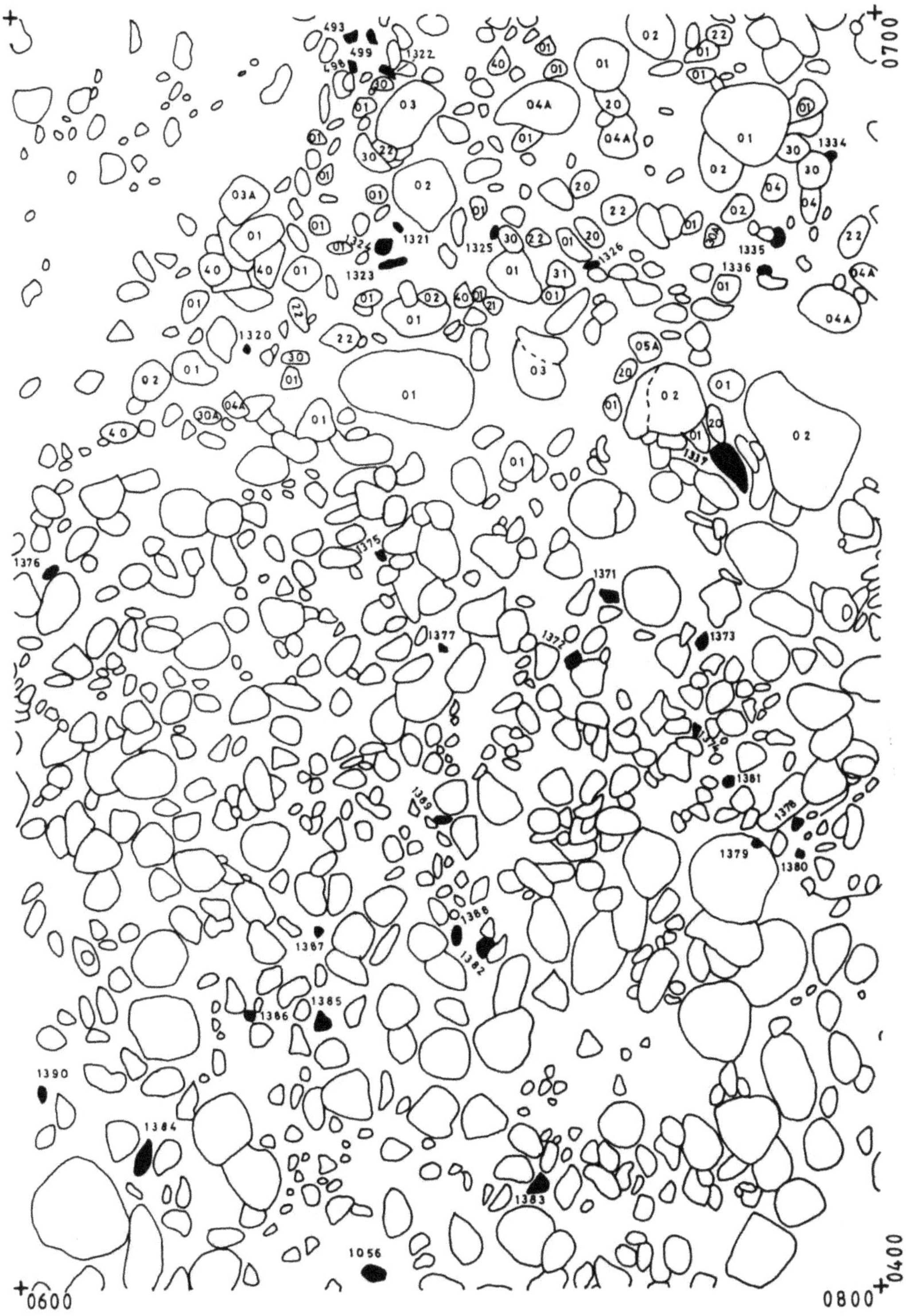

Fig. I.9 Site 55, Riwat: Squares D5-6.

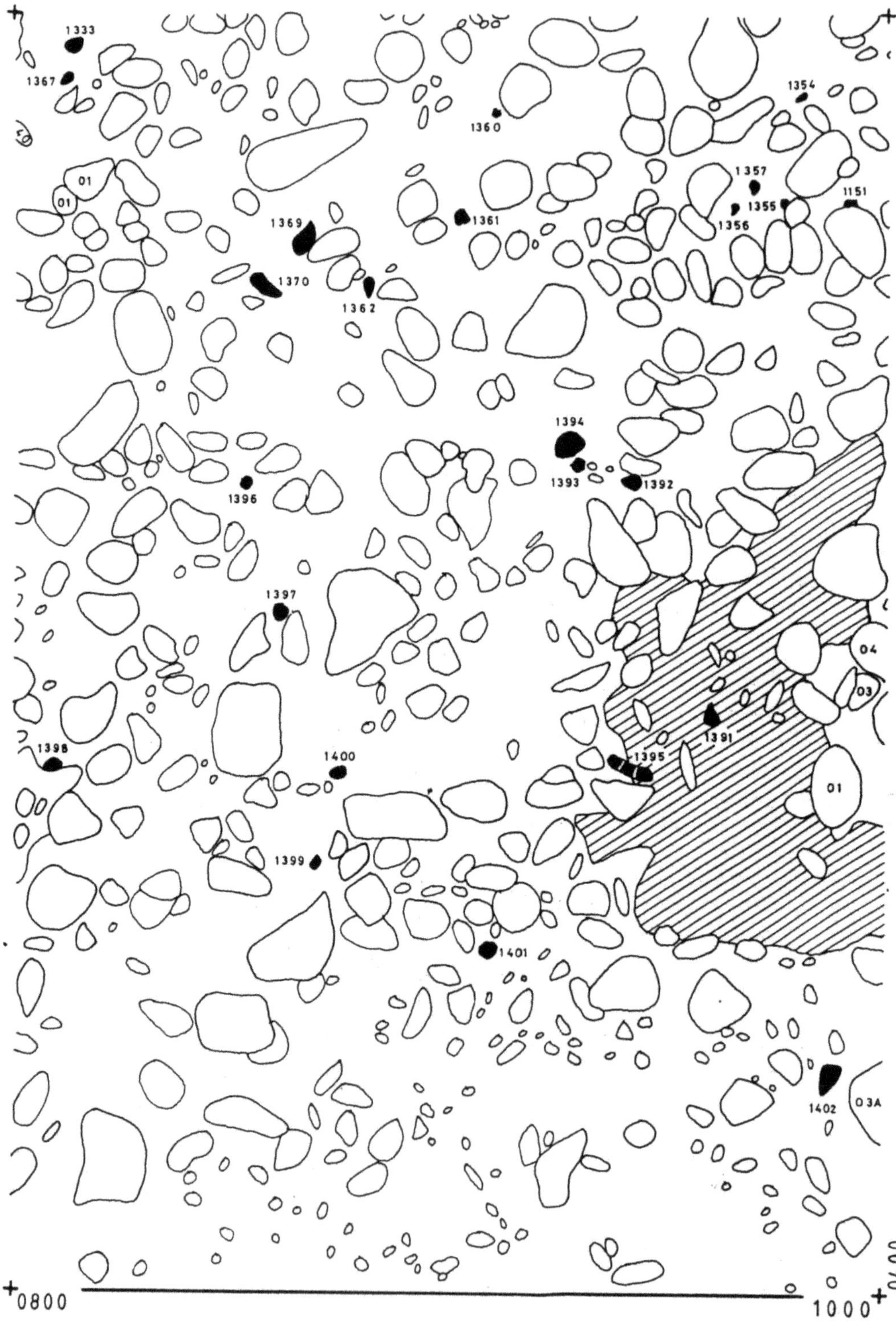

Fig. I.10 Site 55, Riwat: Squares C5-6.

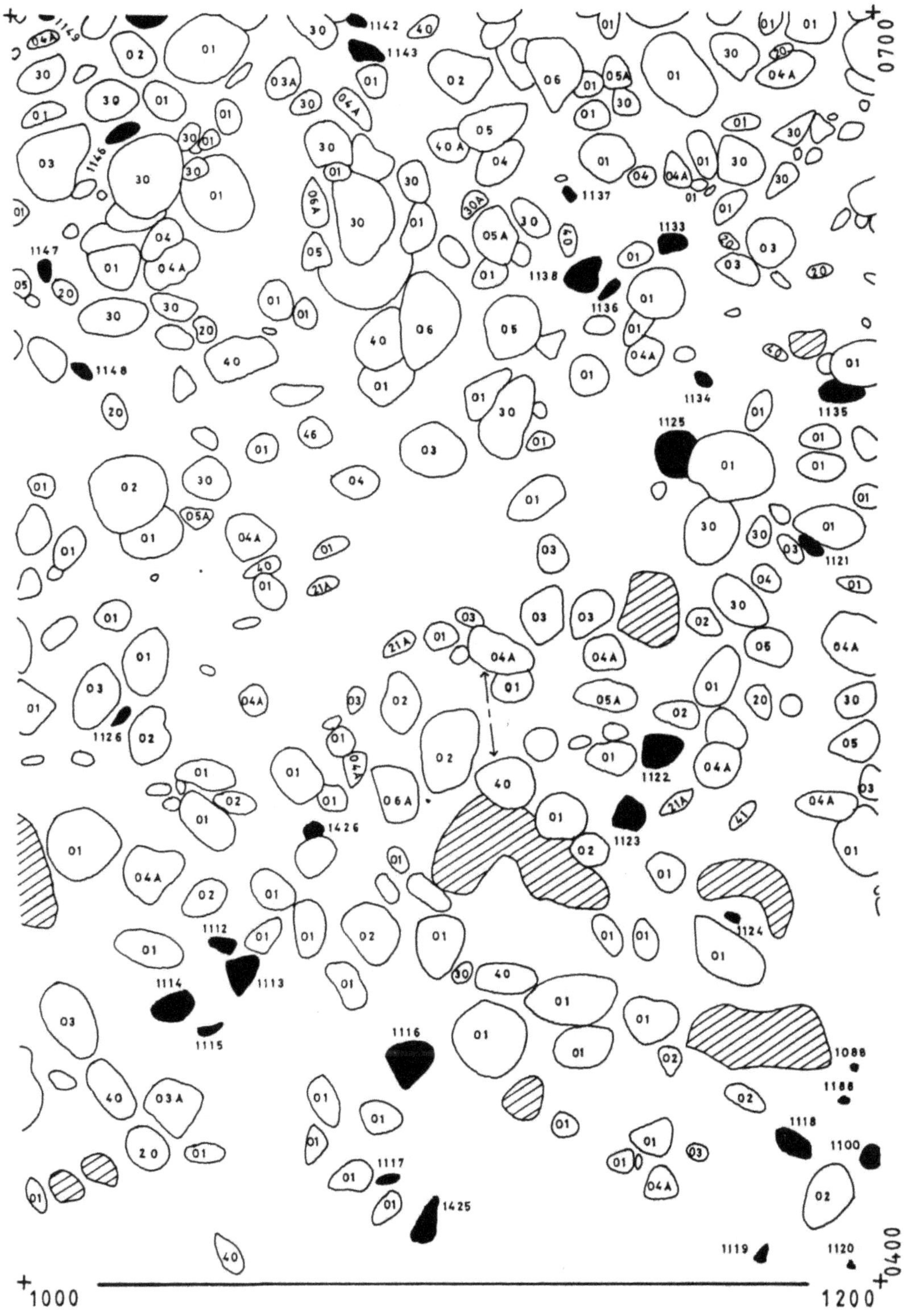

Fig. I.11 Site 55, Riwat: Squares B5-6.

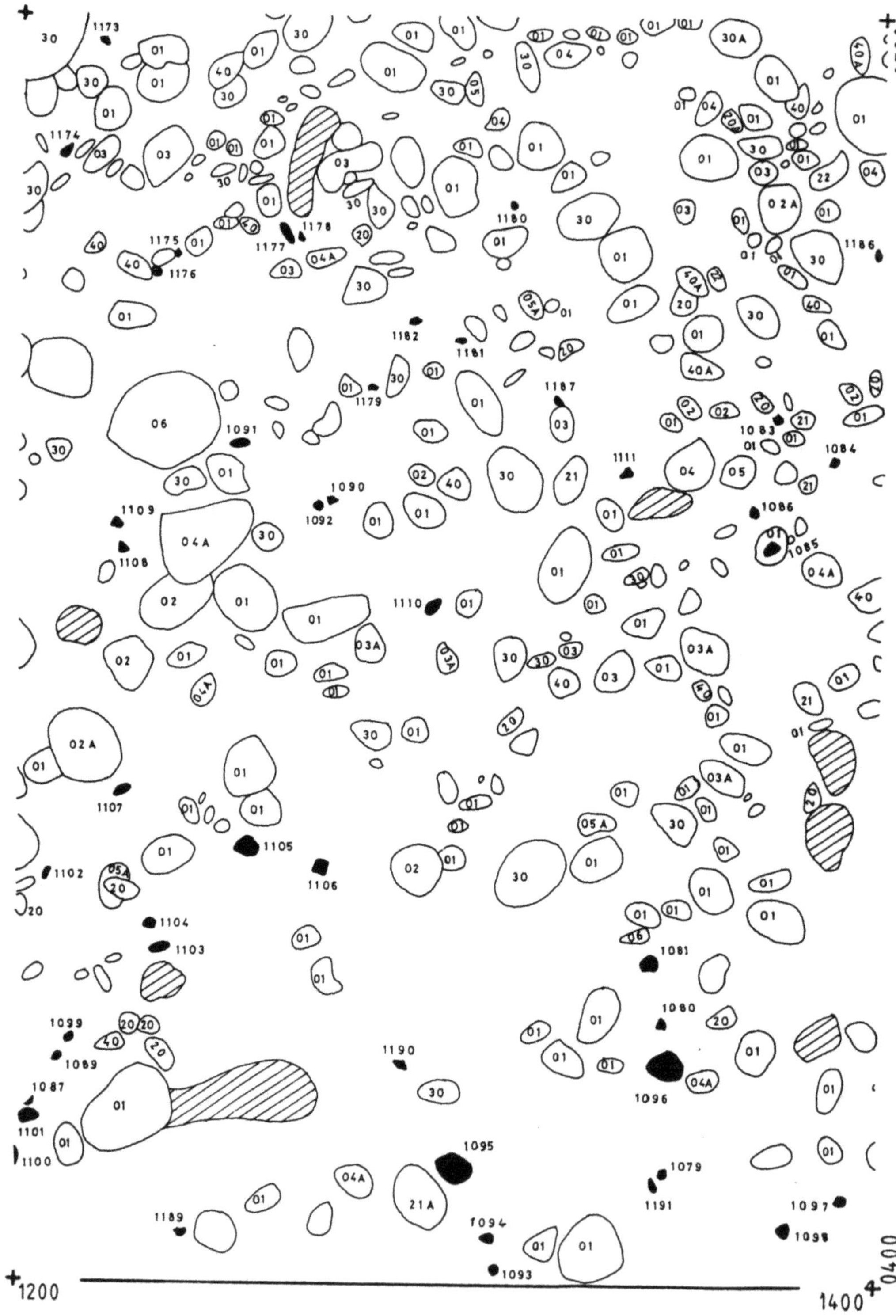

Fig. I.12 Site 55, Riwat: Squares A5-6.

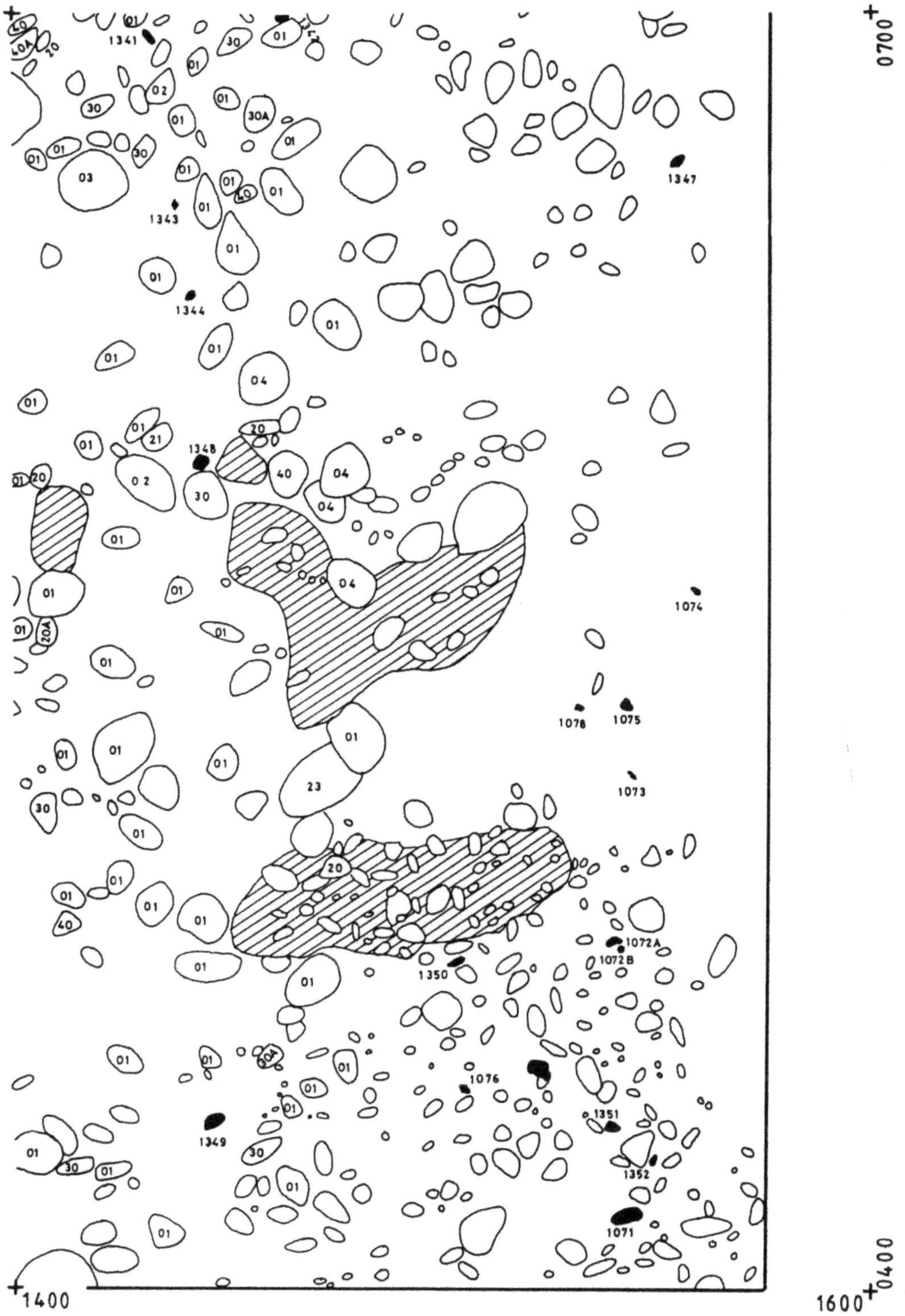

Fig. I.13 Site 55, Riwat: Squares F4-5.

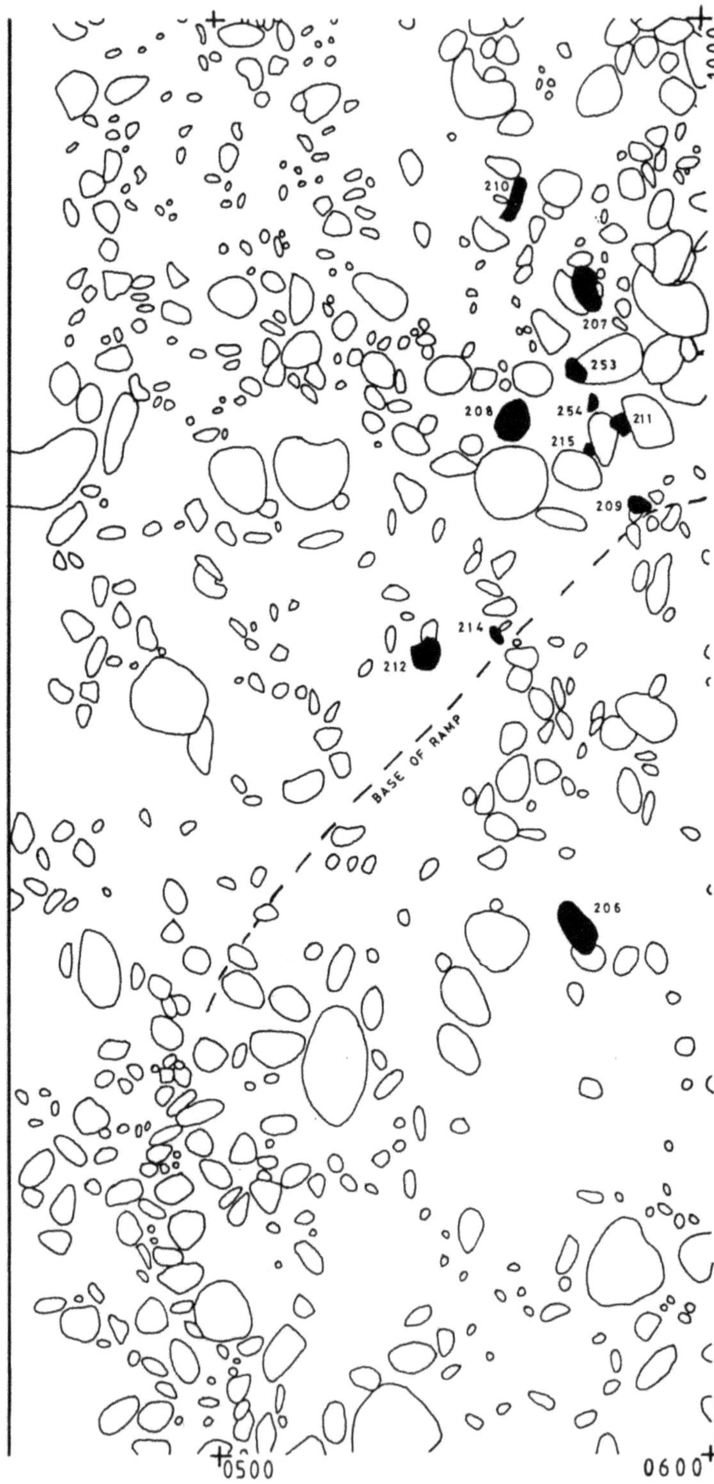

Fig. I.14 Site 55, Riwat: Squares E4-5.

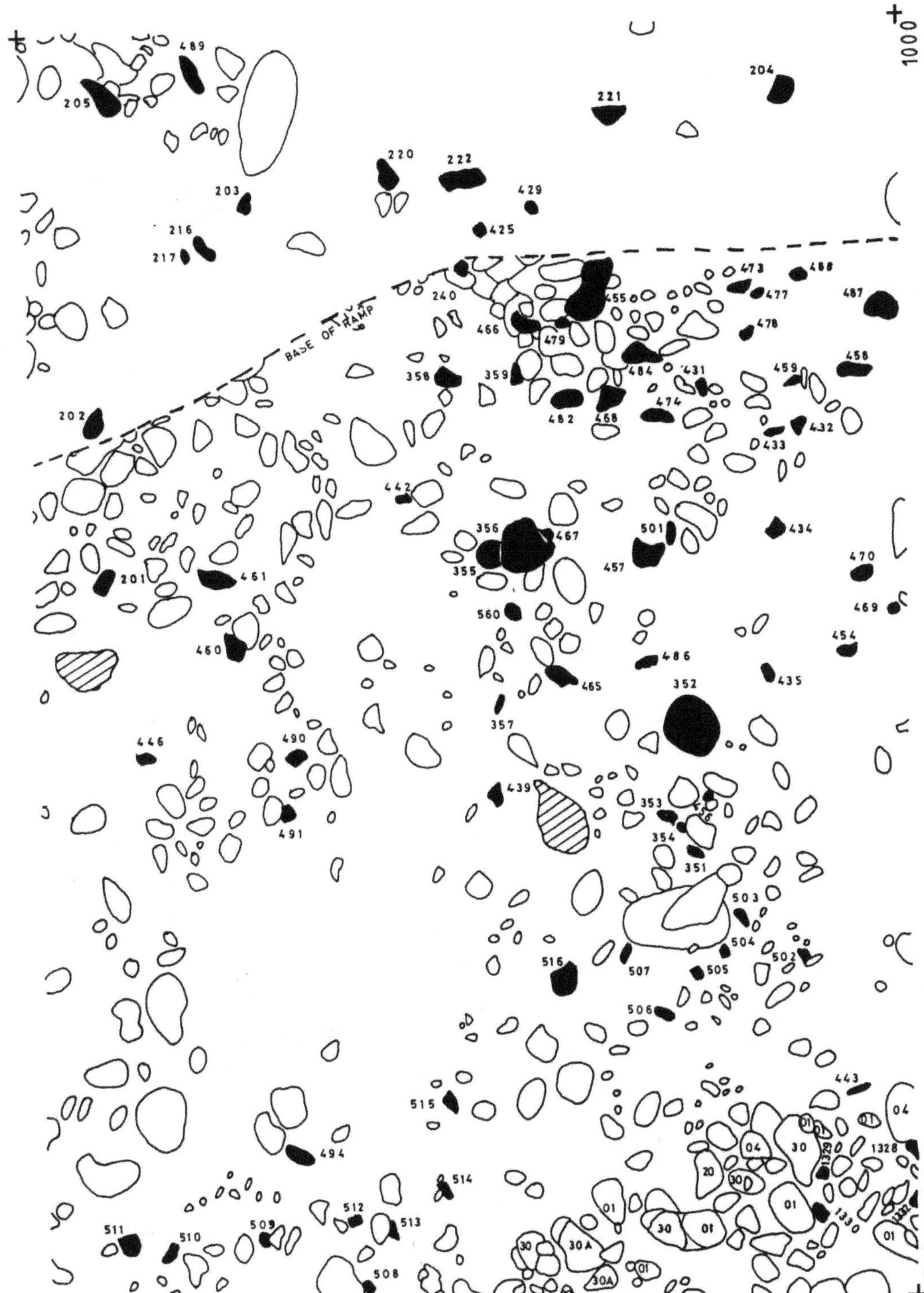

Fig. I.15 Site 55, Riwat: Squares D4-5.

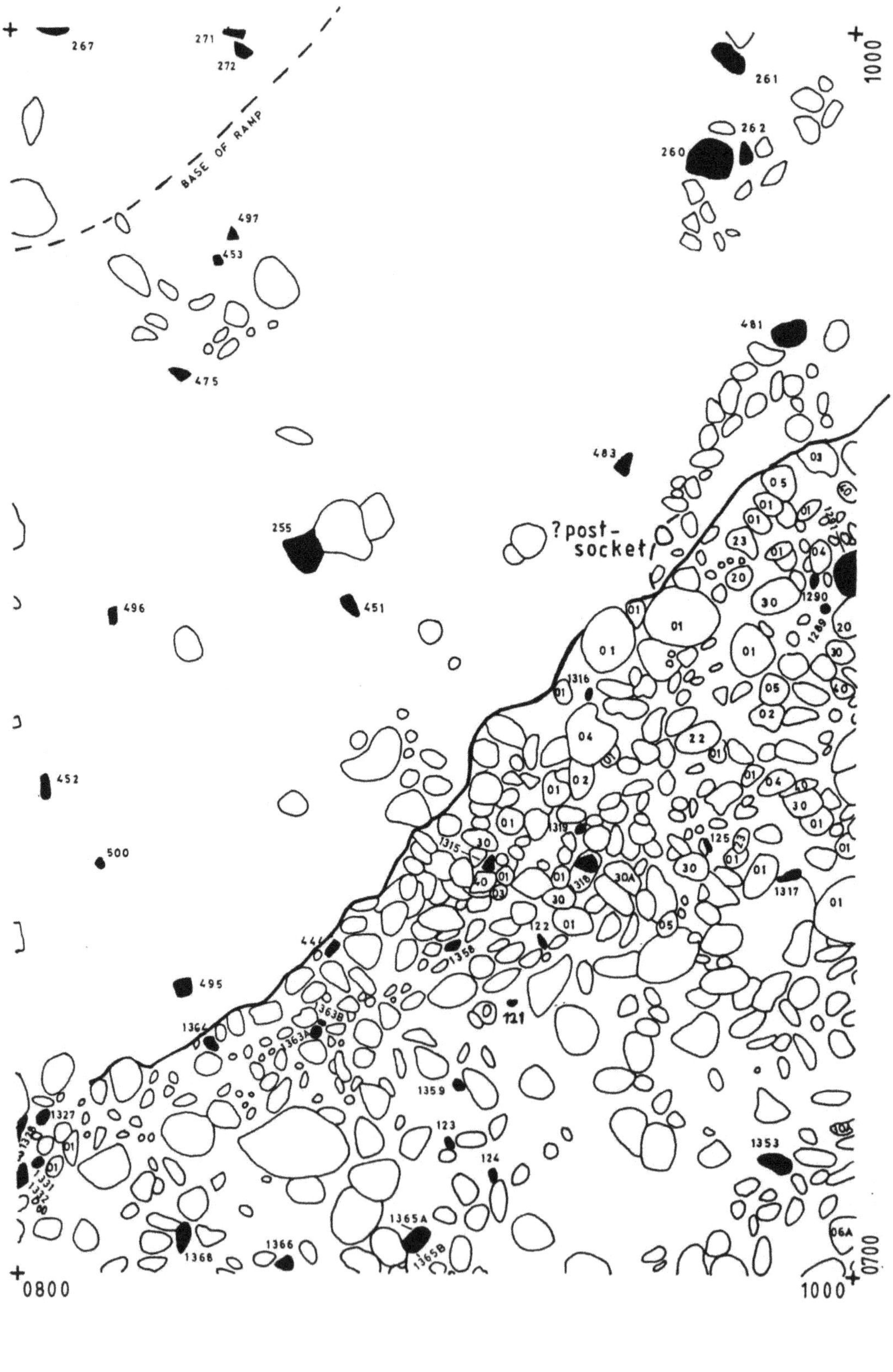

Fig. I.16 Site 55, Riwat: Squares C4-5.

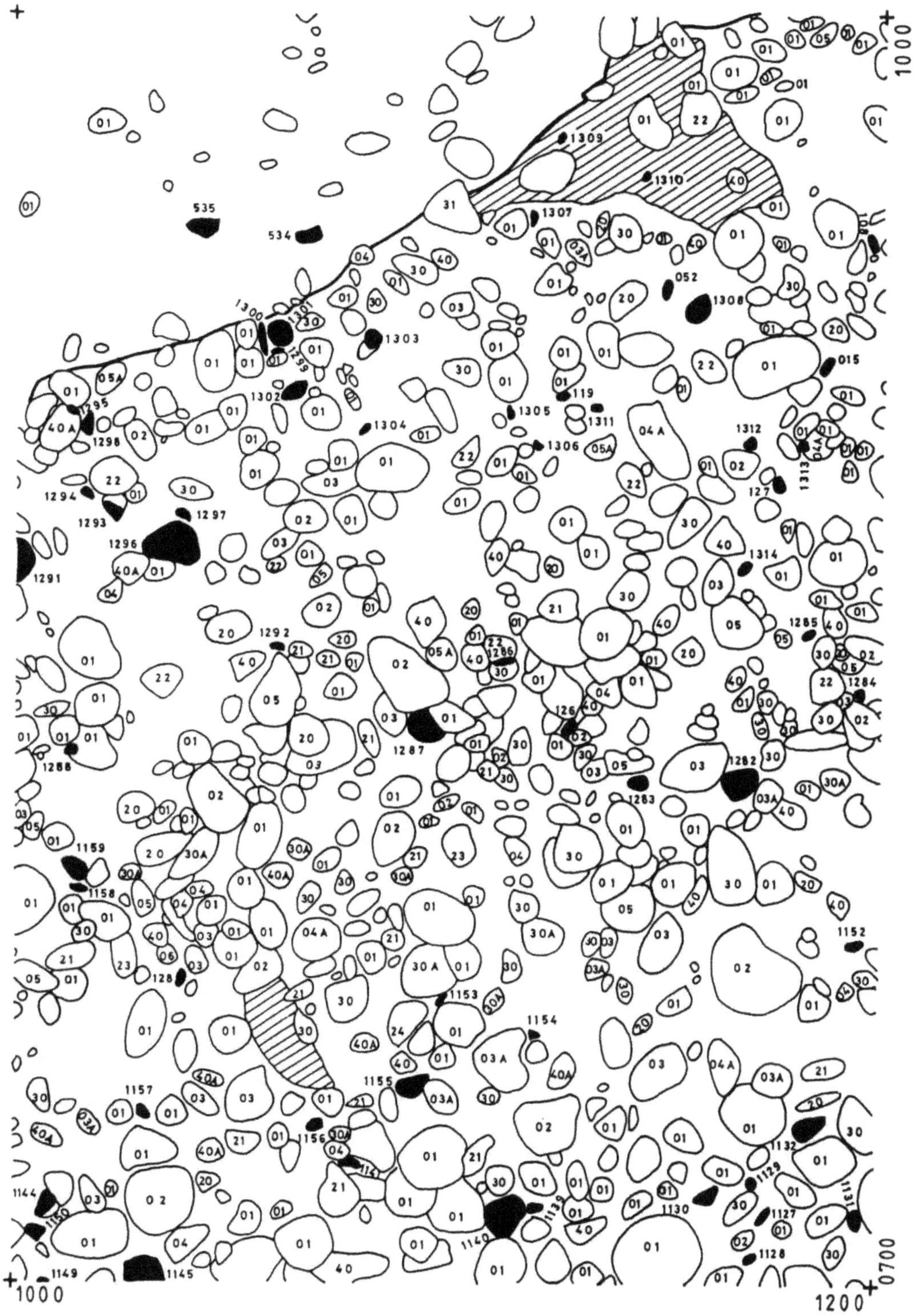

Fig. I.17 Site 55, Riwat: Squares B4-5.

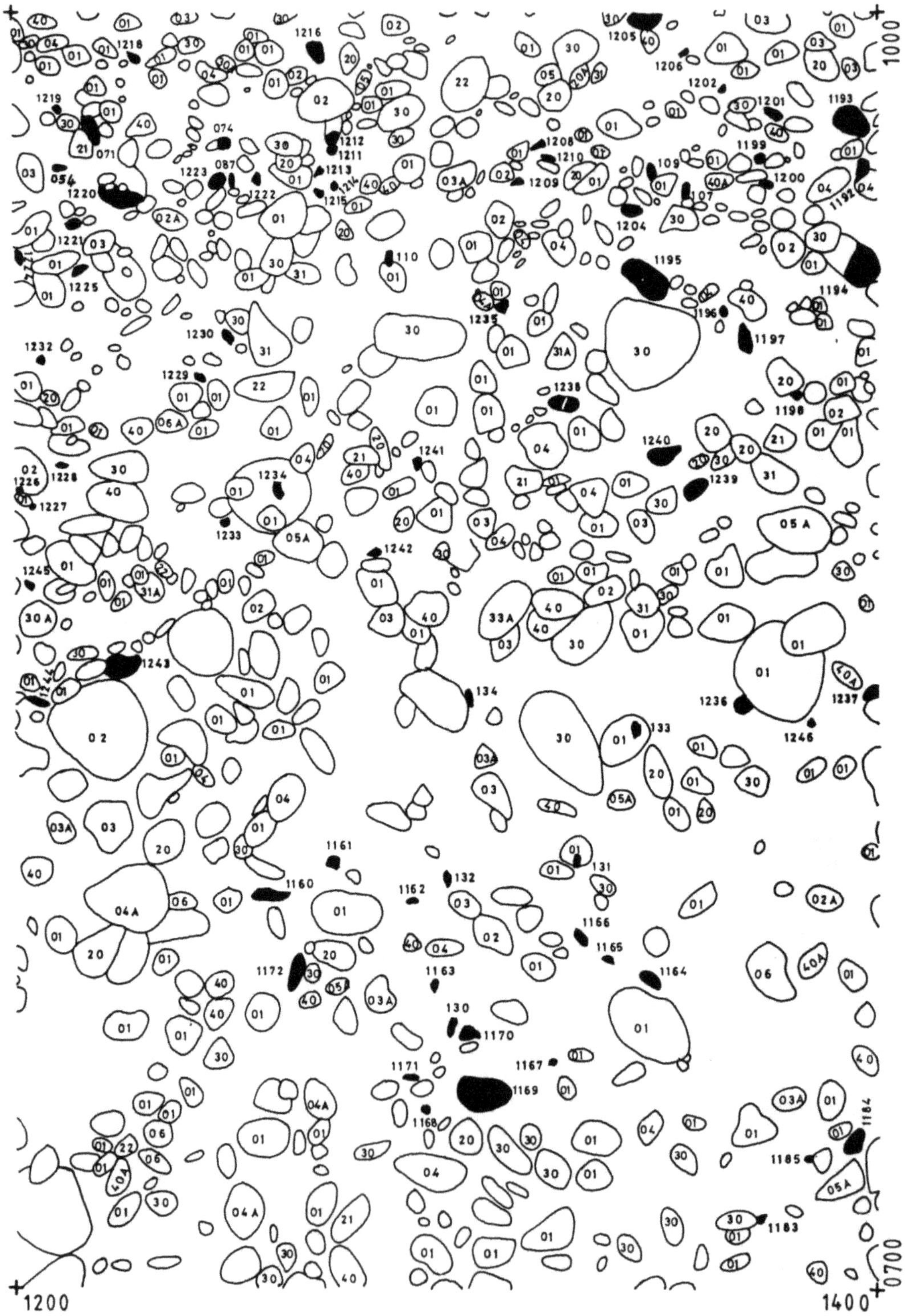

Fig. I.18 Site 55, Riwat: Squares A4-5.

Fig. I.19 Site 55, Riwat: Square F3.

Fig. I.20 Site 55, Riwat: Square E3.

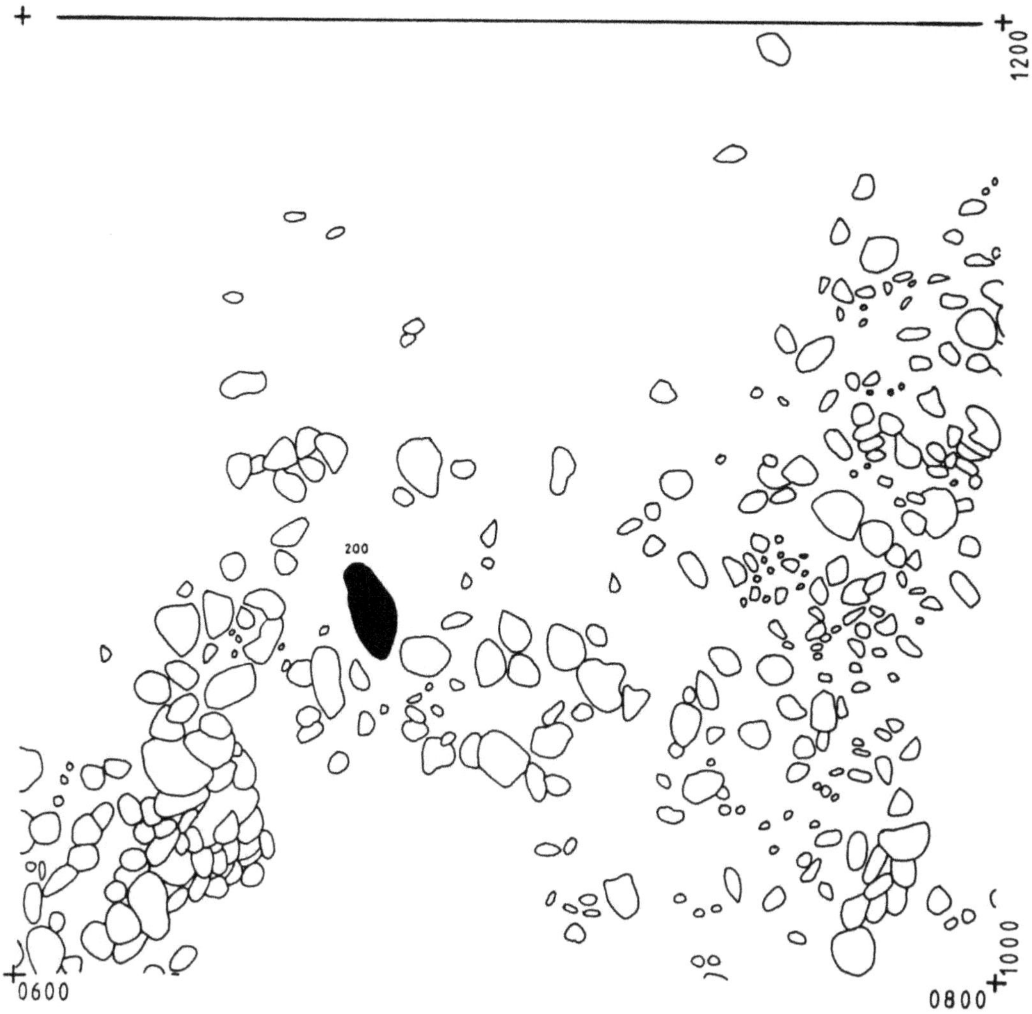

Fig. I.21 Site 55, Riwat: Squares D2-3.

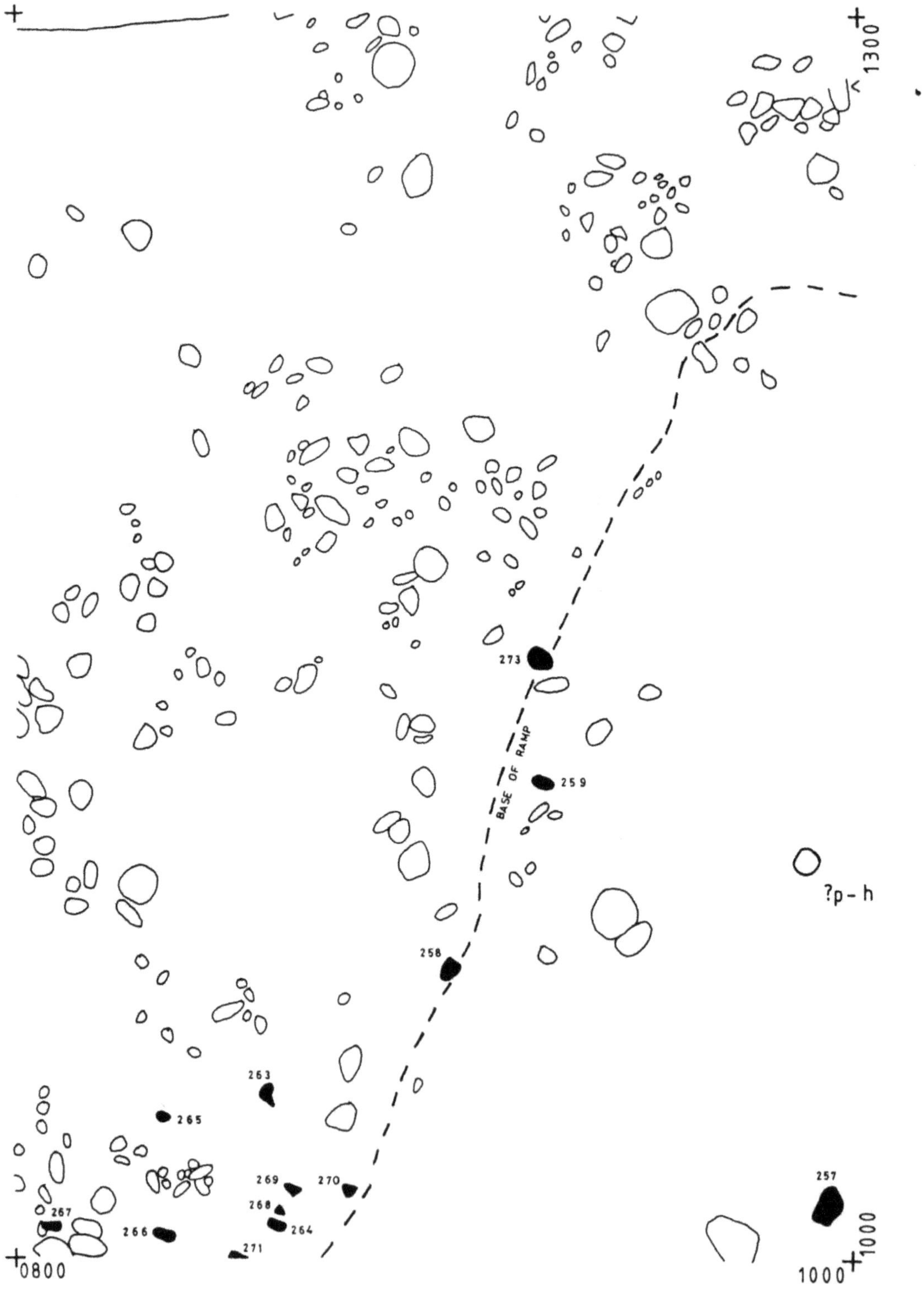

Fig. I.22 Site 55, Riwat: Squares C2-3.

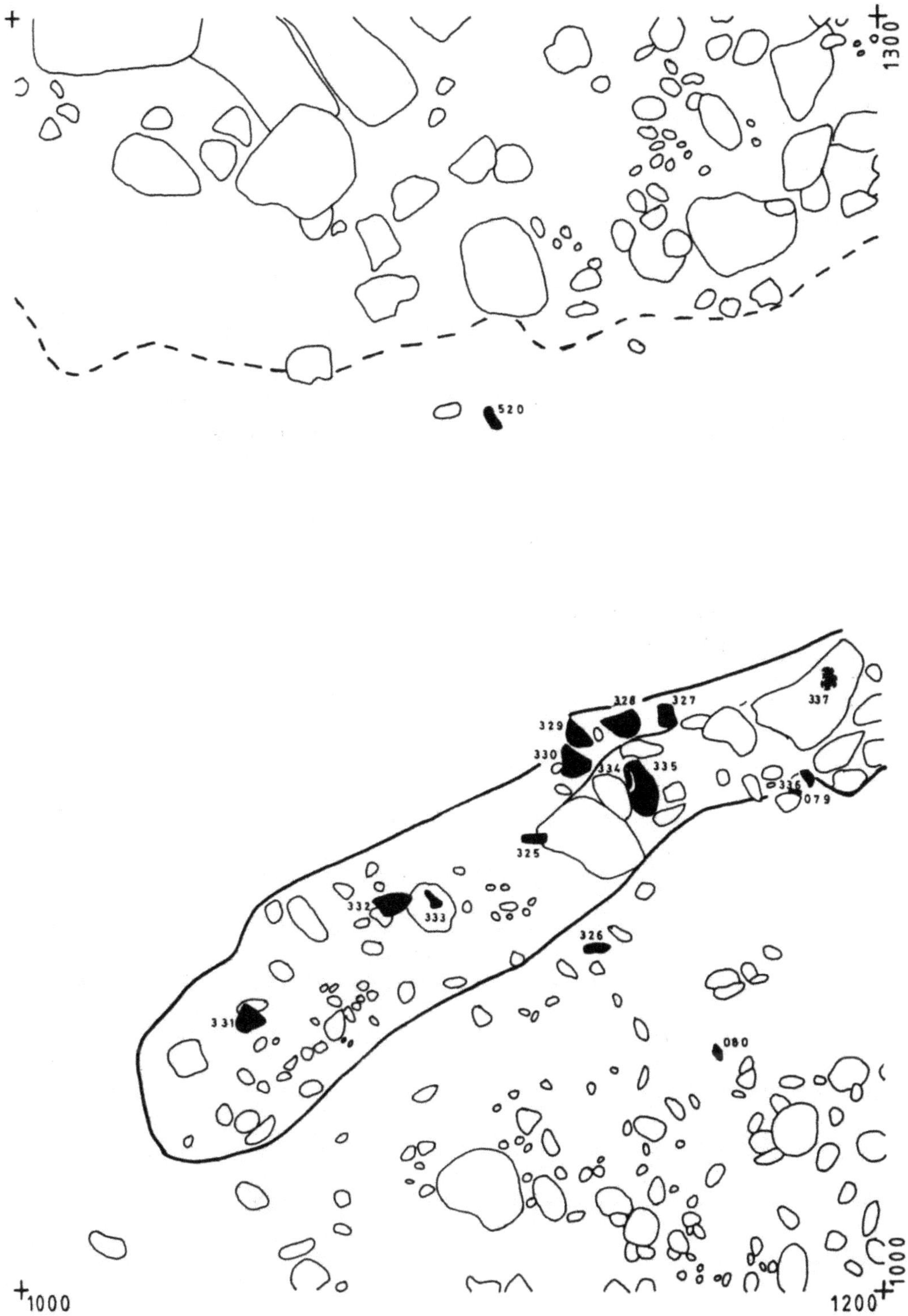

Fig. I.23 Site 55, Riwat: Squares B2-3.

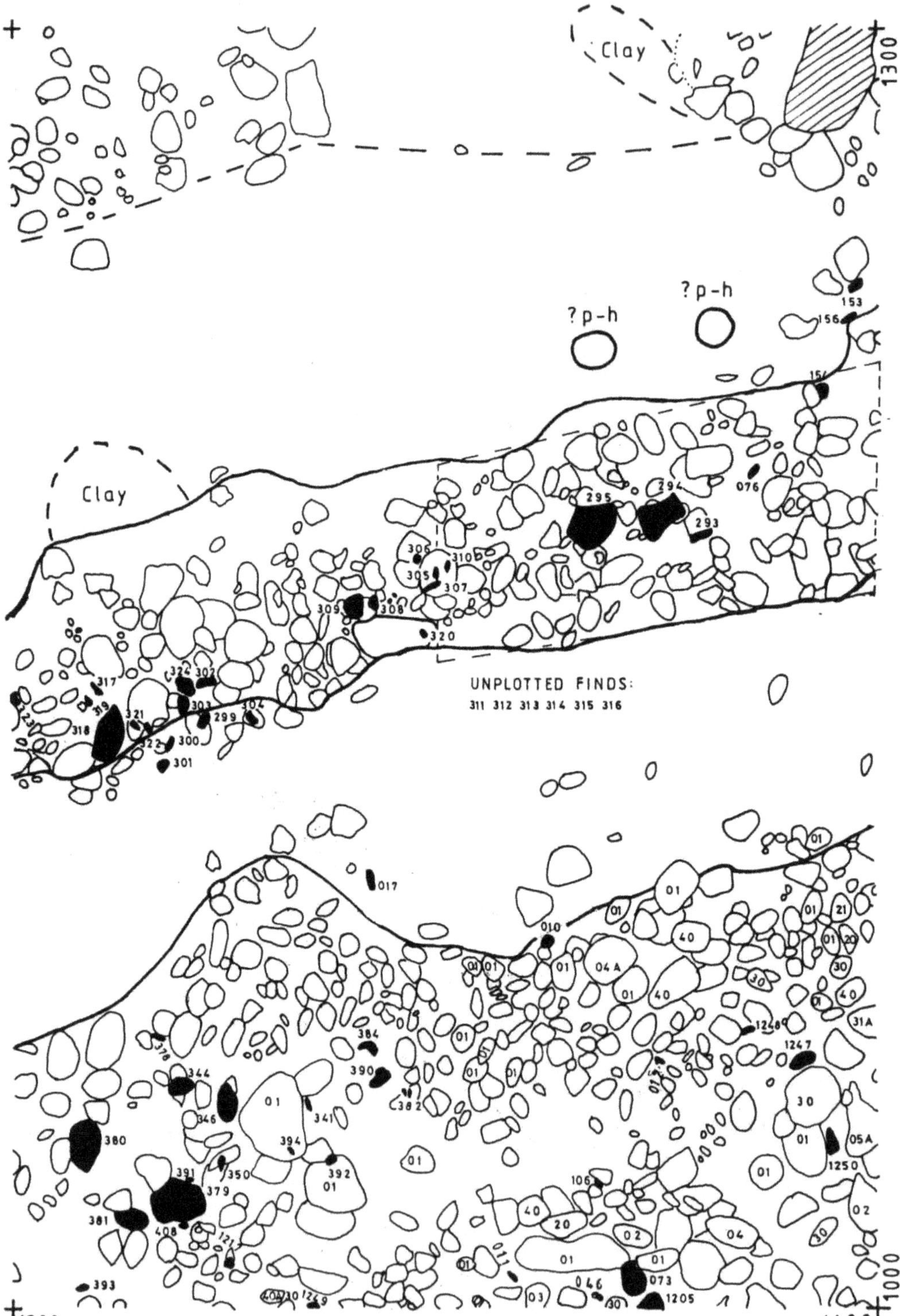

Fig. I.24 Site 55, Riwat: Squares A2-3.

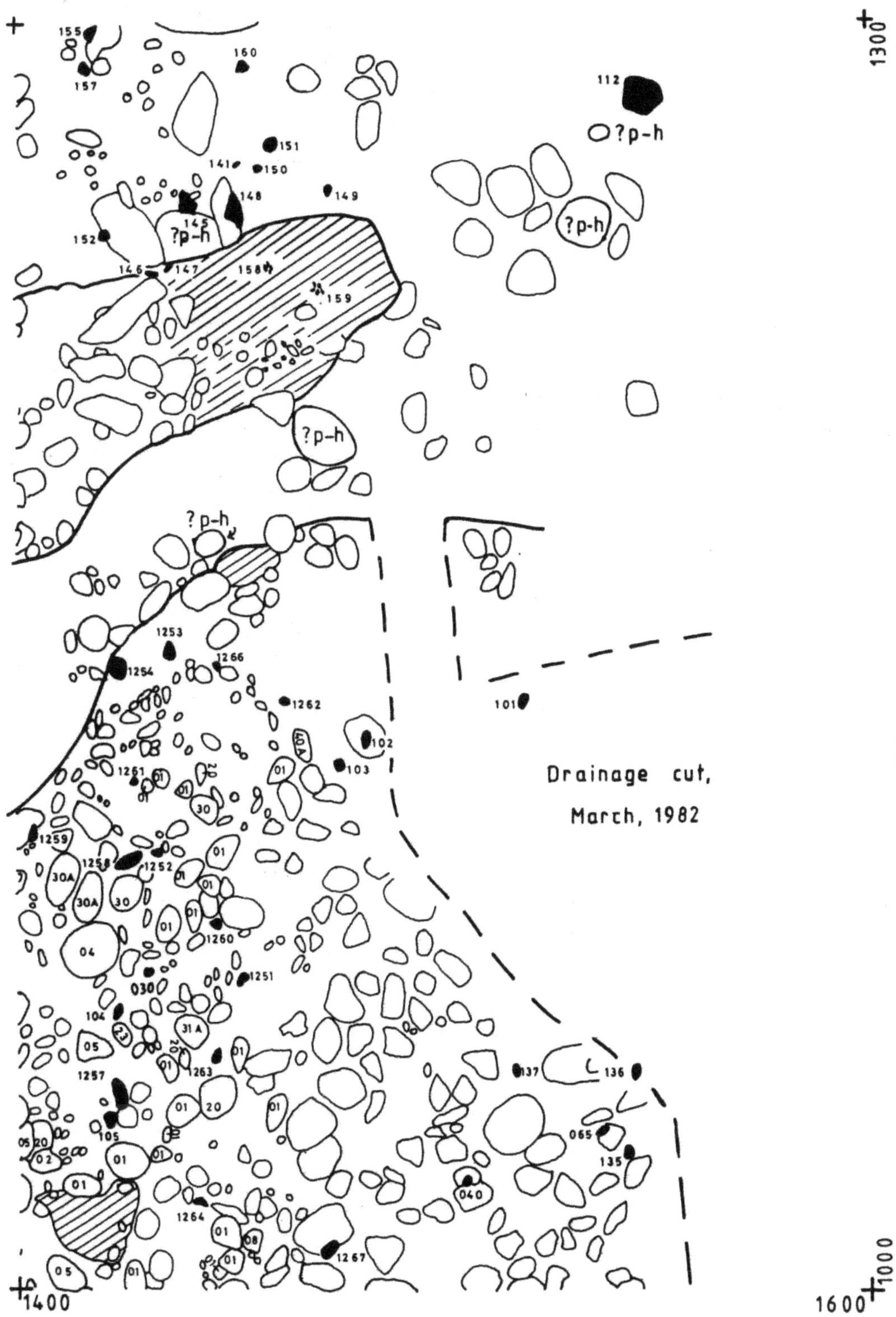

(RIGHT) Fig. I.25 Site 55, Riwat: Square D2. (LEFT) Fig. I.26 Site 55, Riwat: Squares B-C/2

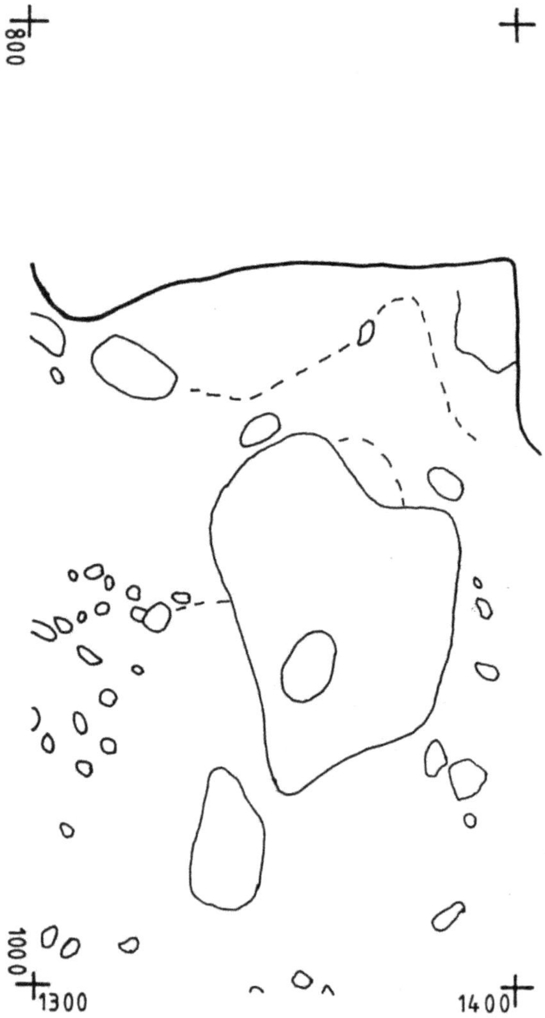

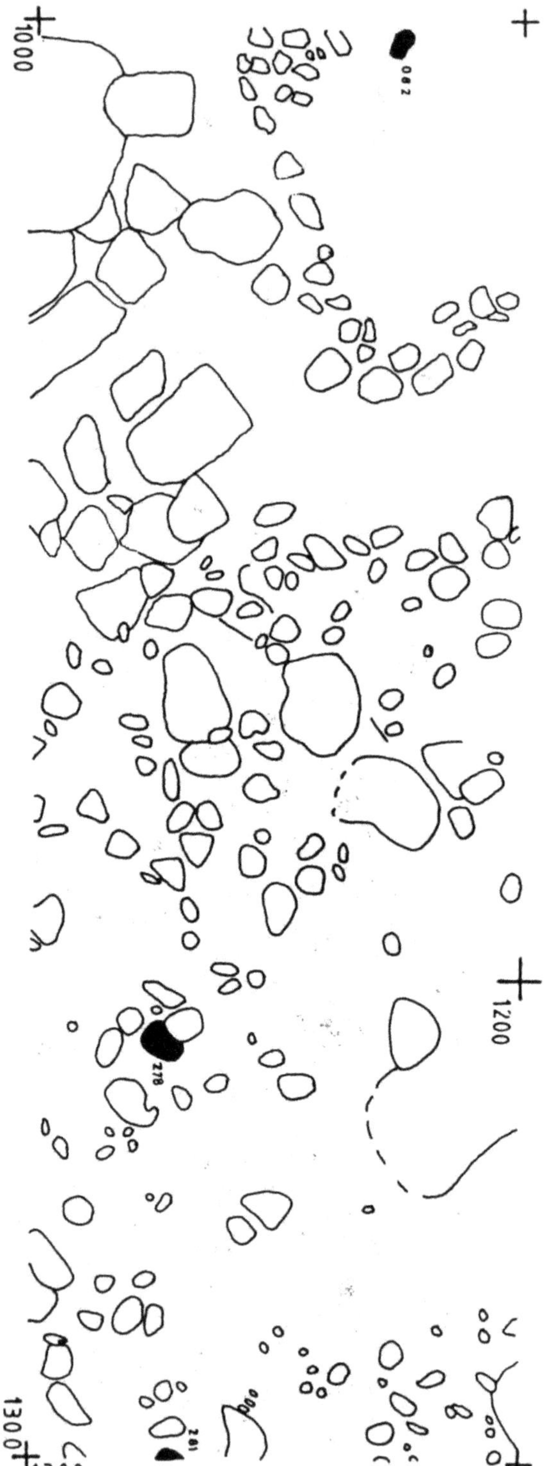

(RIGHT) Fig. I.27 Site 55, Riwat: Square A2. (LEFT) Fig. I.28 Site 55, Riwat: Square C1.

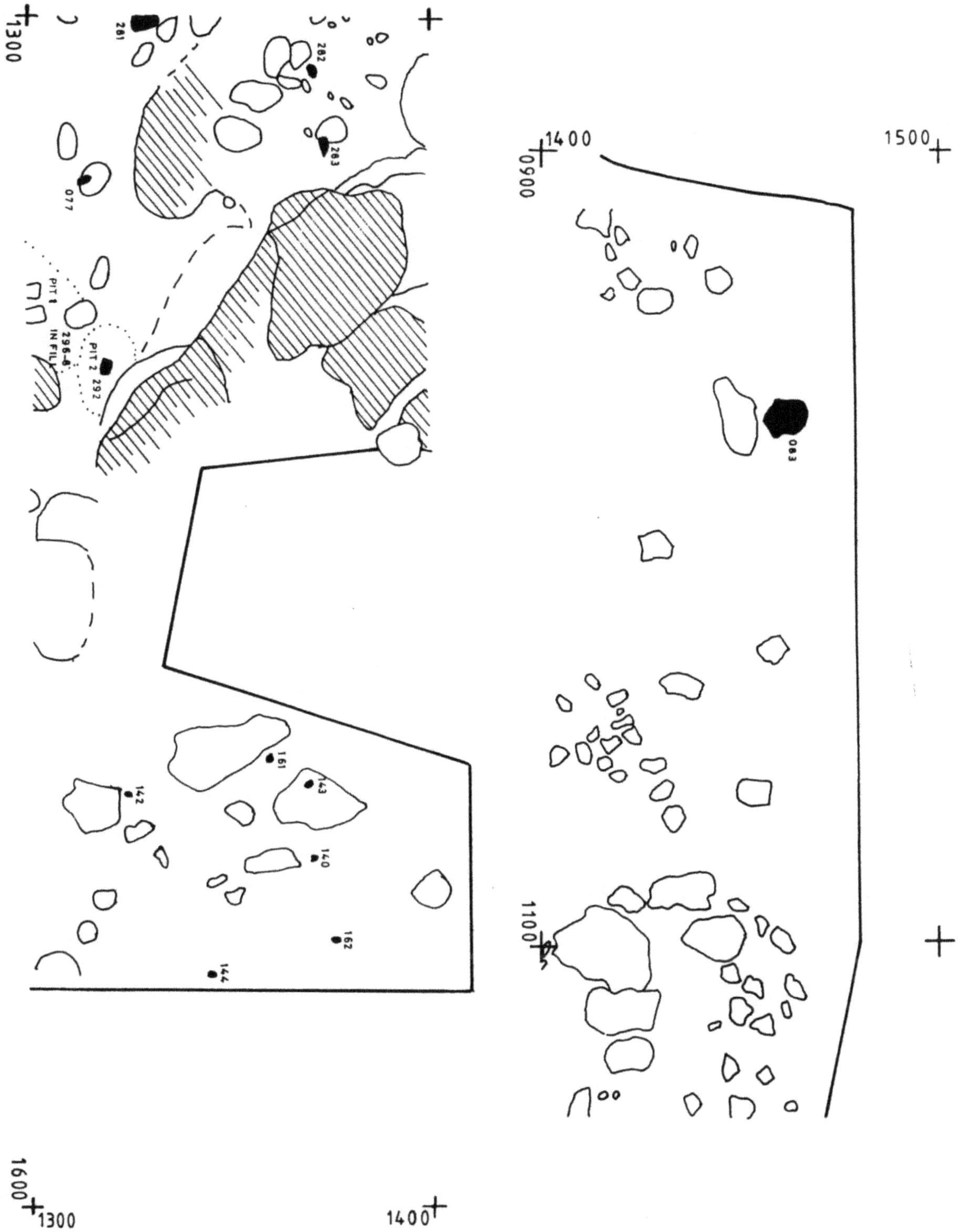

Fig. I.29 Site 55, Riwat: Squares A-B/1.

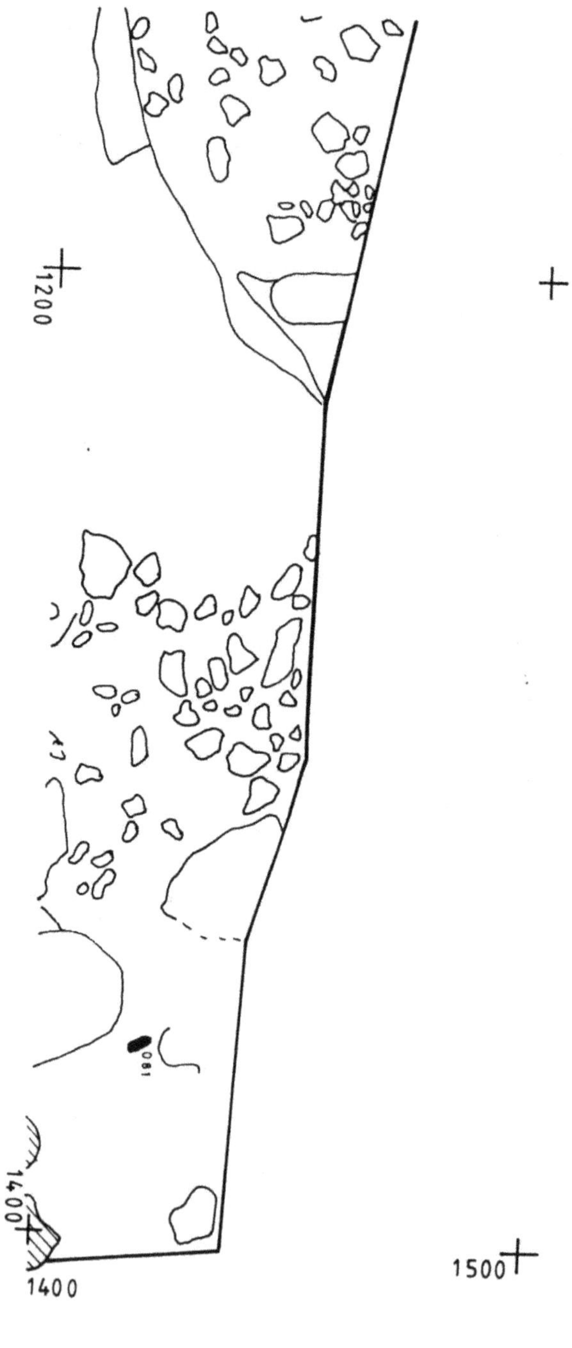

Fig. I.30 Site 55, Riwat: Squares E4 SE and E4 SW.

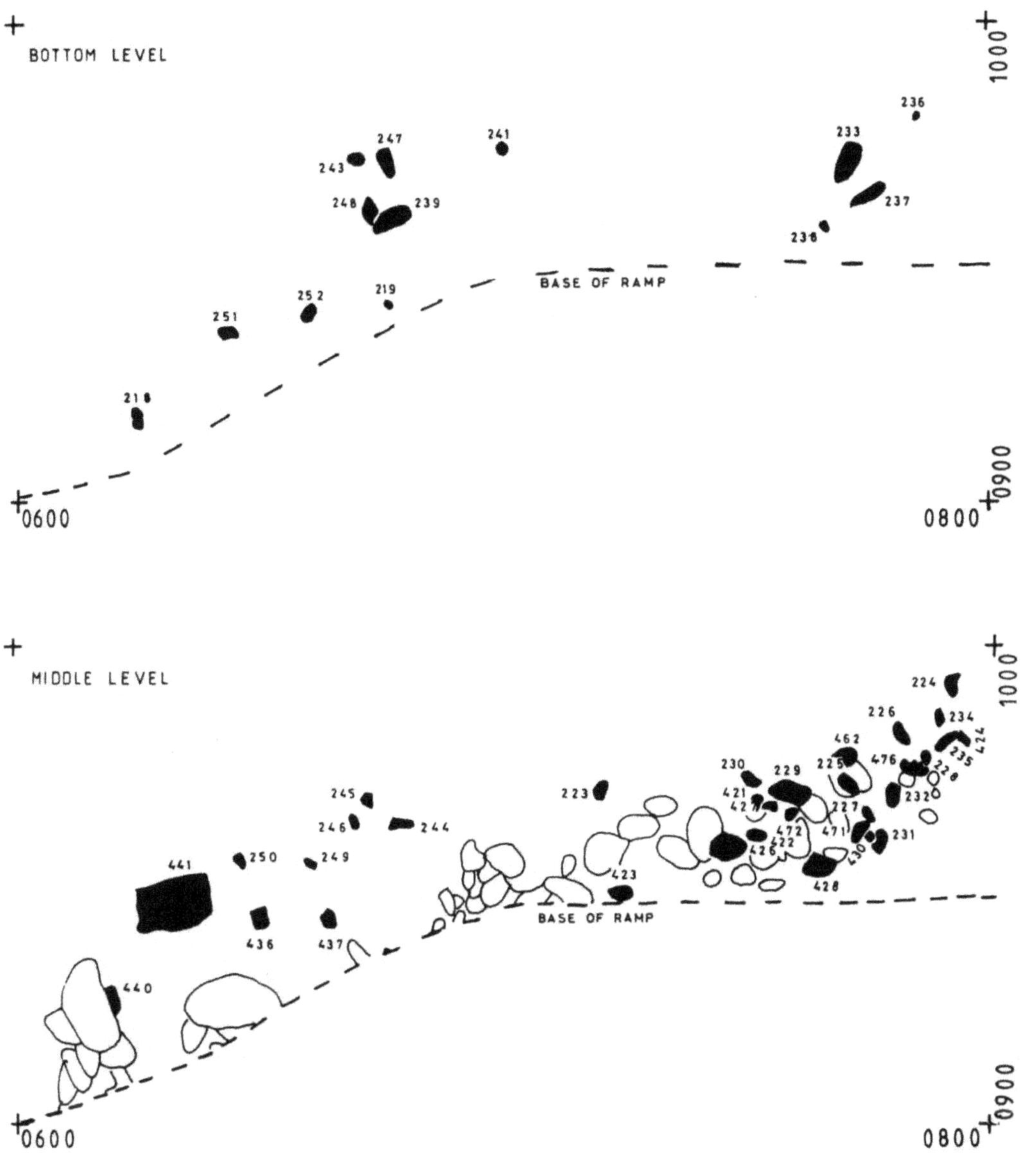

Fig. I.31 Site 55, Riwat: Square C3 NE: level II.

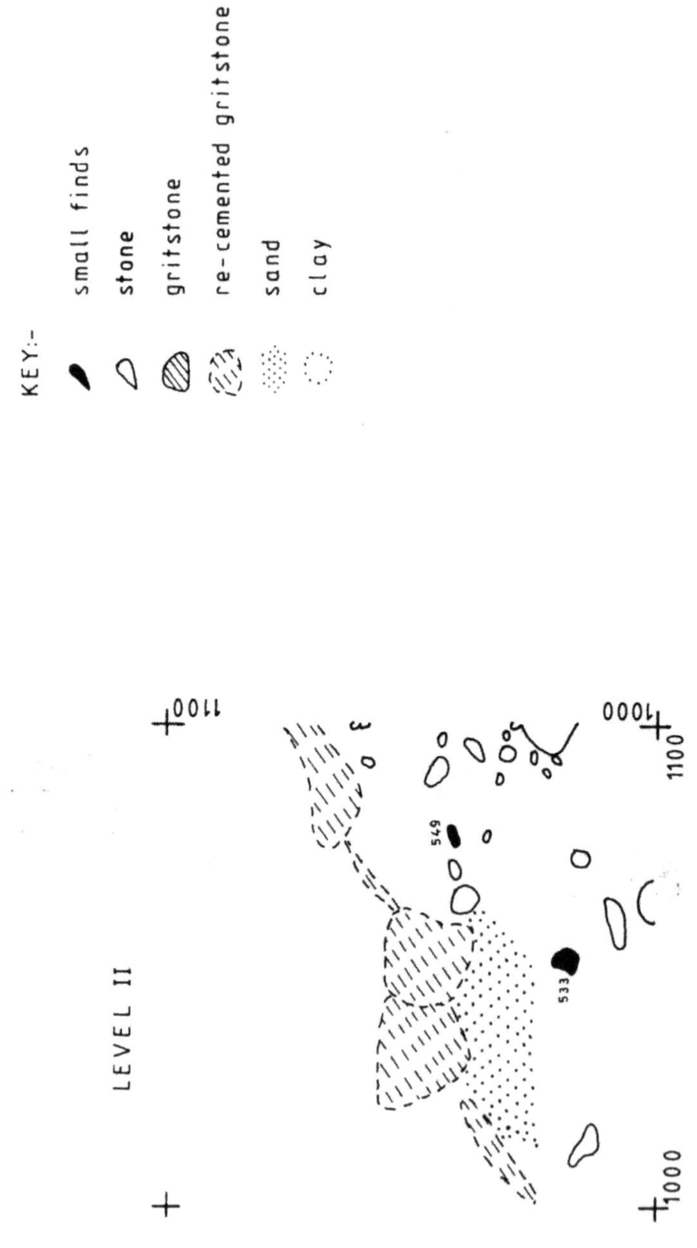

Fig. I.32 Site 55, Riwat: Squares A-B/3: level II.

Fig. I.33 Site 55, Riwat: Squares A-B/2: level II.

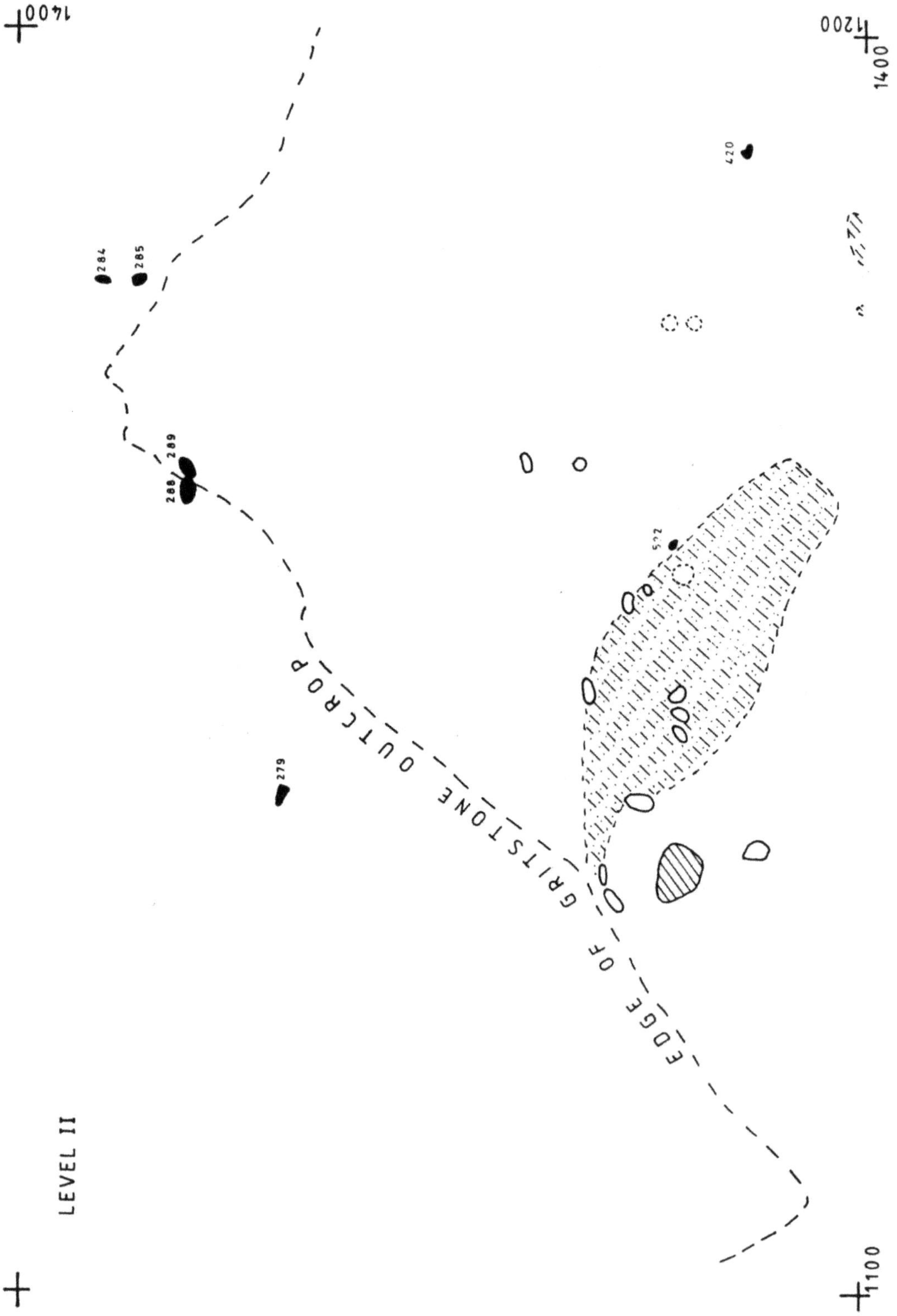

Fig. I.34 Site 55, Riwat: Square C3: level III.

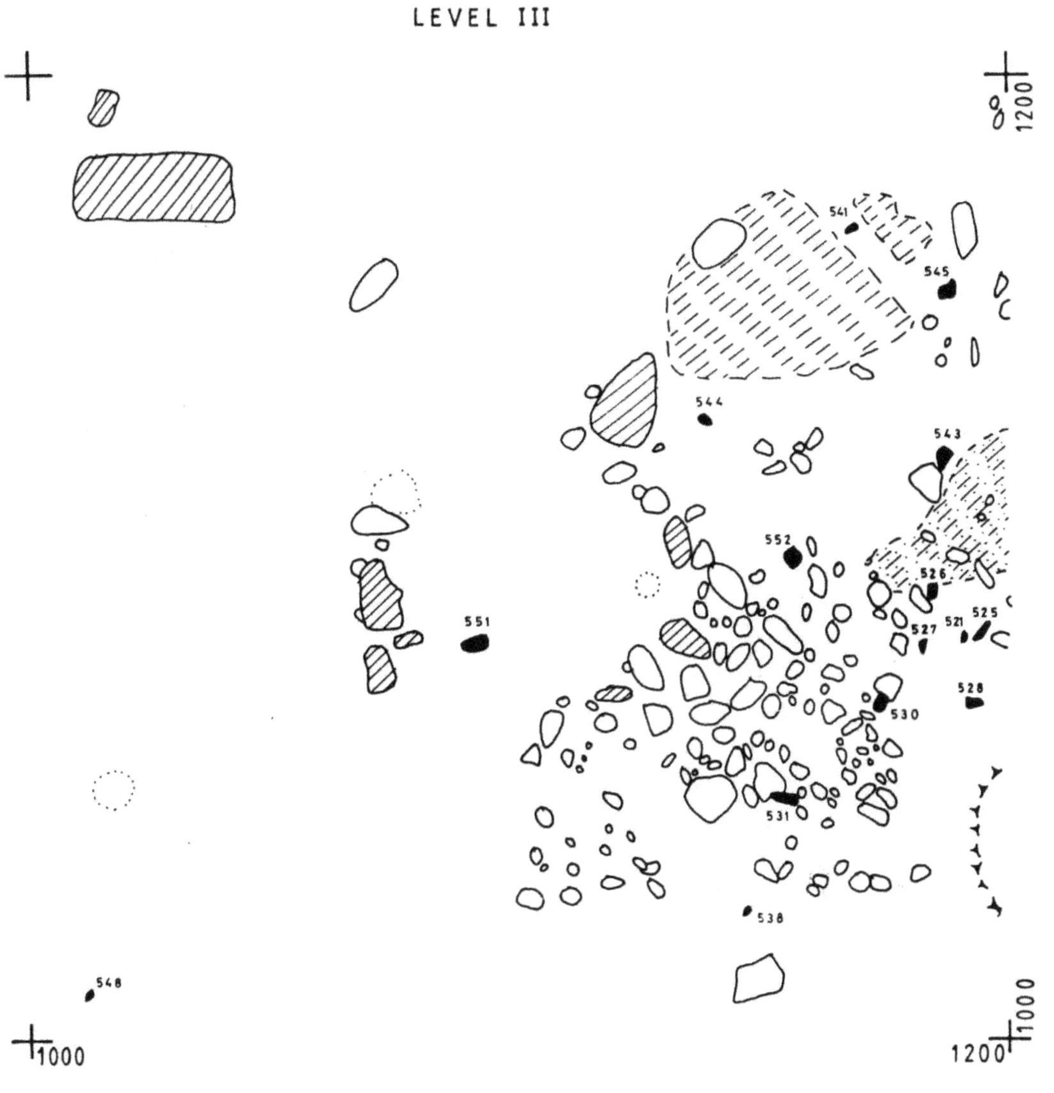

Fig. I.35 Site 55, Riwat: Squares A-B/2: level III.

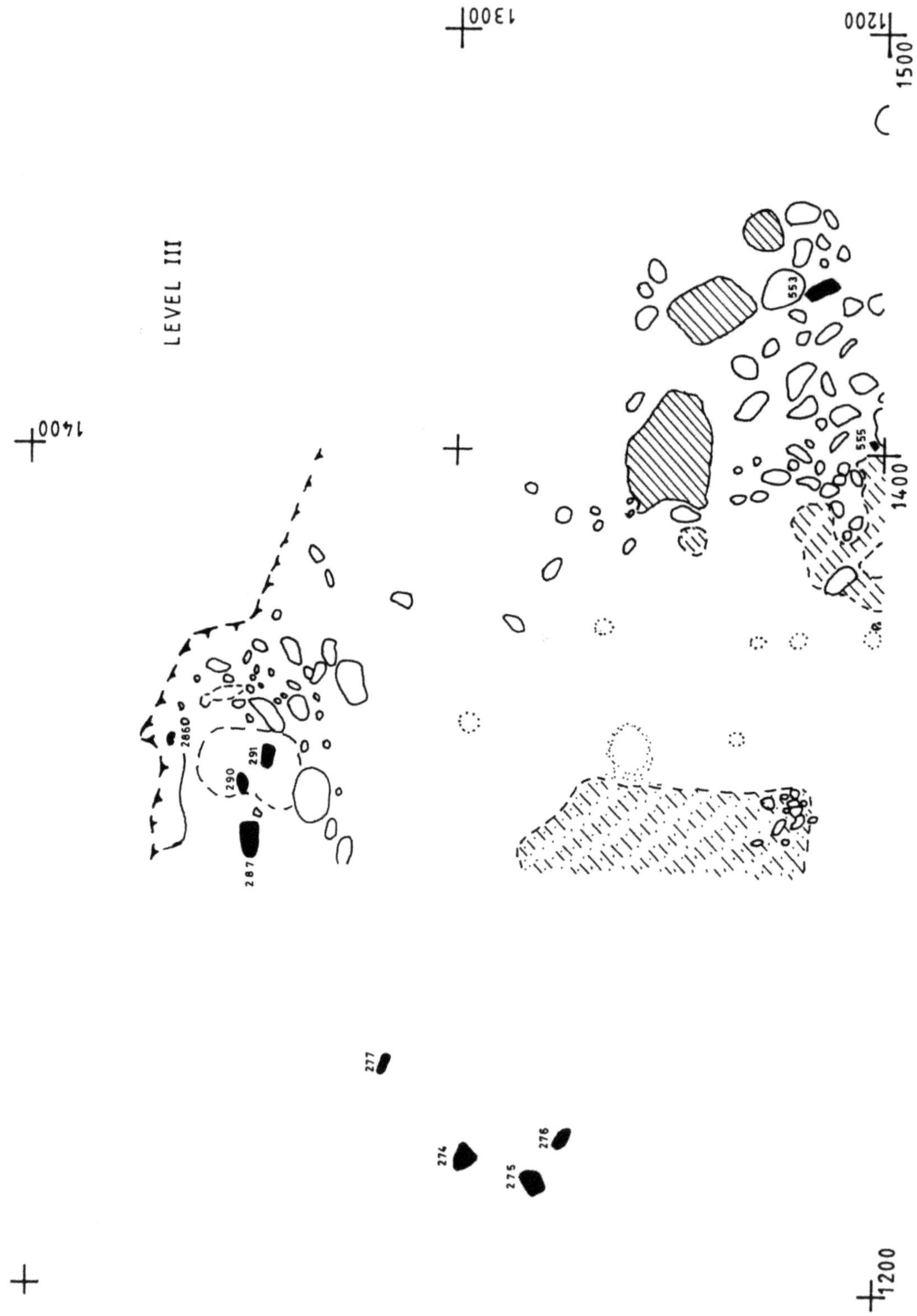

Fig. I.36 Site 55, Riwat: Squares A-B/3: level III.

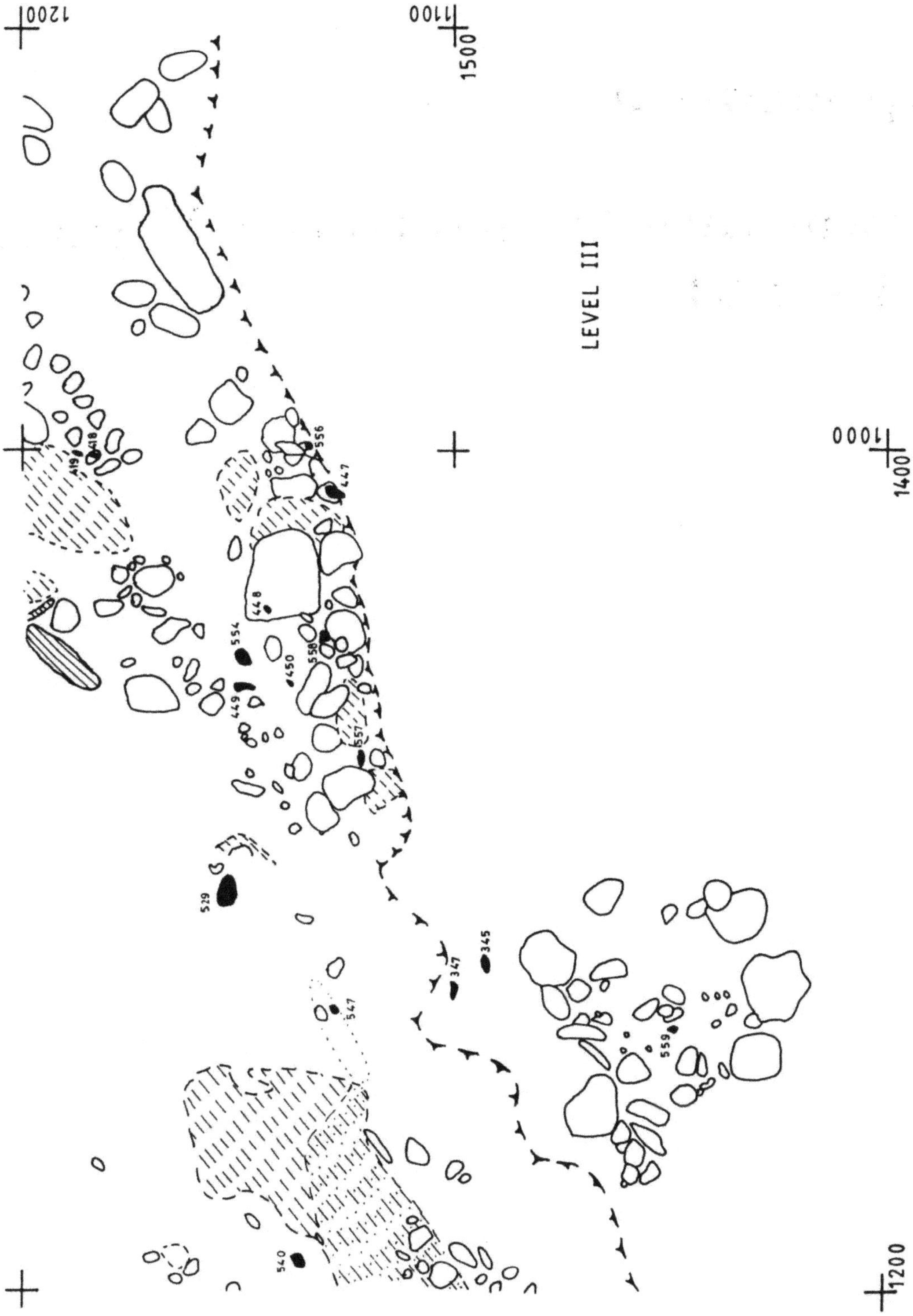

Appendix J

Riwat Site 55 artefacts : Figures J.1 - J.31

Fig. J.1 : Blades and blade-like flakes from site 55, Riwat

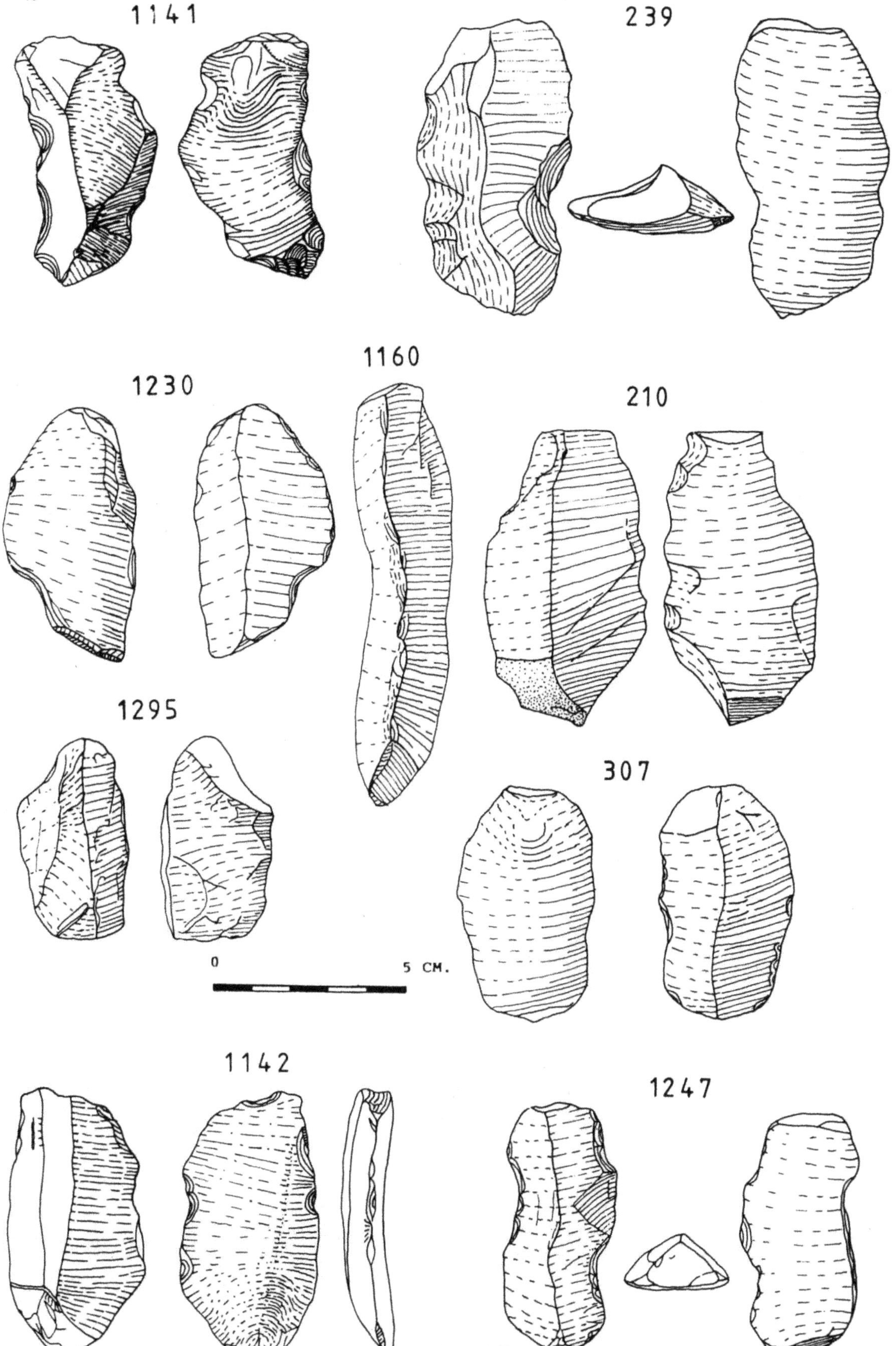

Fig. J.2 : Blades and blade-like flakes from site 55, Riwat

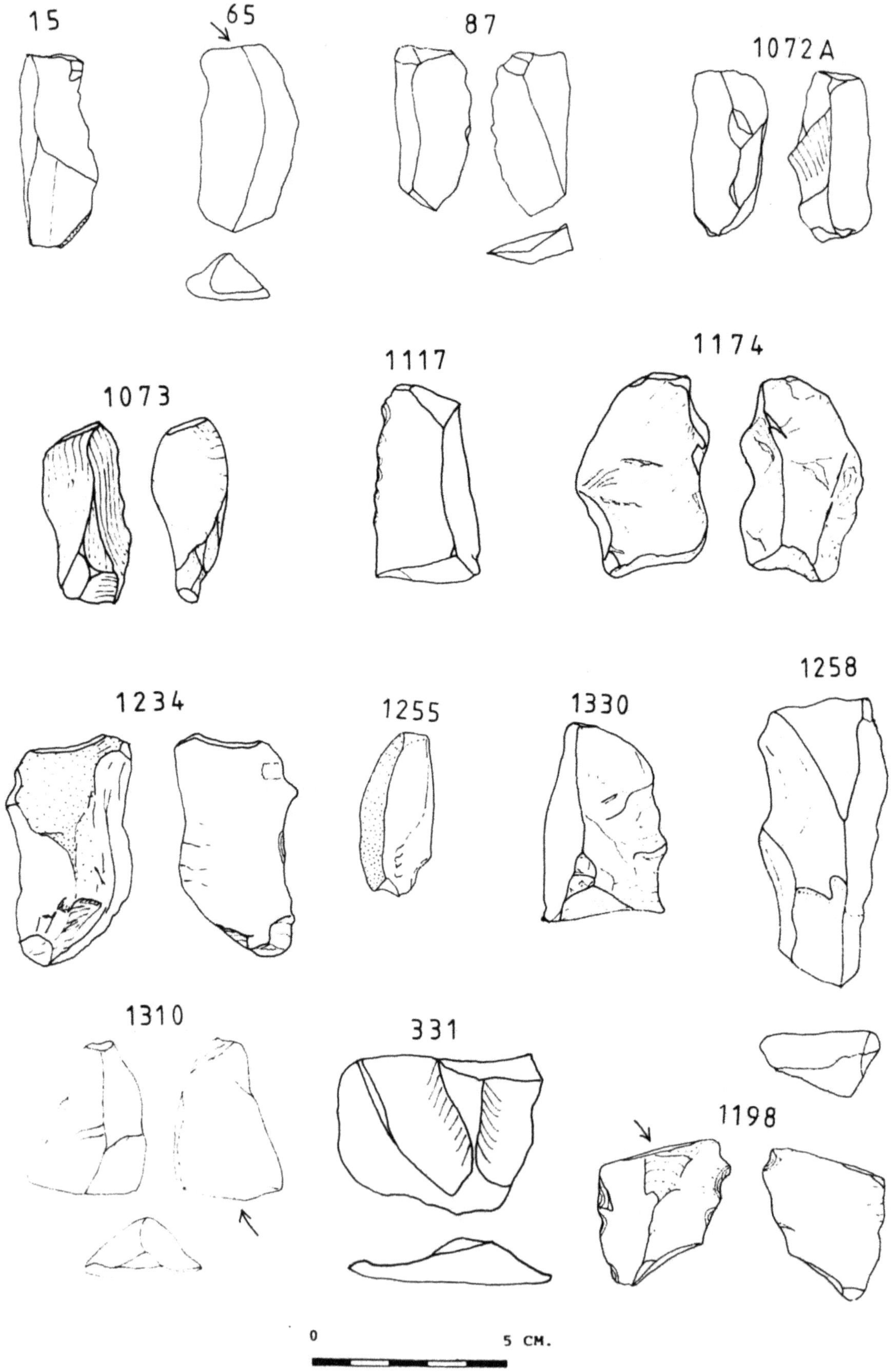

Fig. J.3 : Blades and blade-like flakes from site 55, Riwat

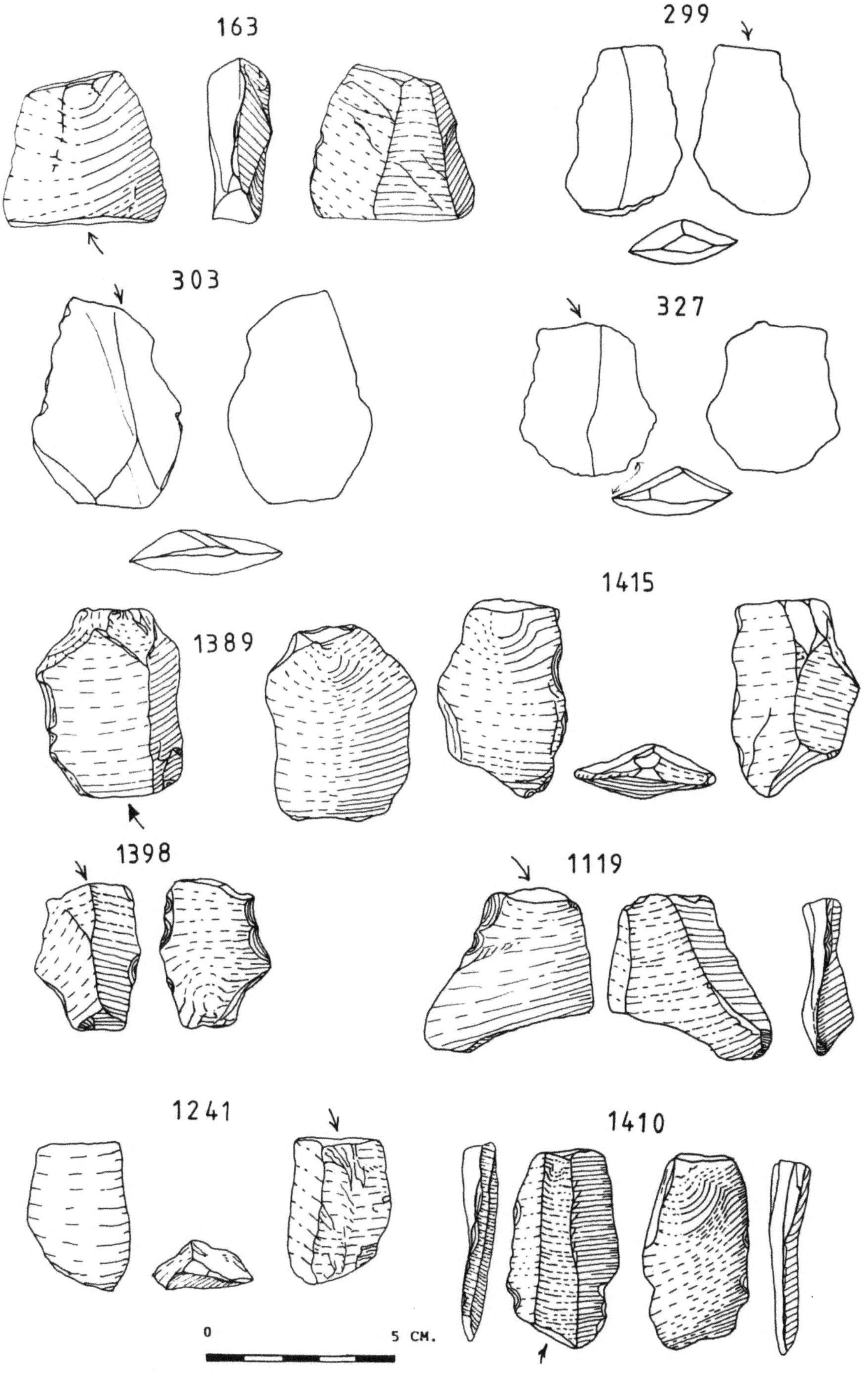

Fig. J.4 : Blades and blade-like flakes from site 55, Riwat

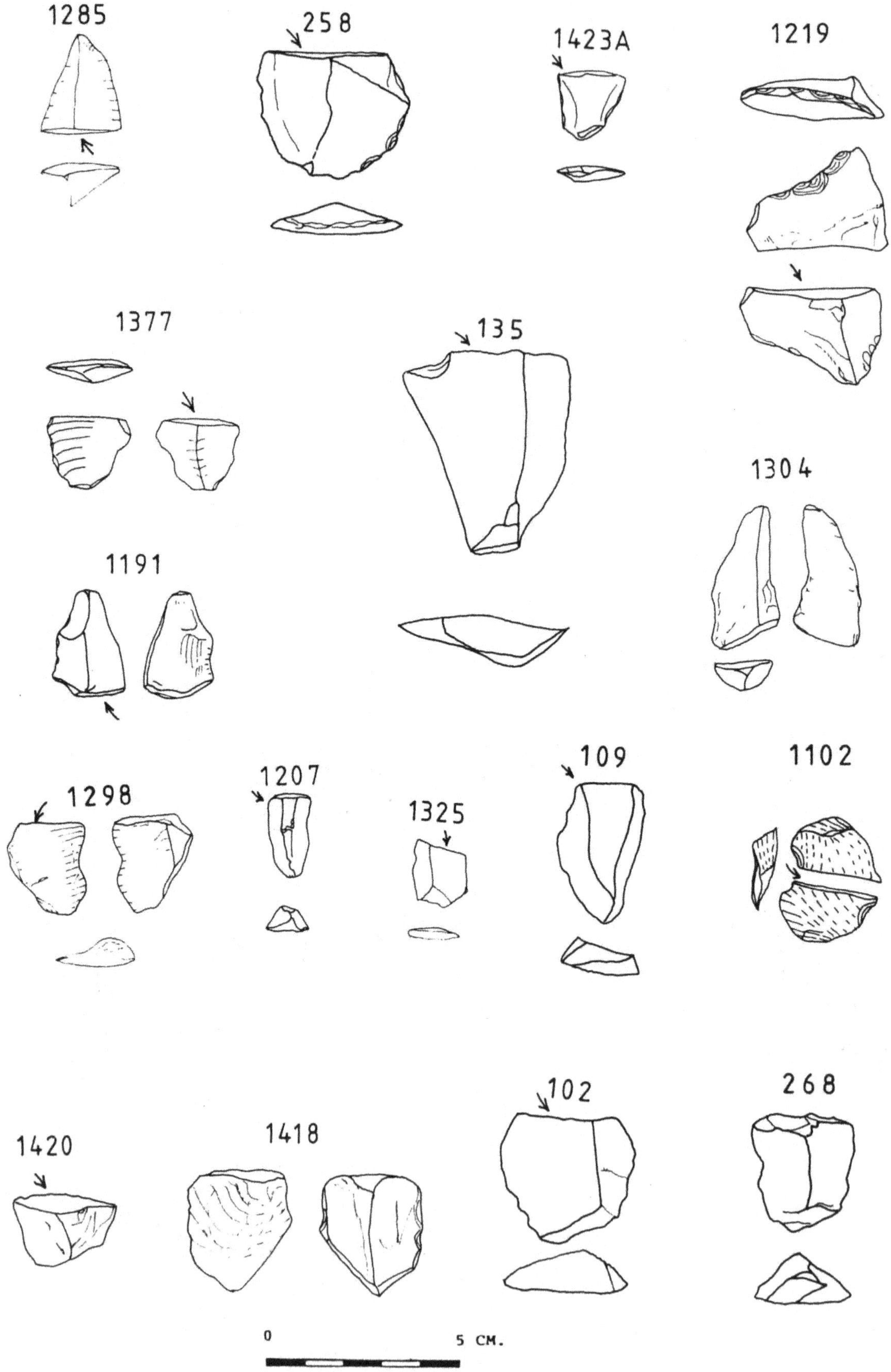

Fig. J.5 : Blades and blade-like flakes from site 55, Riwat

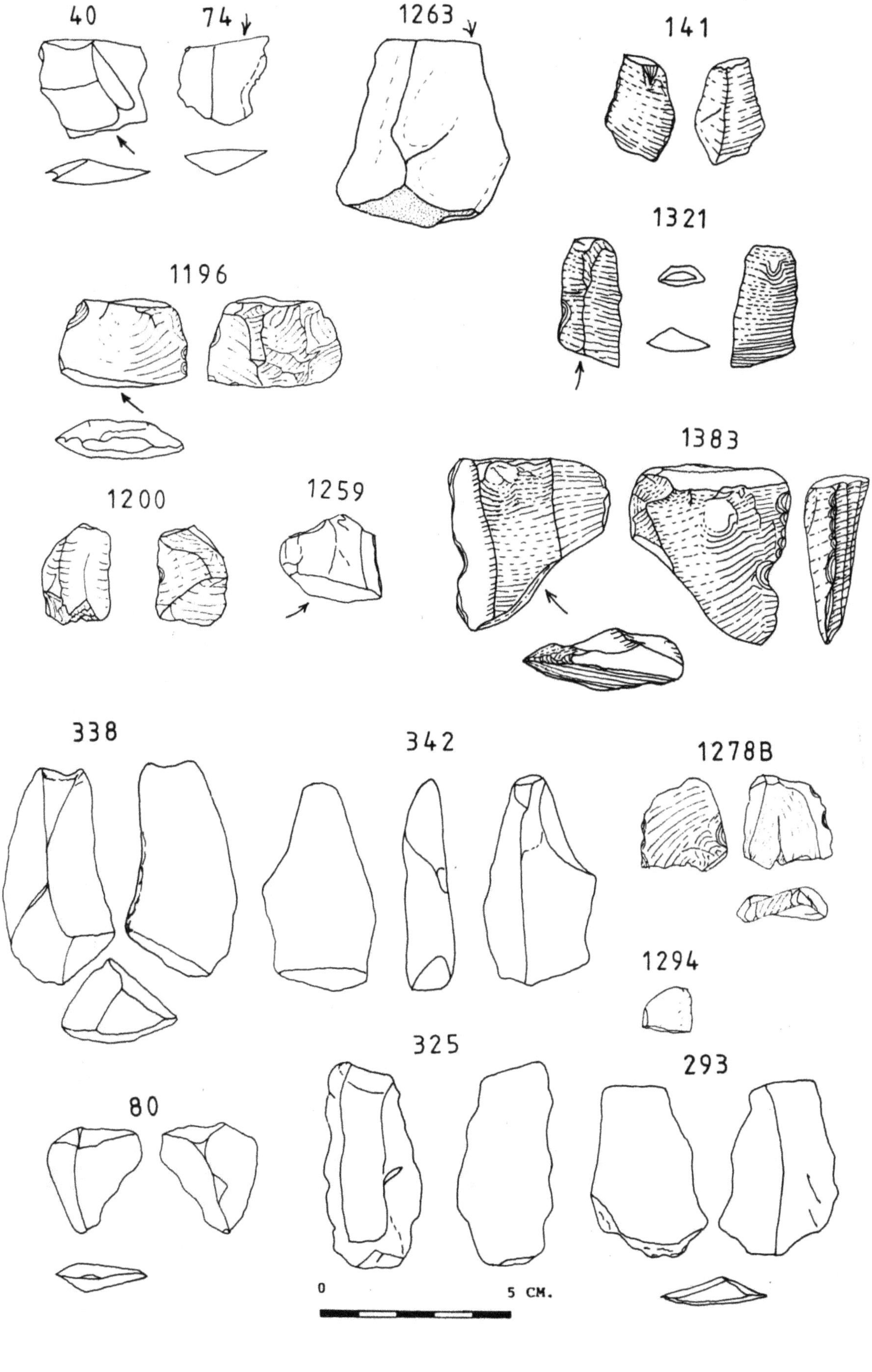

Fig. J.6 : Blades and blade-like flakes from site 55, Riwat

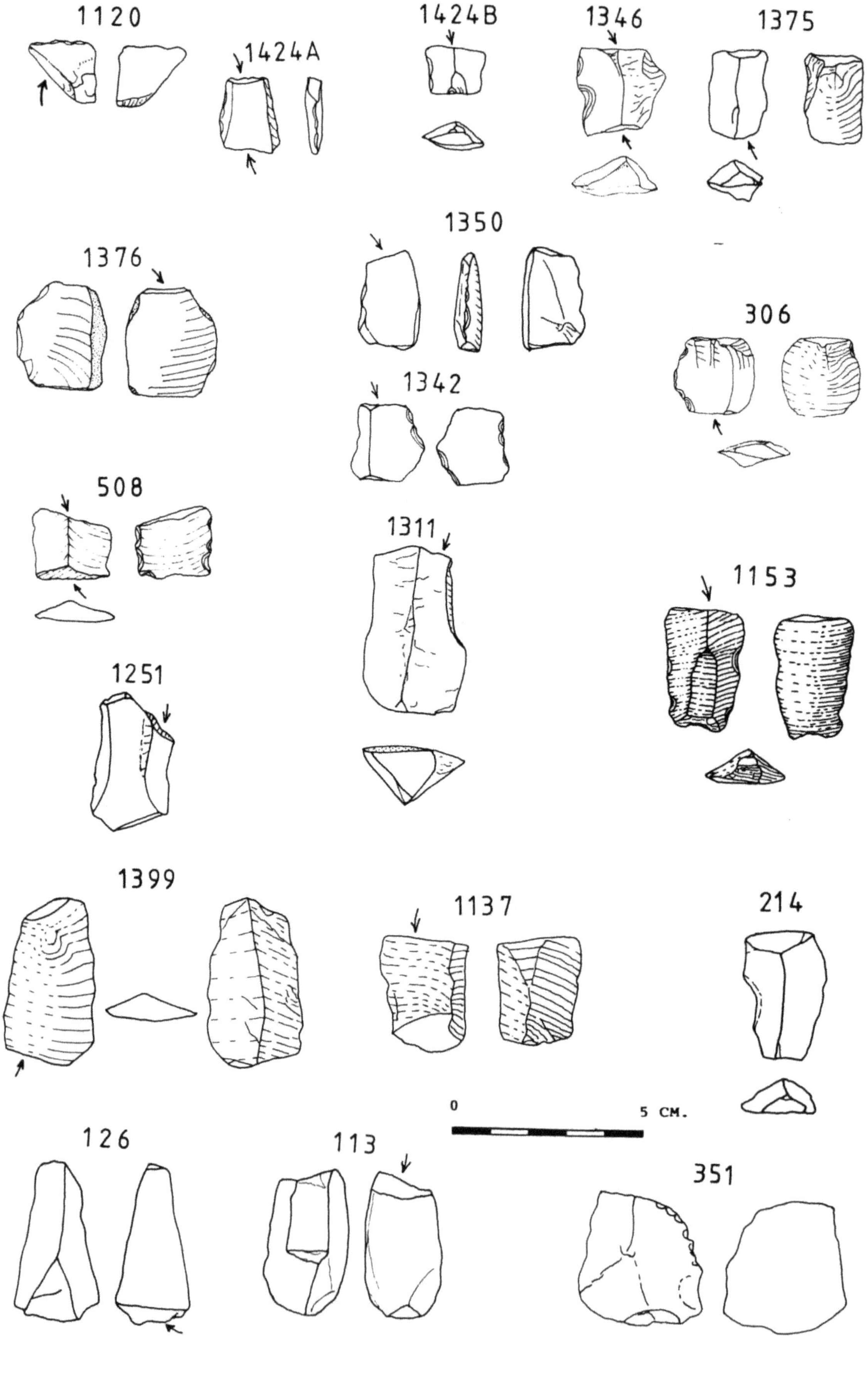

Fig. J.7 : Blades and blade-like flakes from site 55, Riwat

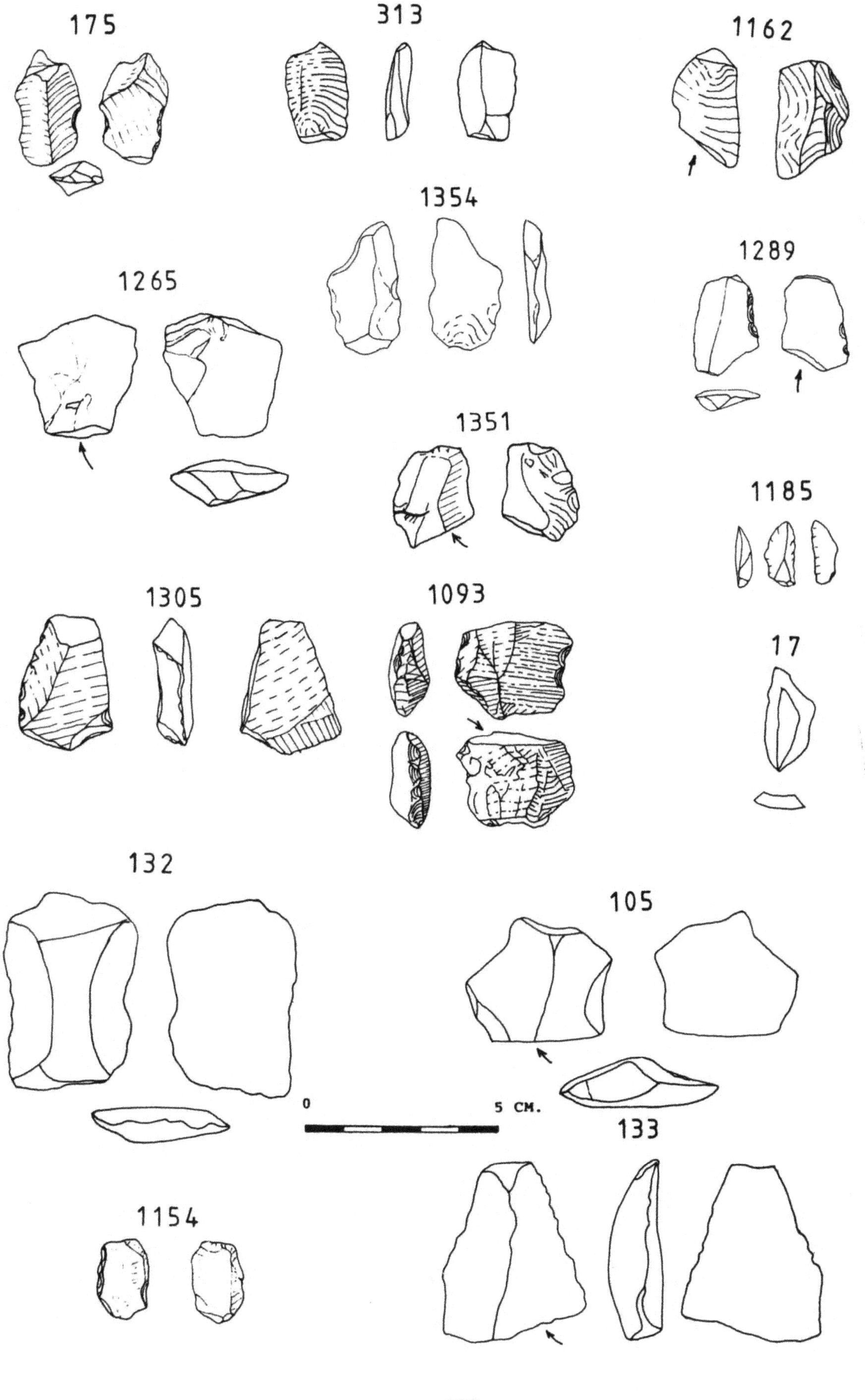

Fig. J.8 : Flakes and flake-like blades from site 55, Riwat

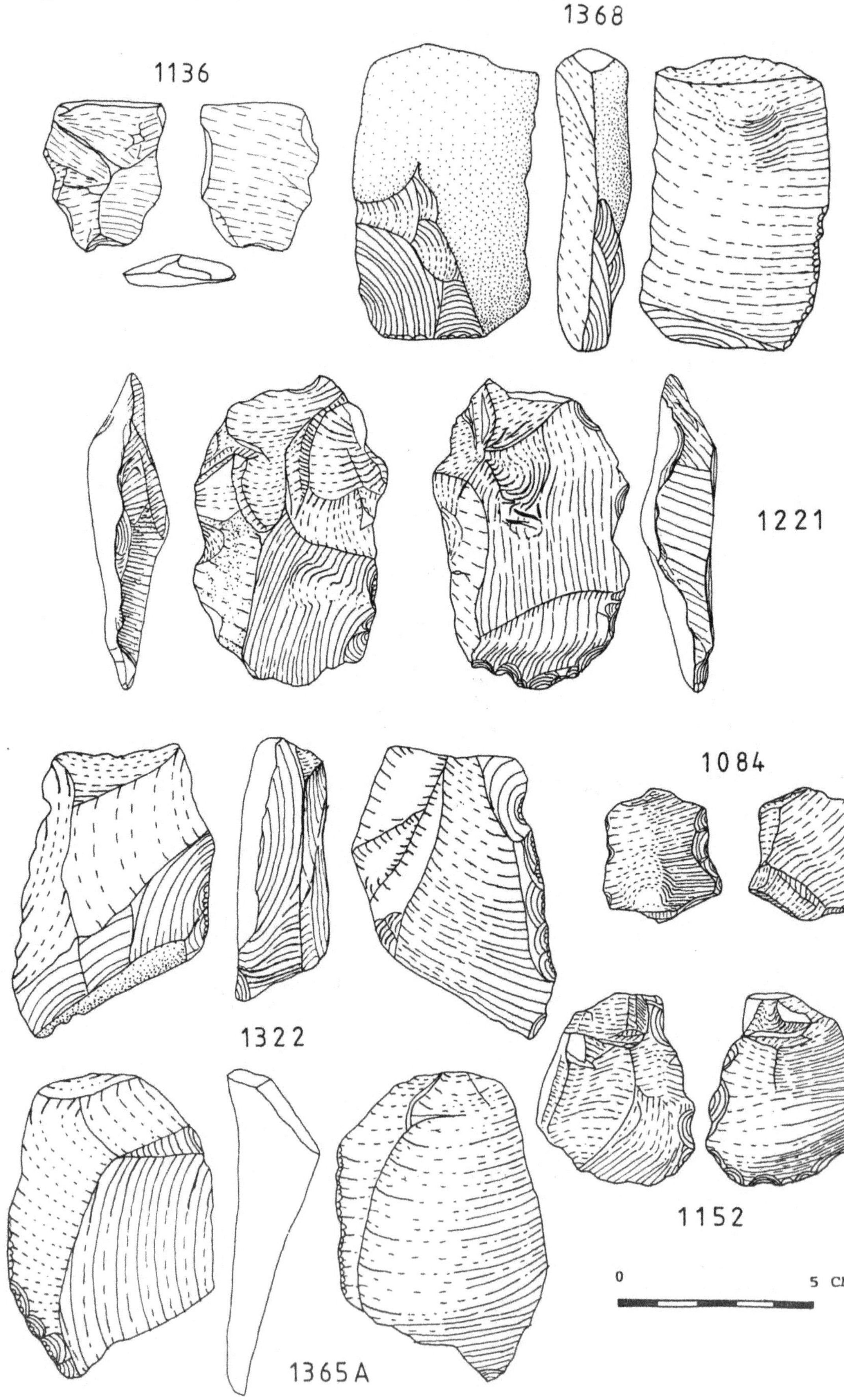

Fig. J.9 : Flakes and flake-like blades from site 55, Riwat

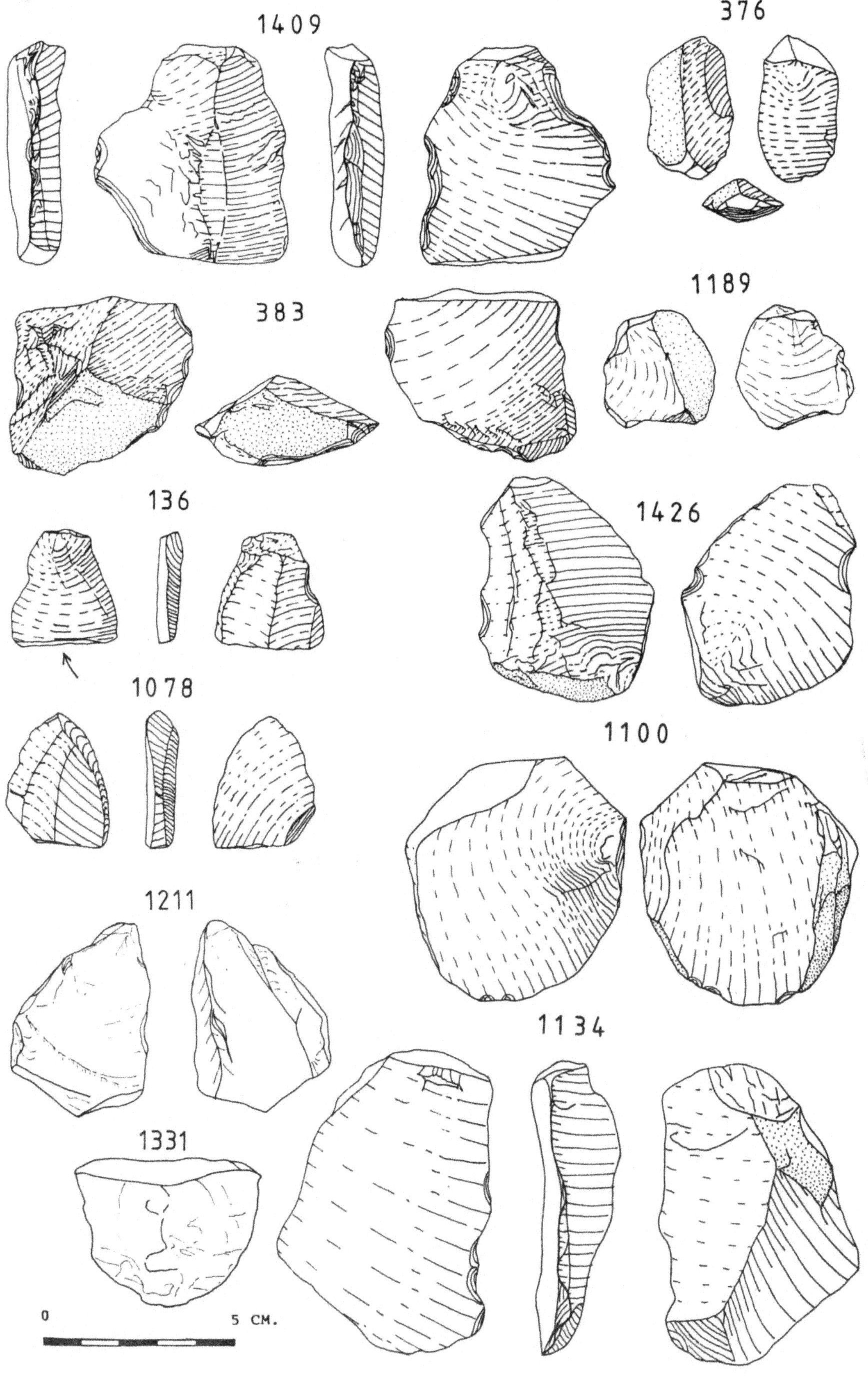

Fig. J.10 : Flakes and flake-like blades from site 55, Riwat

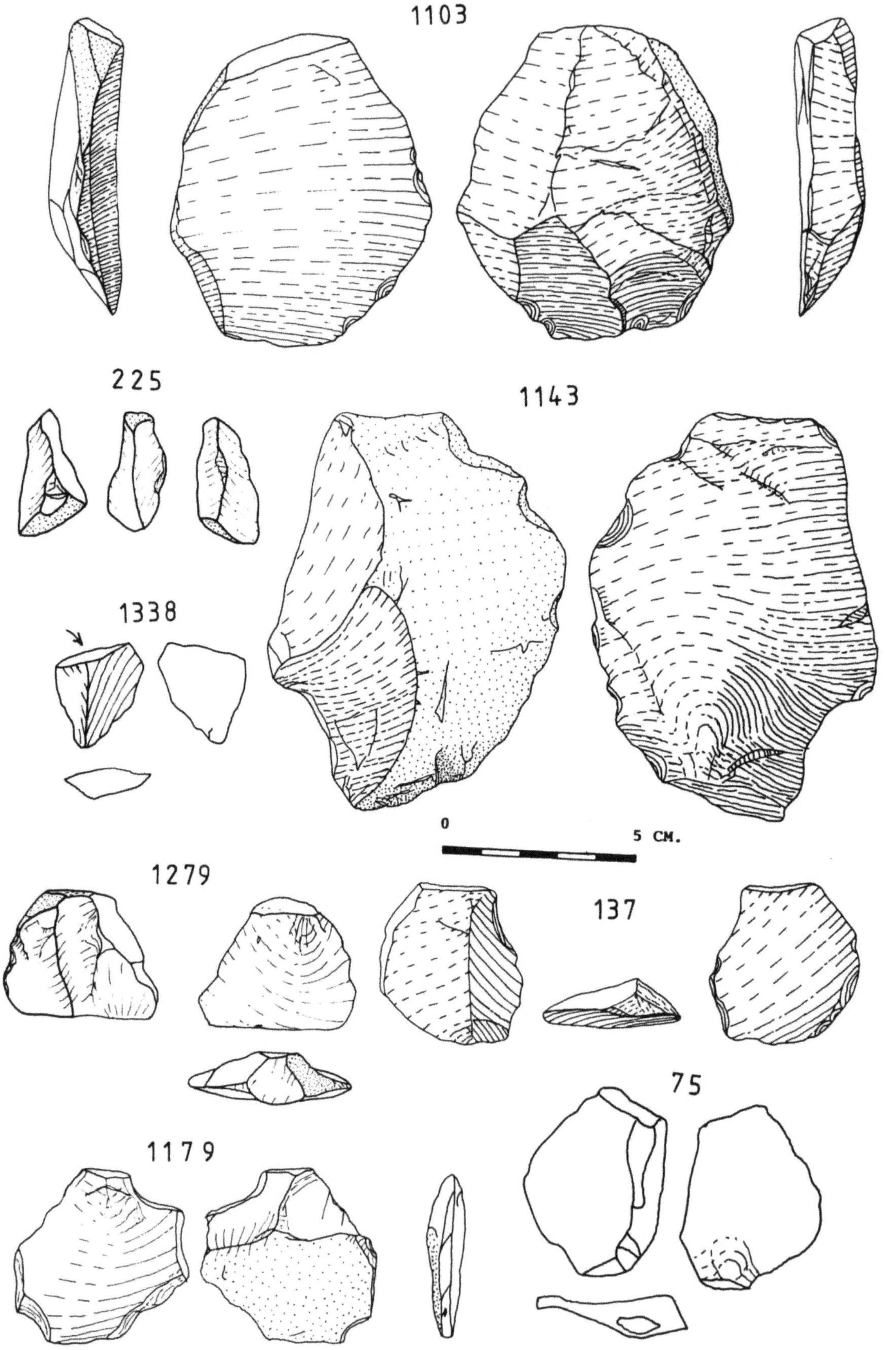

Fig. J.11 : Flakes and flake-like blades from site 55, Riwat

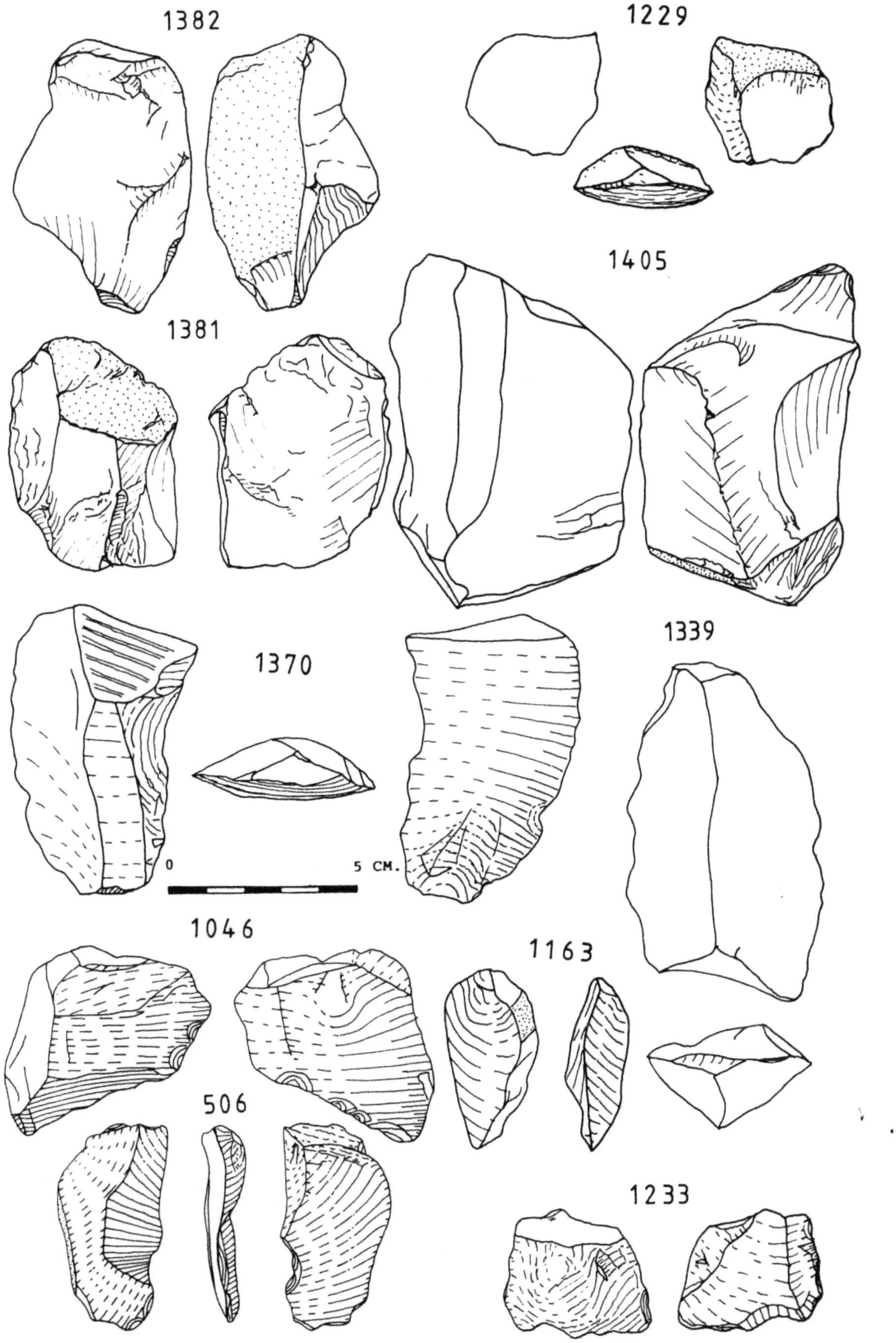

Fig. J.12 : Flakes and flake-like blades from site 55, Riwat

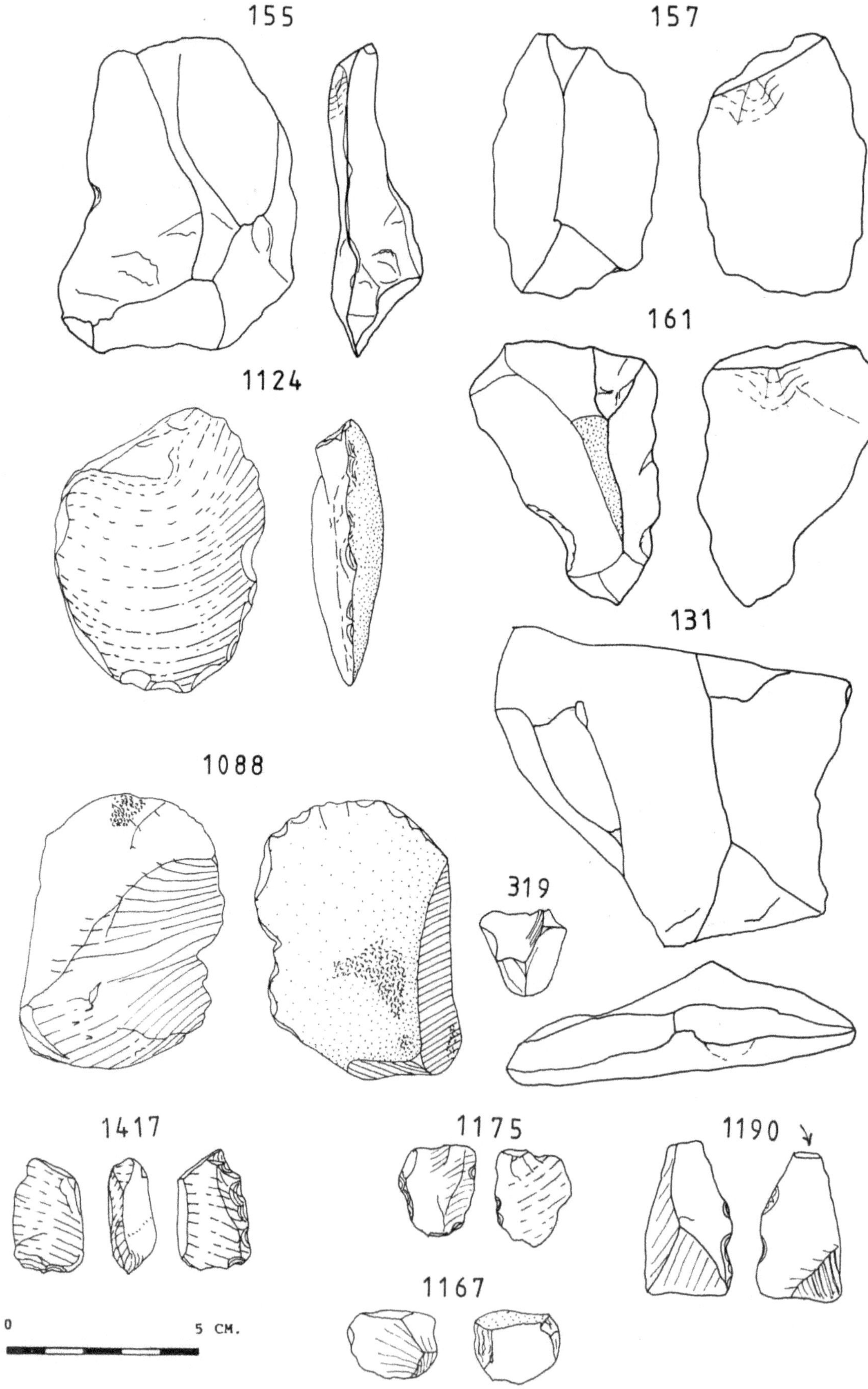

Fig. J.13 : Flakes and flake-like blades from site 55, Riwat

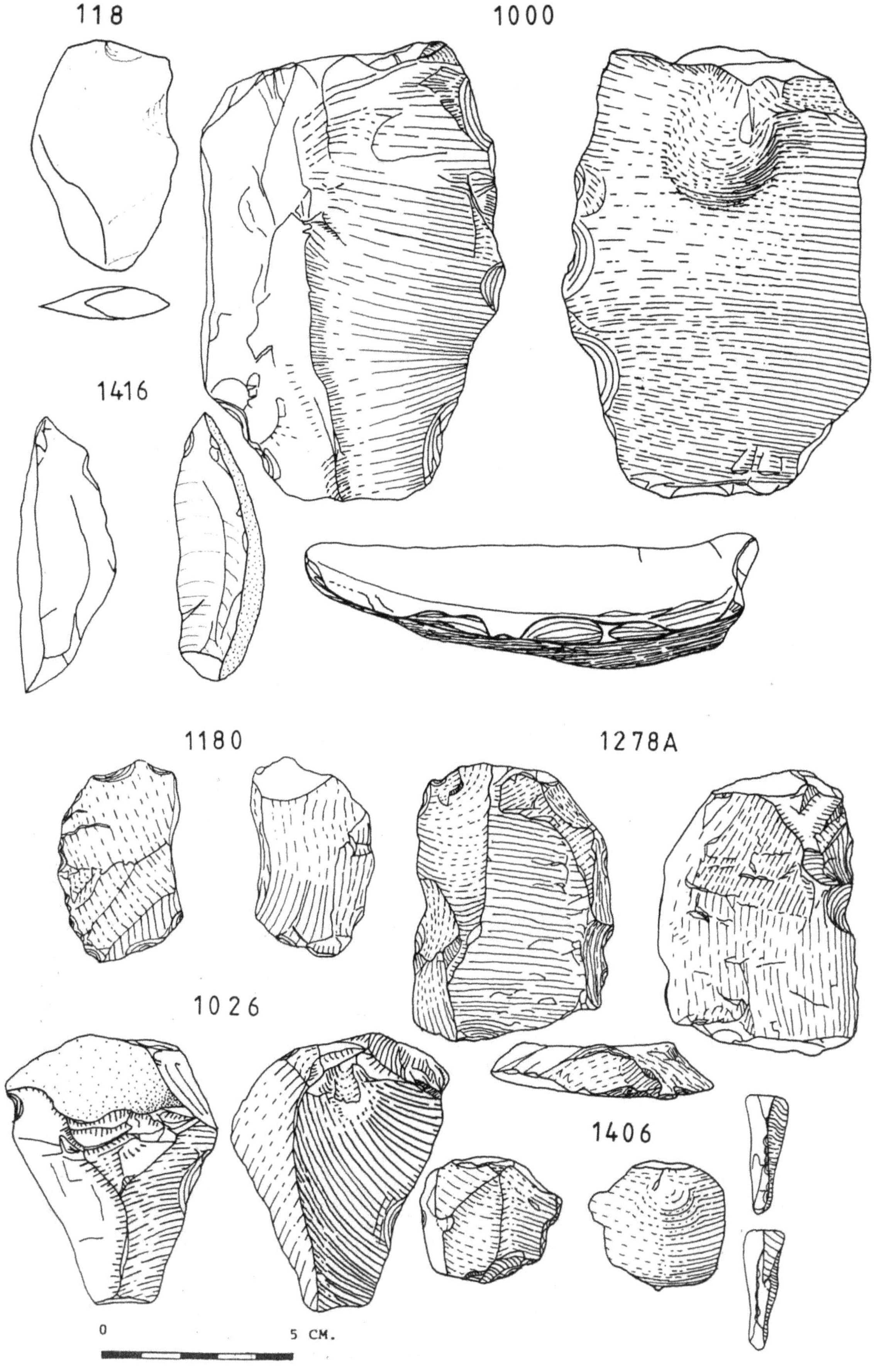

Fig. J.14 : Flakes and flake-like blades from site 55, Riwat

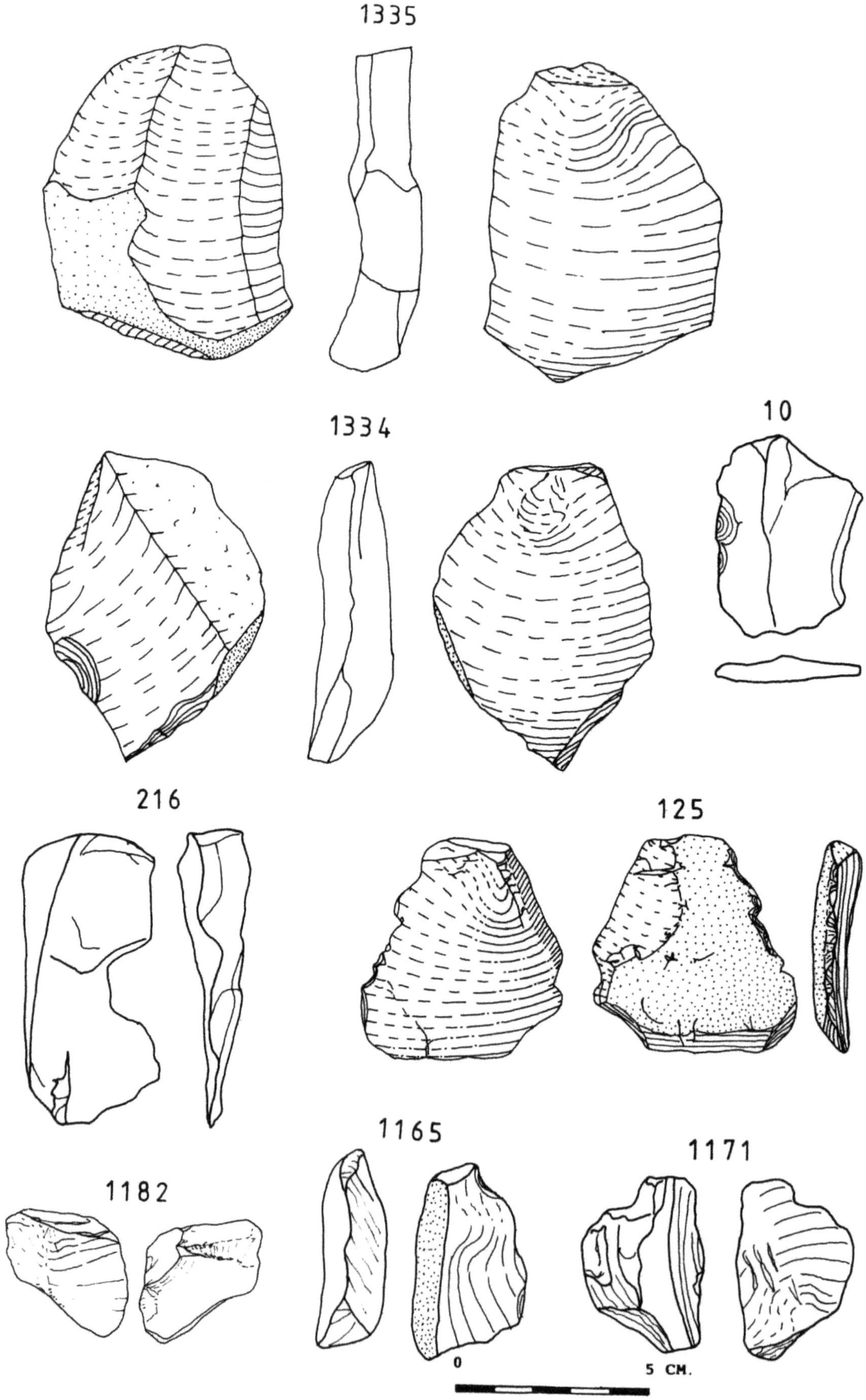

Fig. J.15 : Flakes and flake-like blades from site 55, Riwat

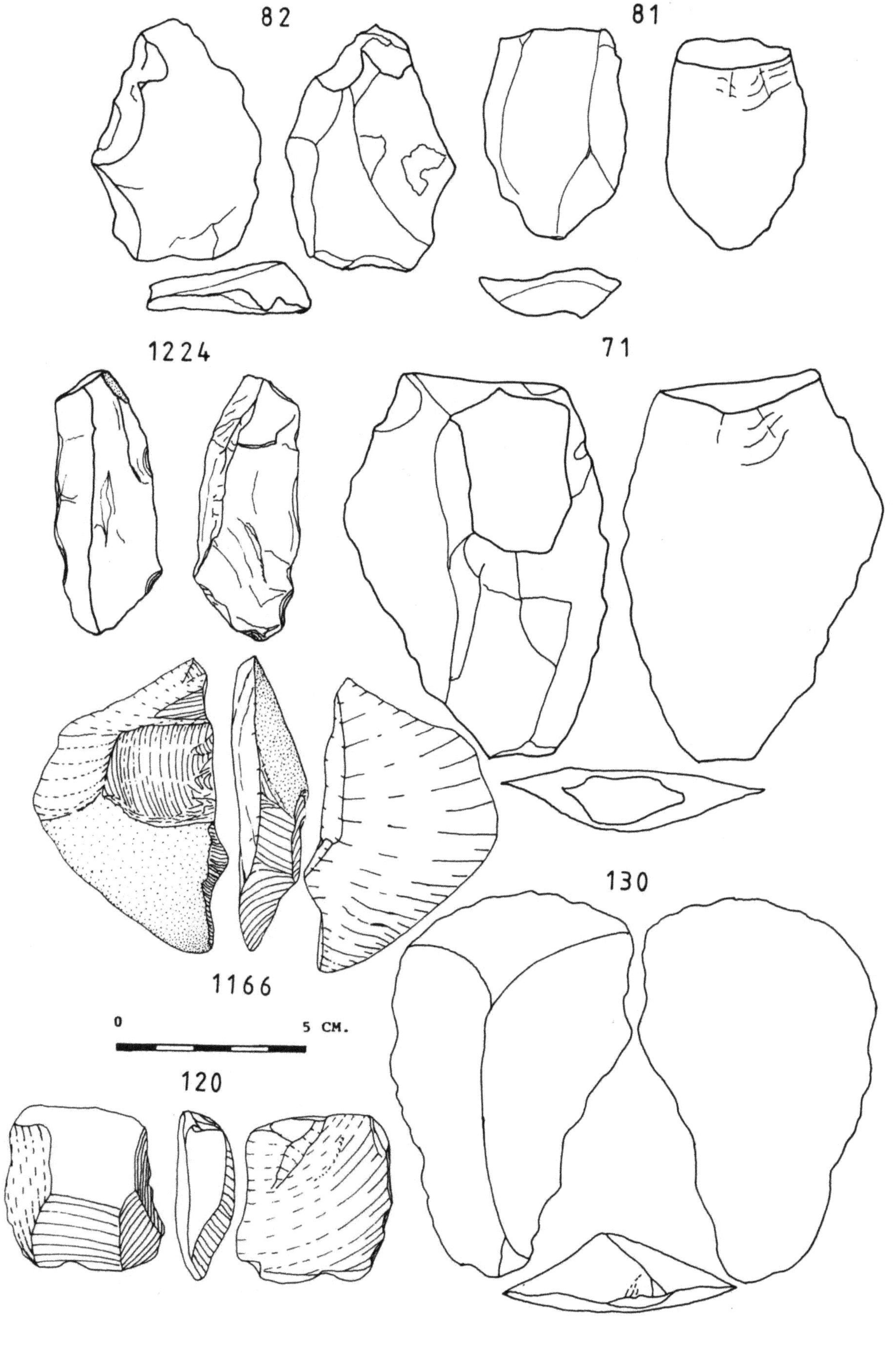

Fig. J.16 : Flakes and flake-like blades from site 55, Riwat

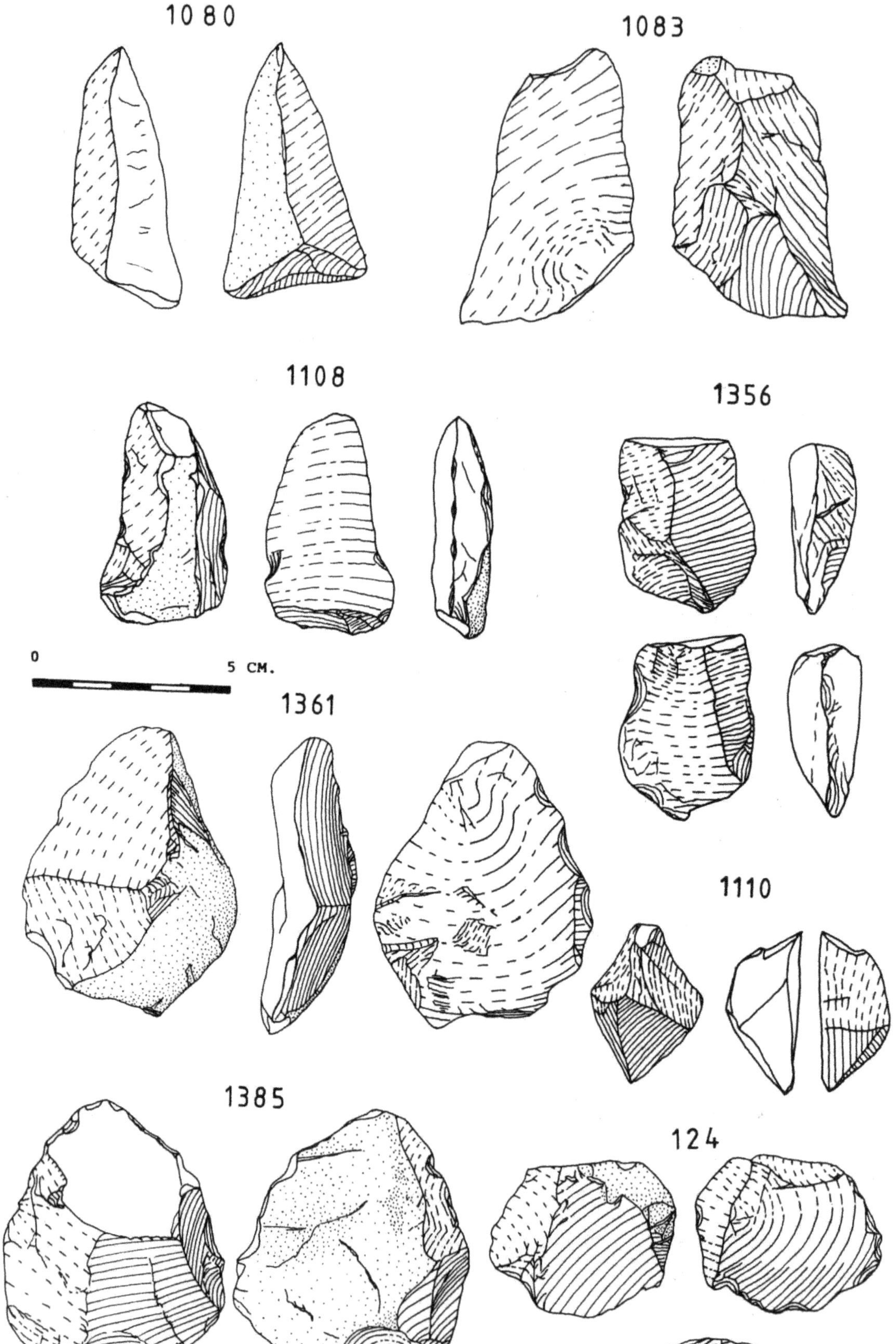

Fig. J.17 : Flakes and flake-like blades from site 55, Riwat

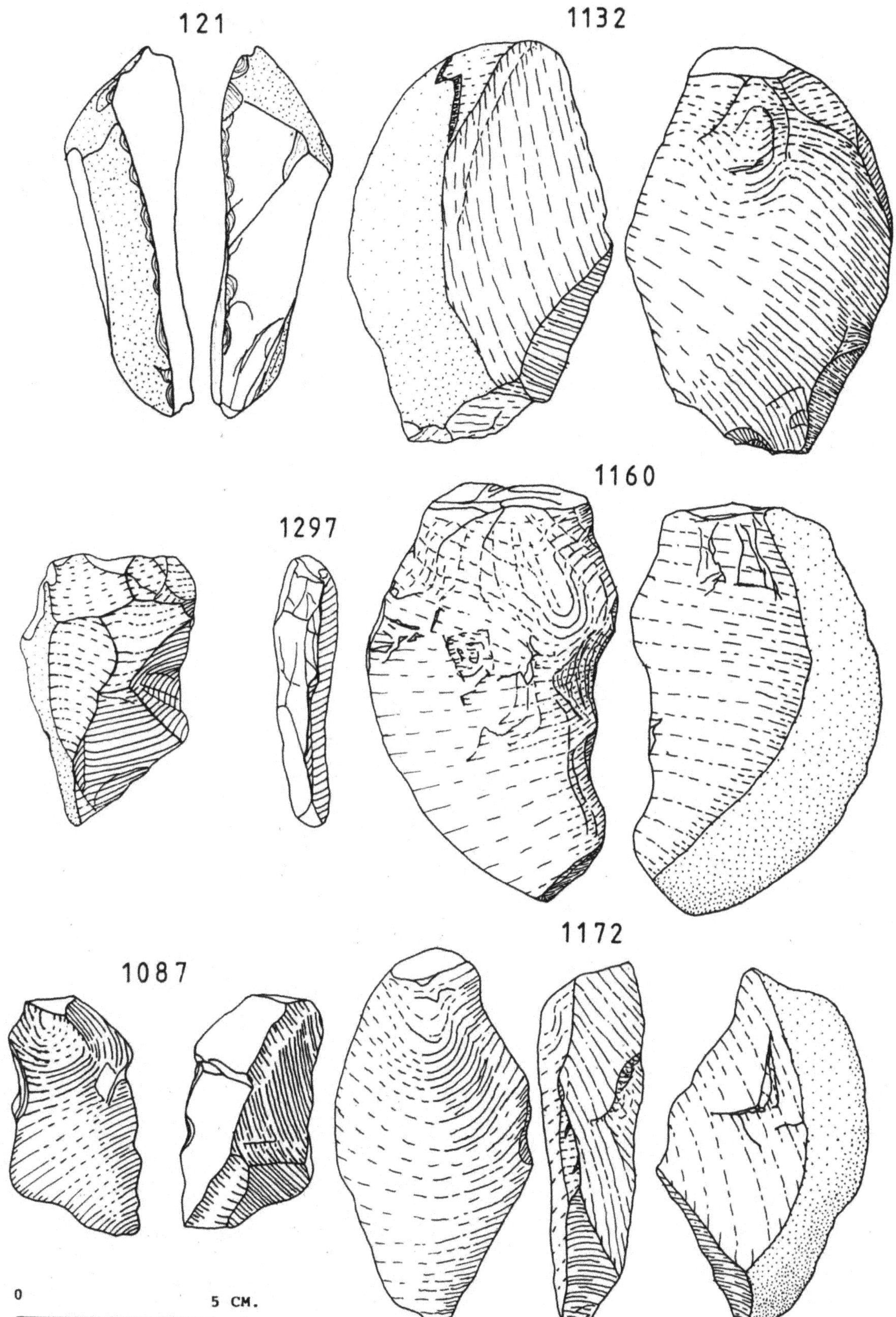

Fig. J.18 : Flakes and flake-like blades from site 55, Riwat

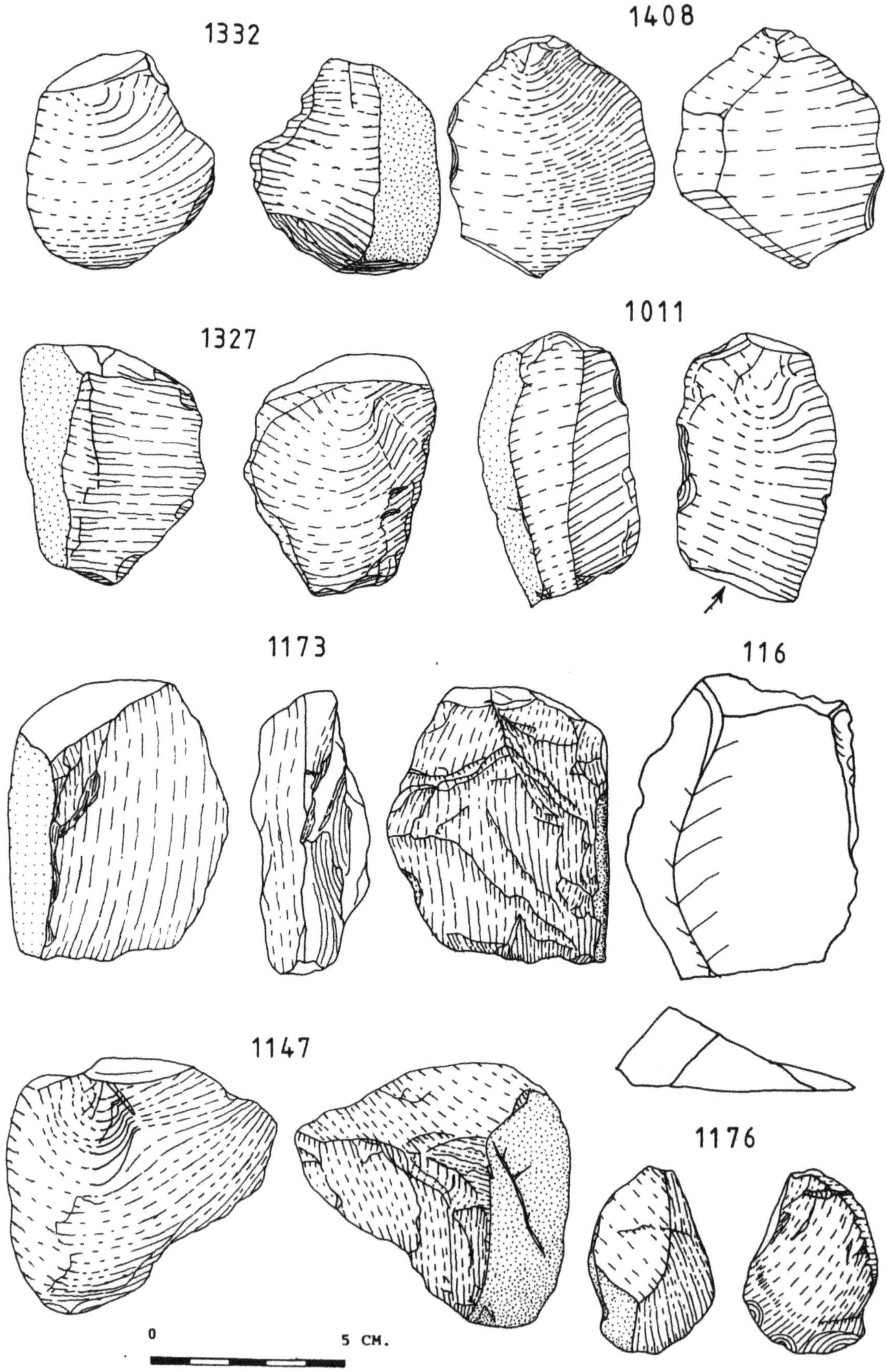

Fig. J.19 : Flakes and flake-like blades from site 55, Riwat

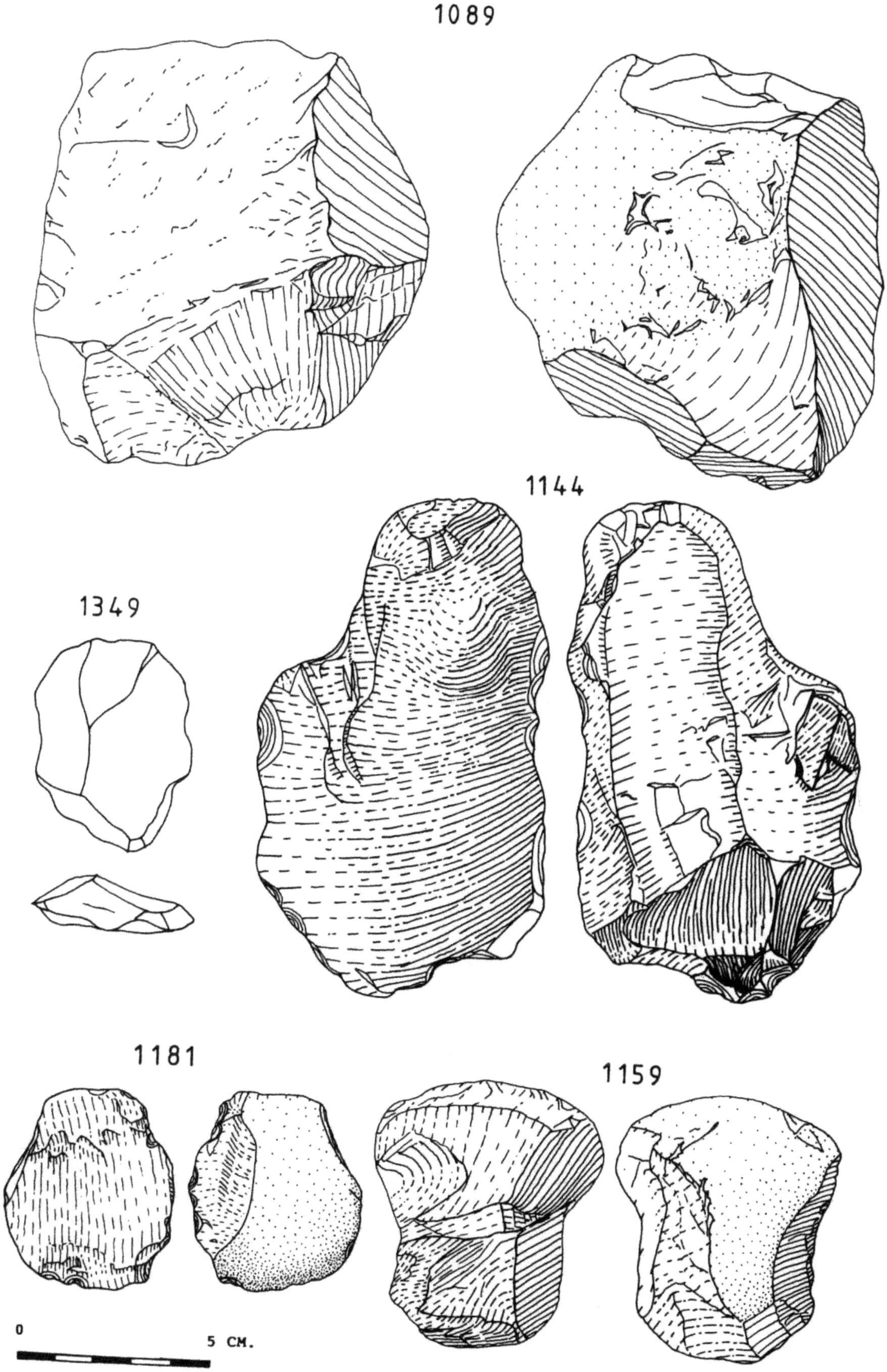

Fig. J.20 : Flakes and flake-like blades from site 55, Riwat

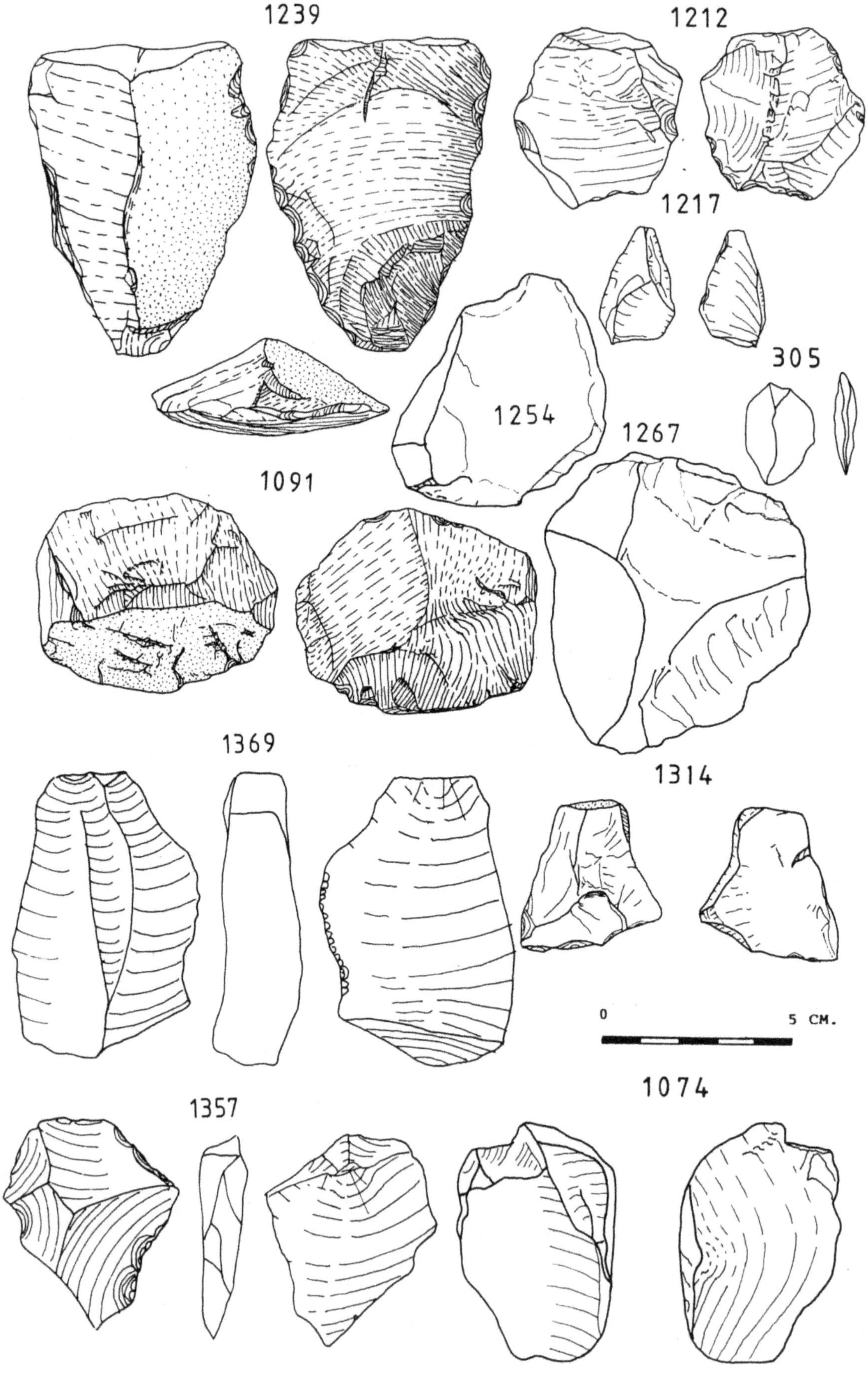

Fig. J.21 : Cores and core fragments from site 55, Riwat

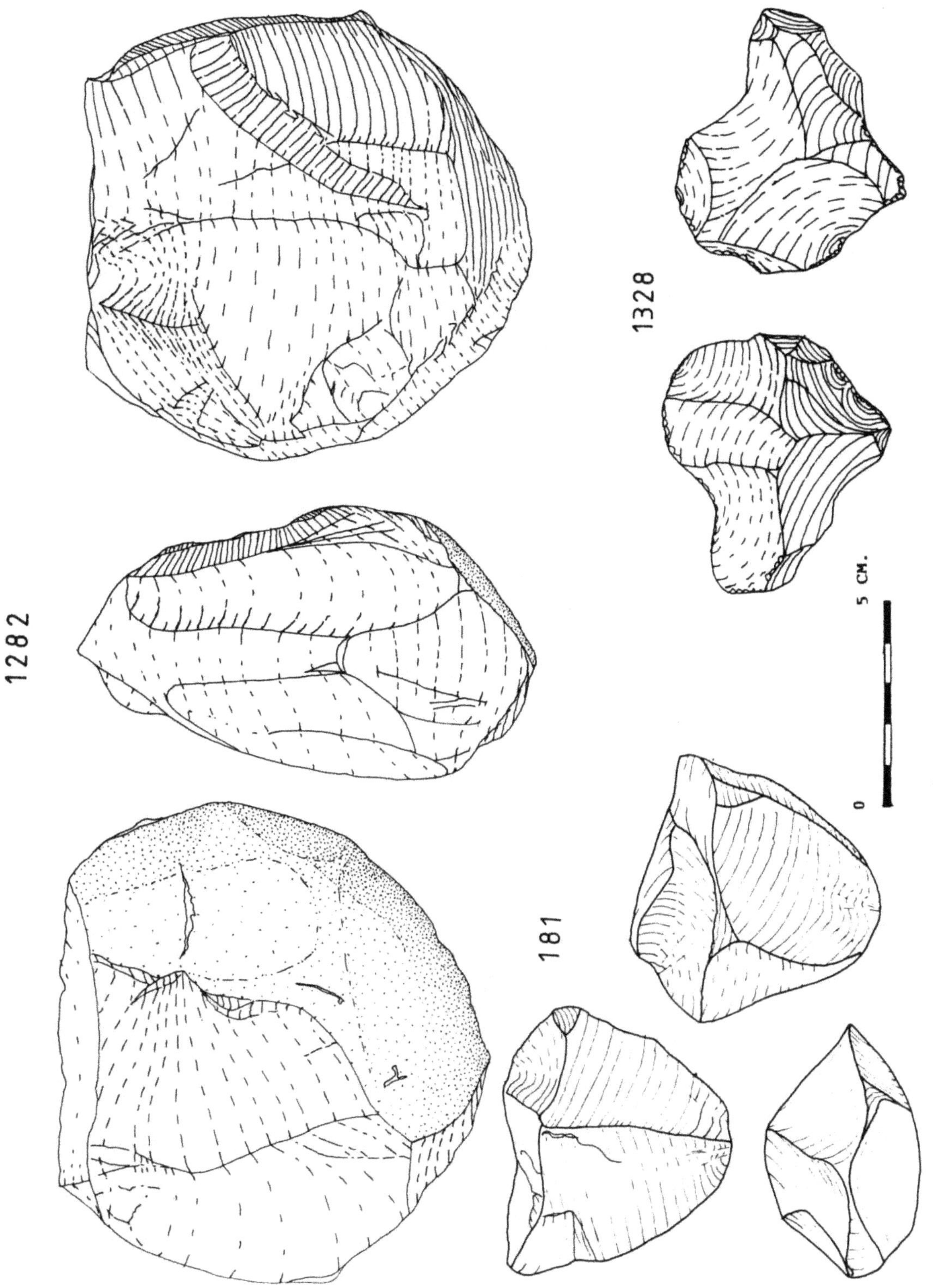

Fig. J.22 : Cores and core fragments from site 55, Riwat

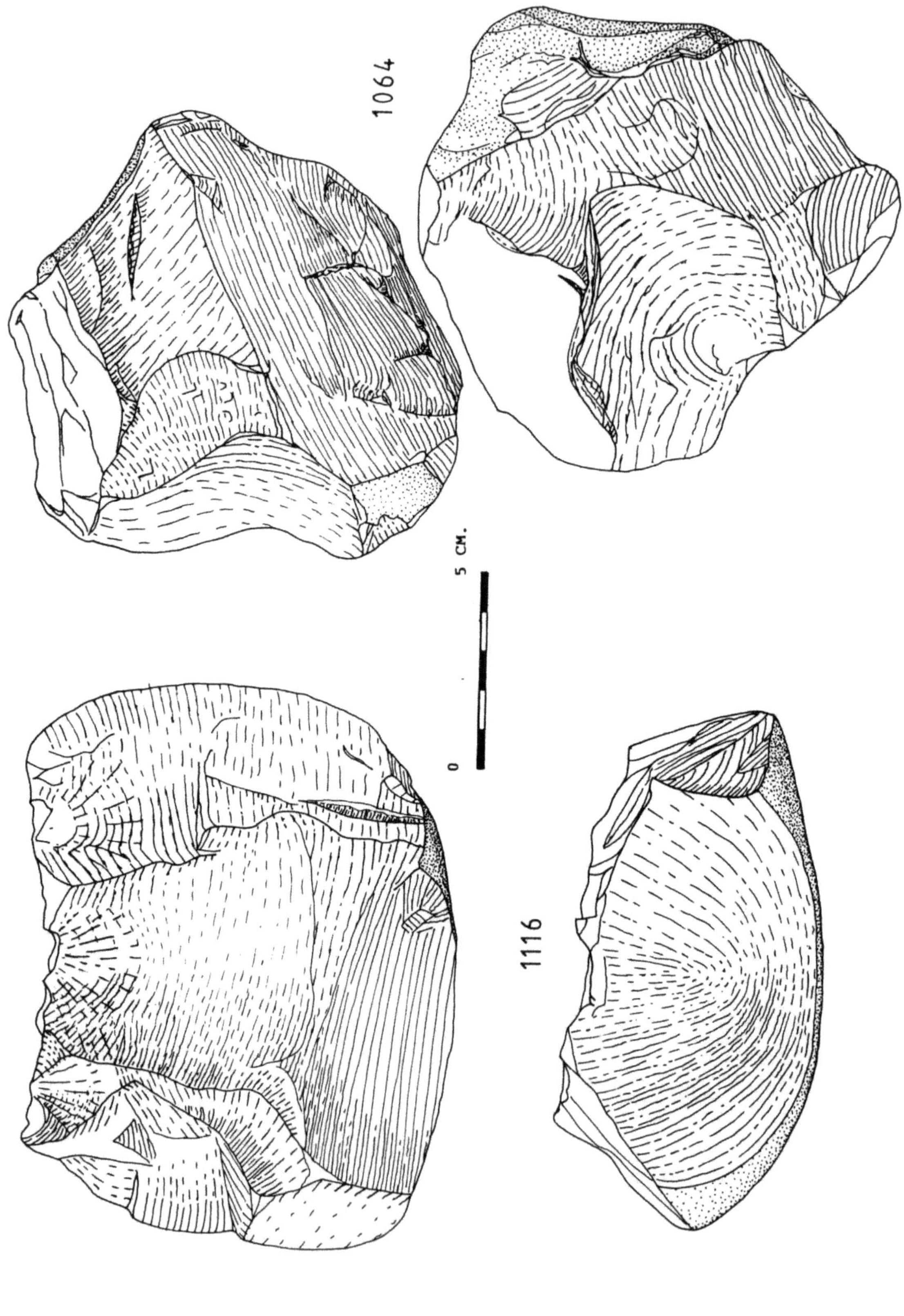

Fig. J.23 : Cores and core fragments from site 55, Riwat

1291

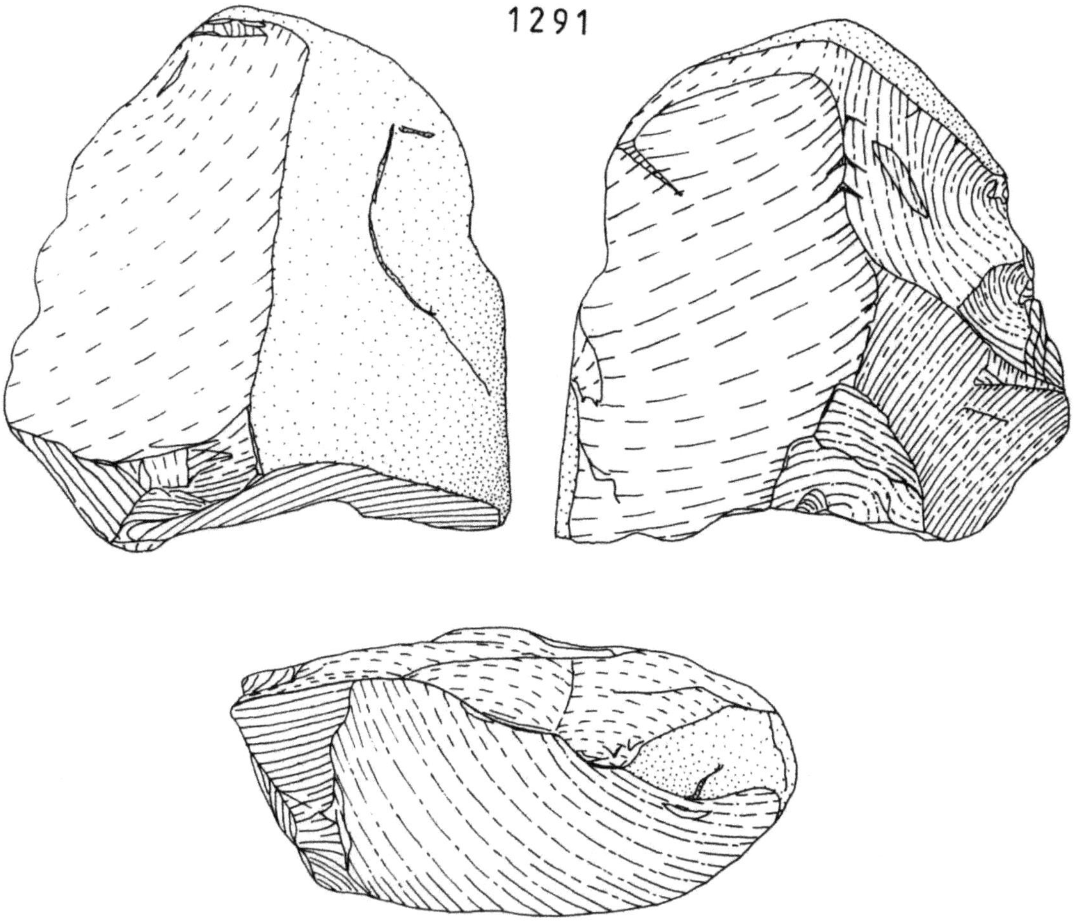

0 5 CM.

169

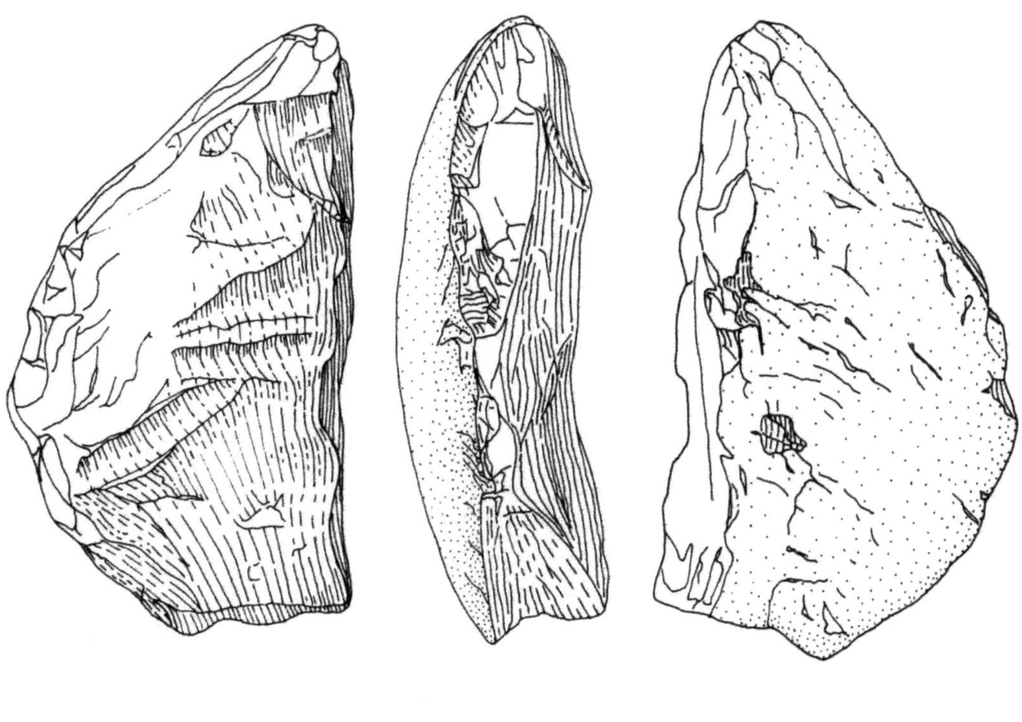

Fig. J.24 : Cores and core fragments from site 55, Riwat

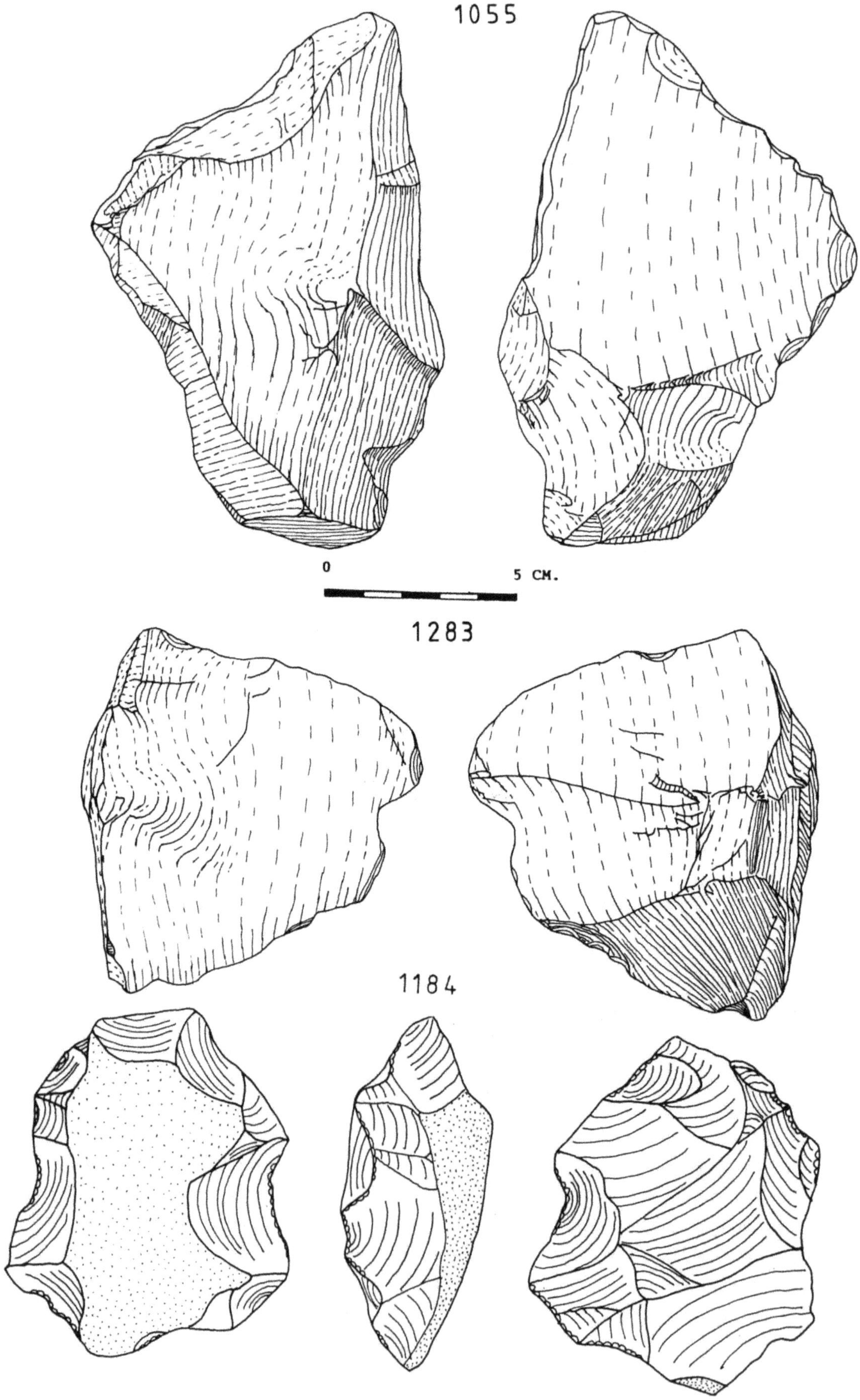

Fig. J.25 : Cores and core fragments from site 55, Riwat

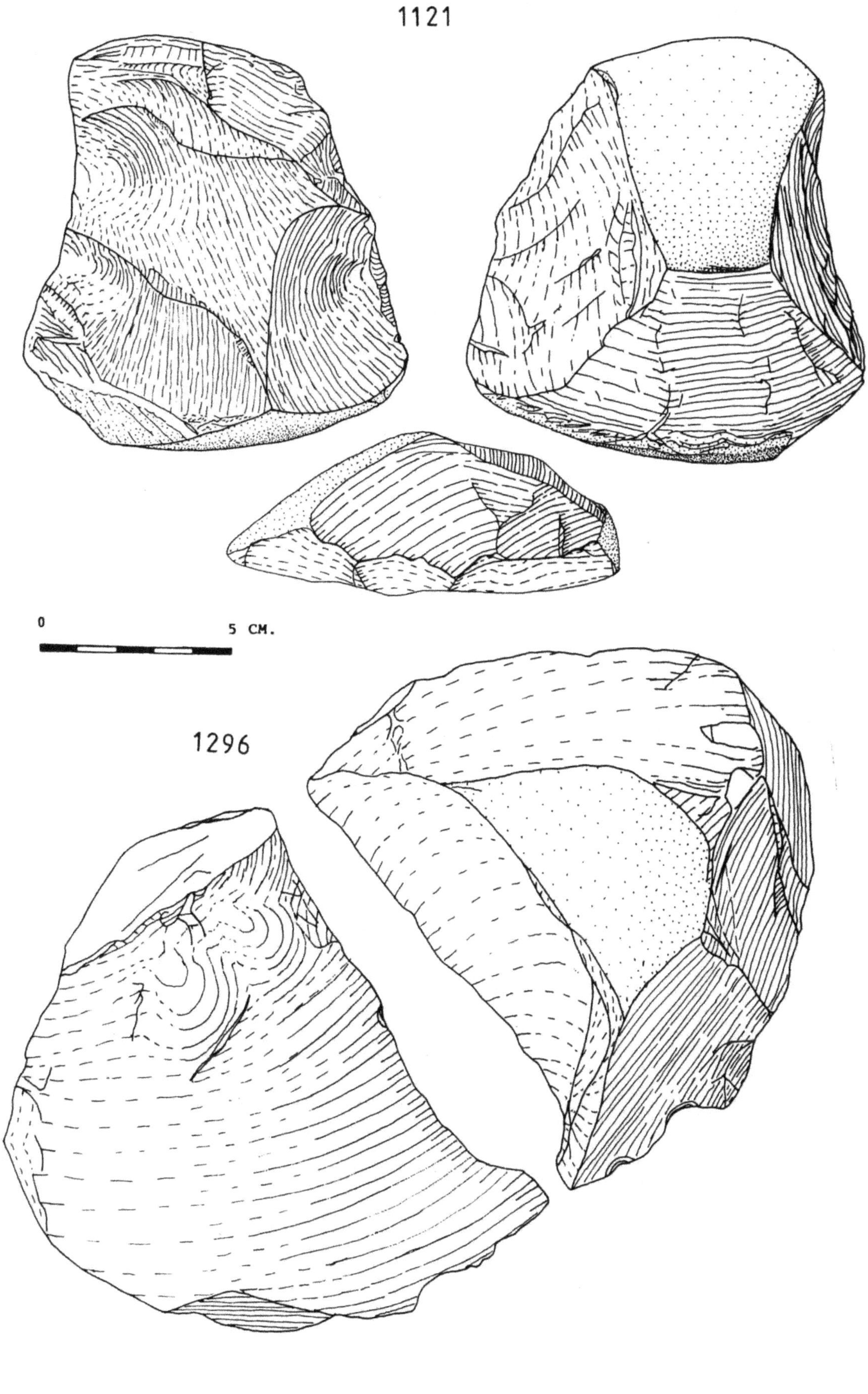

Fig. J.26 : Cores and core fragments from site 55, Riwat

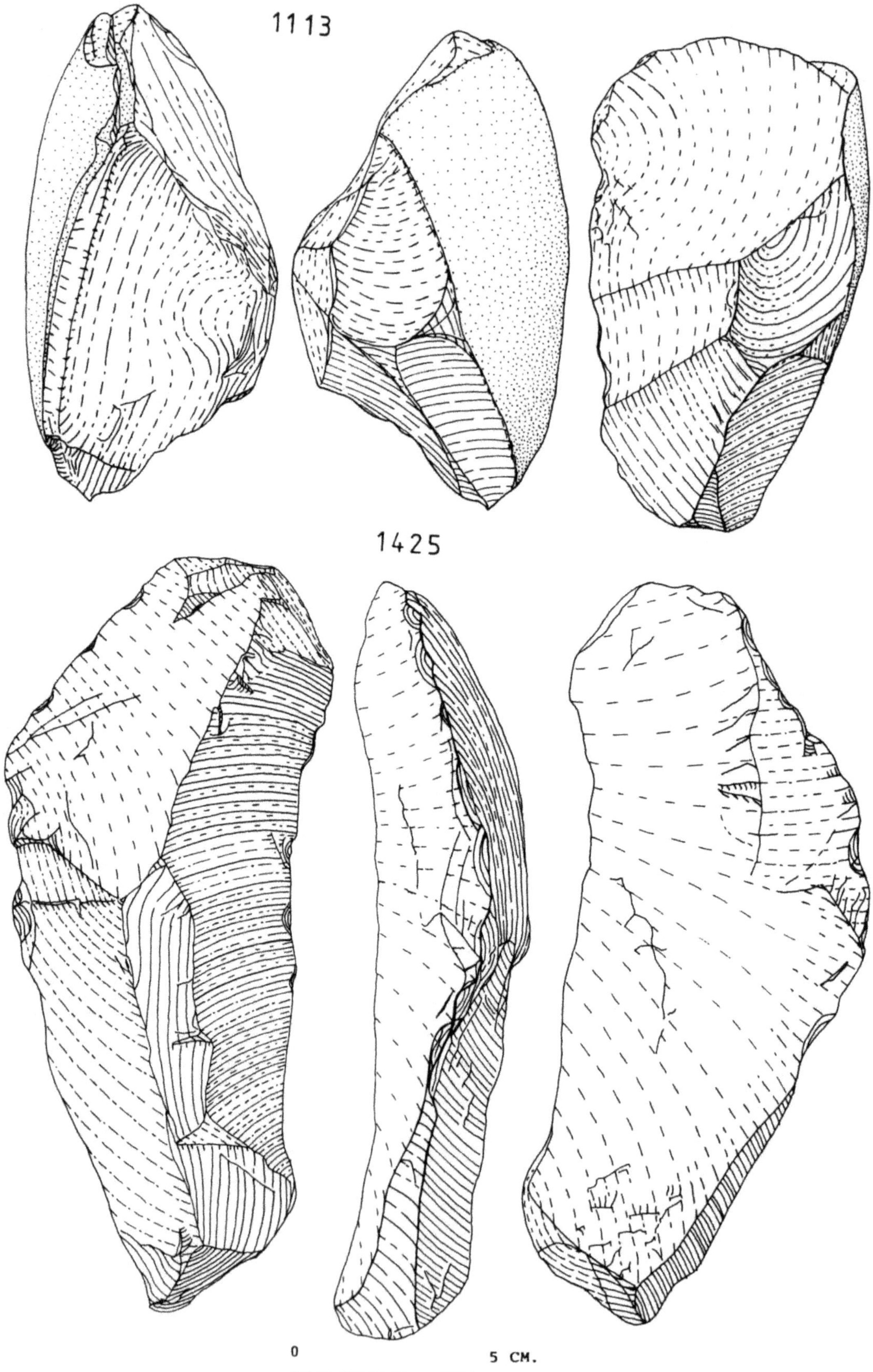

Fig. J.27 : Cores and core fragments from site 55, Riwat

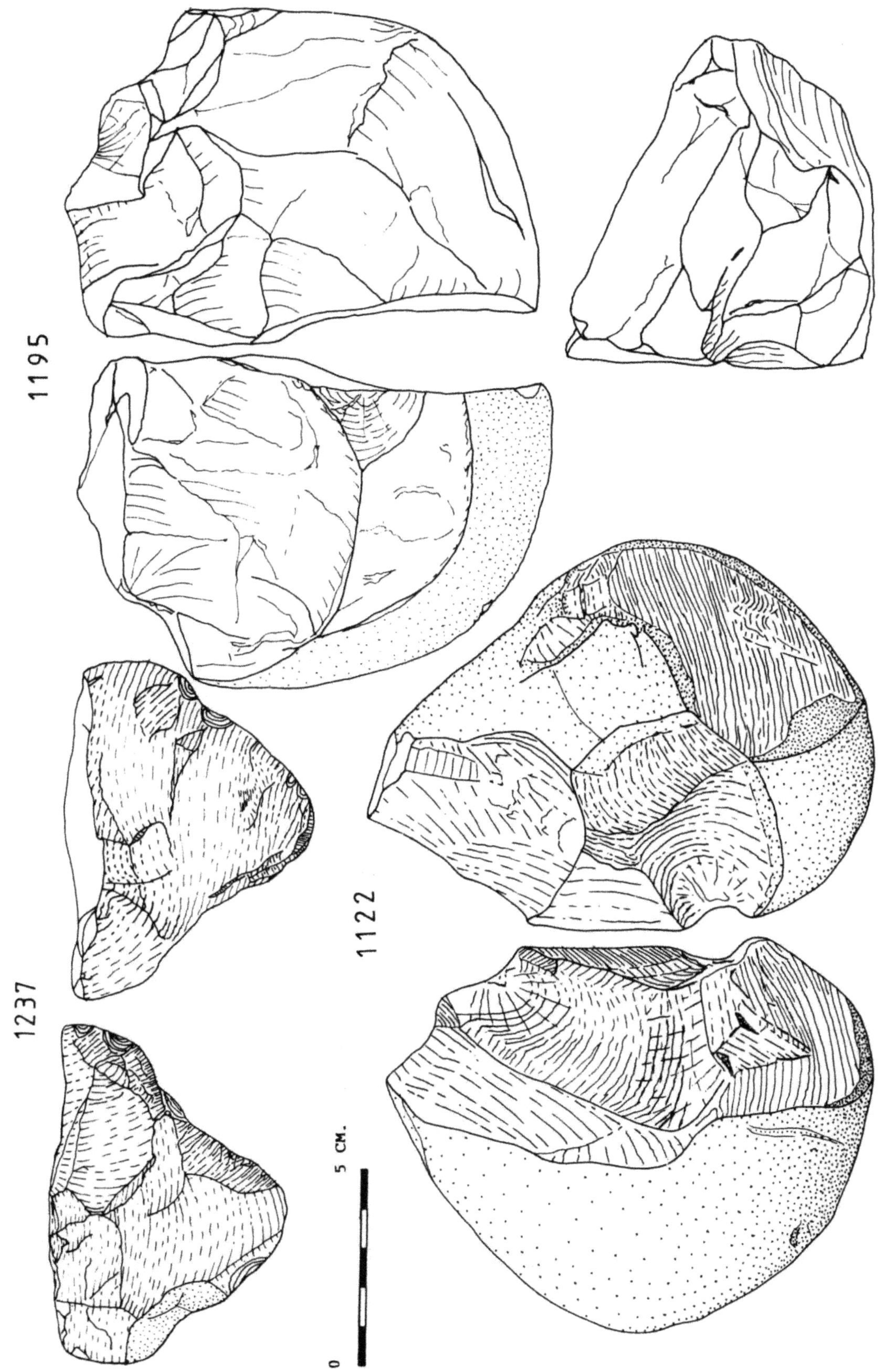

Fig. J.28 : Cores and core fragments from site 55, Riwat

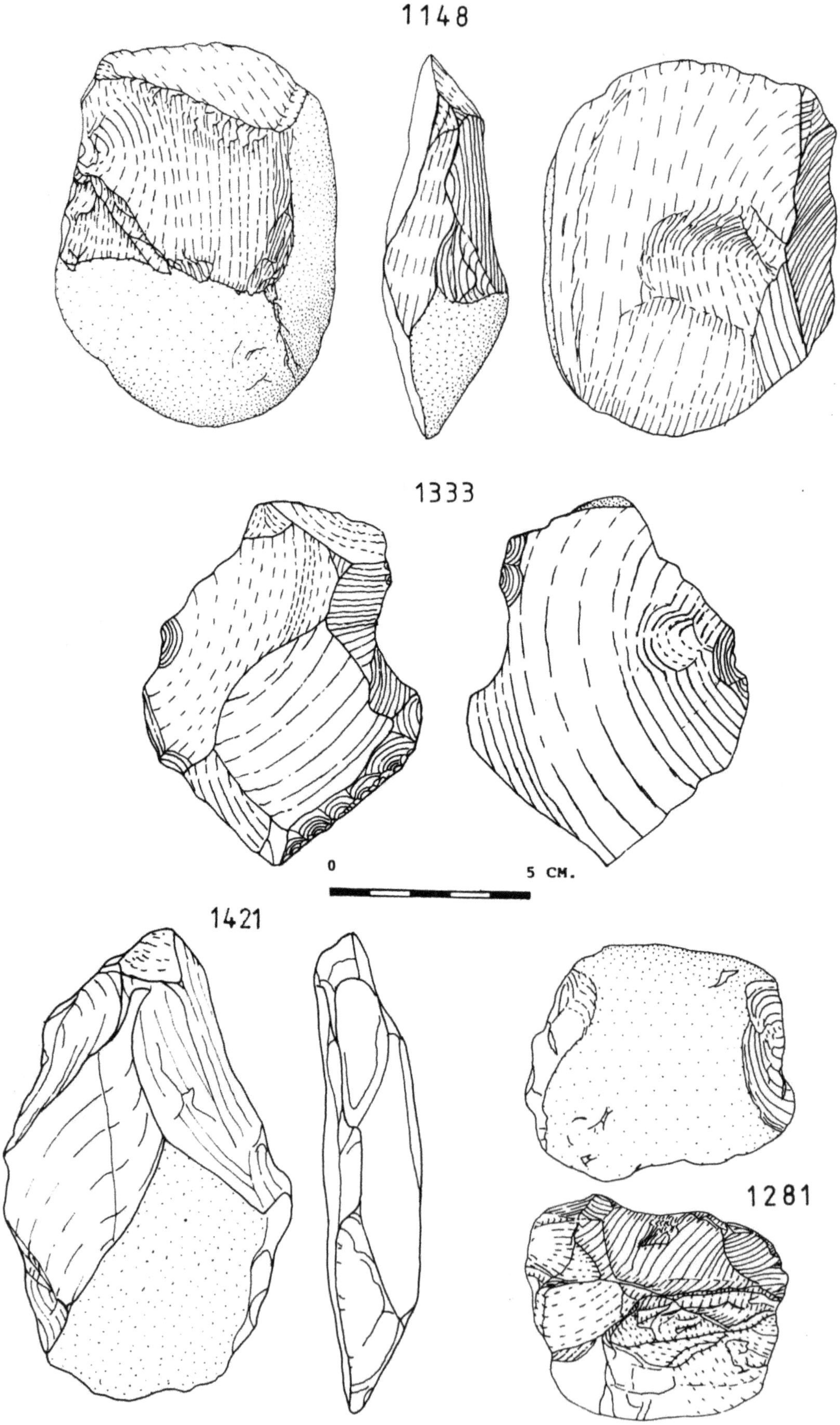

Fig. J.29 : Cores and core fragments from site 55, Riwat

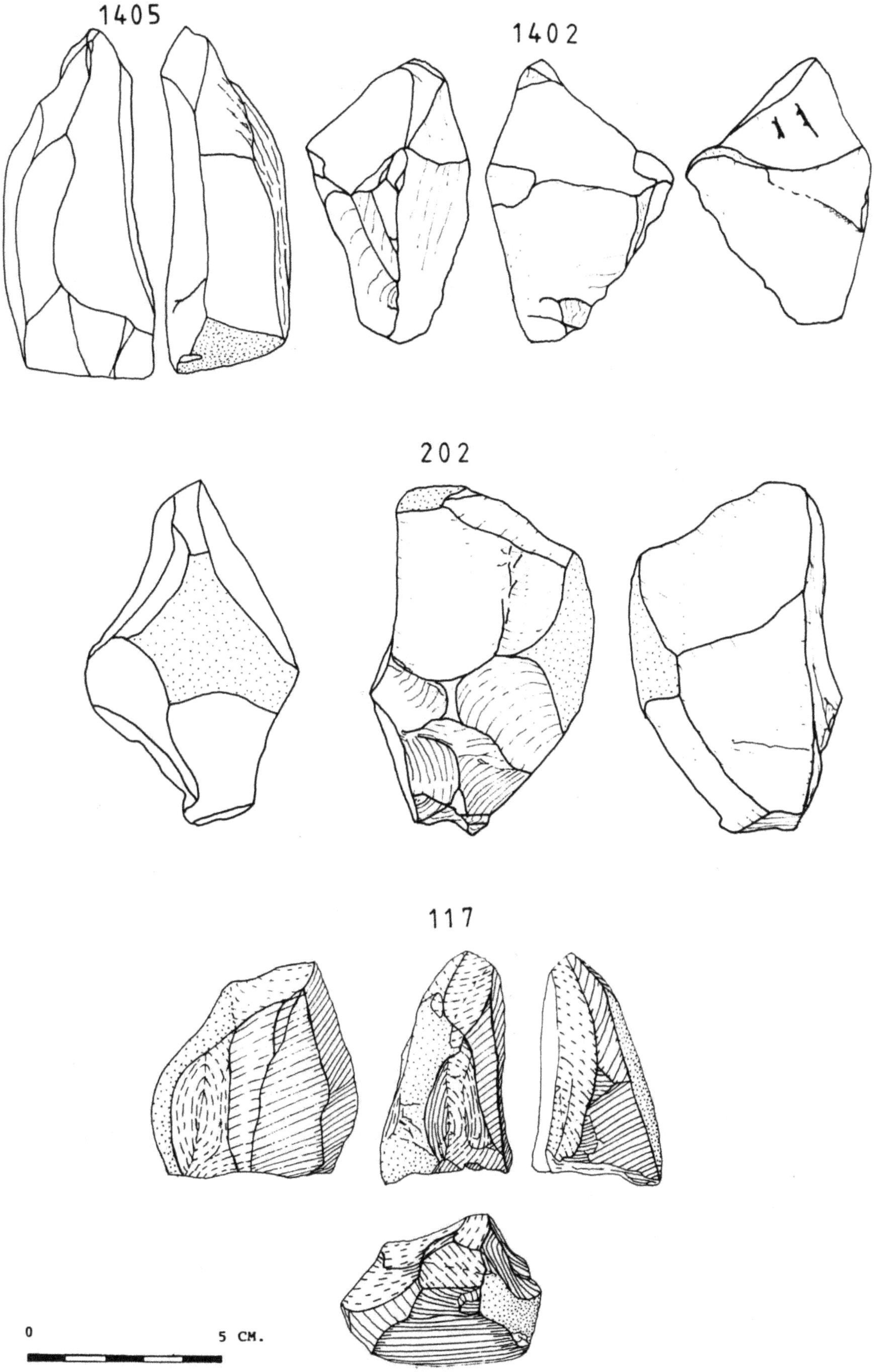

Fig. J.30 : Bifaces from site 55, Riwat

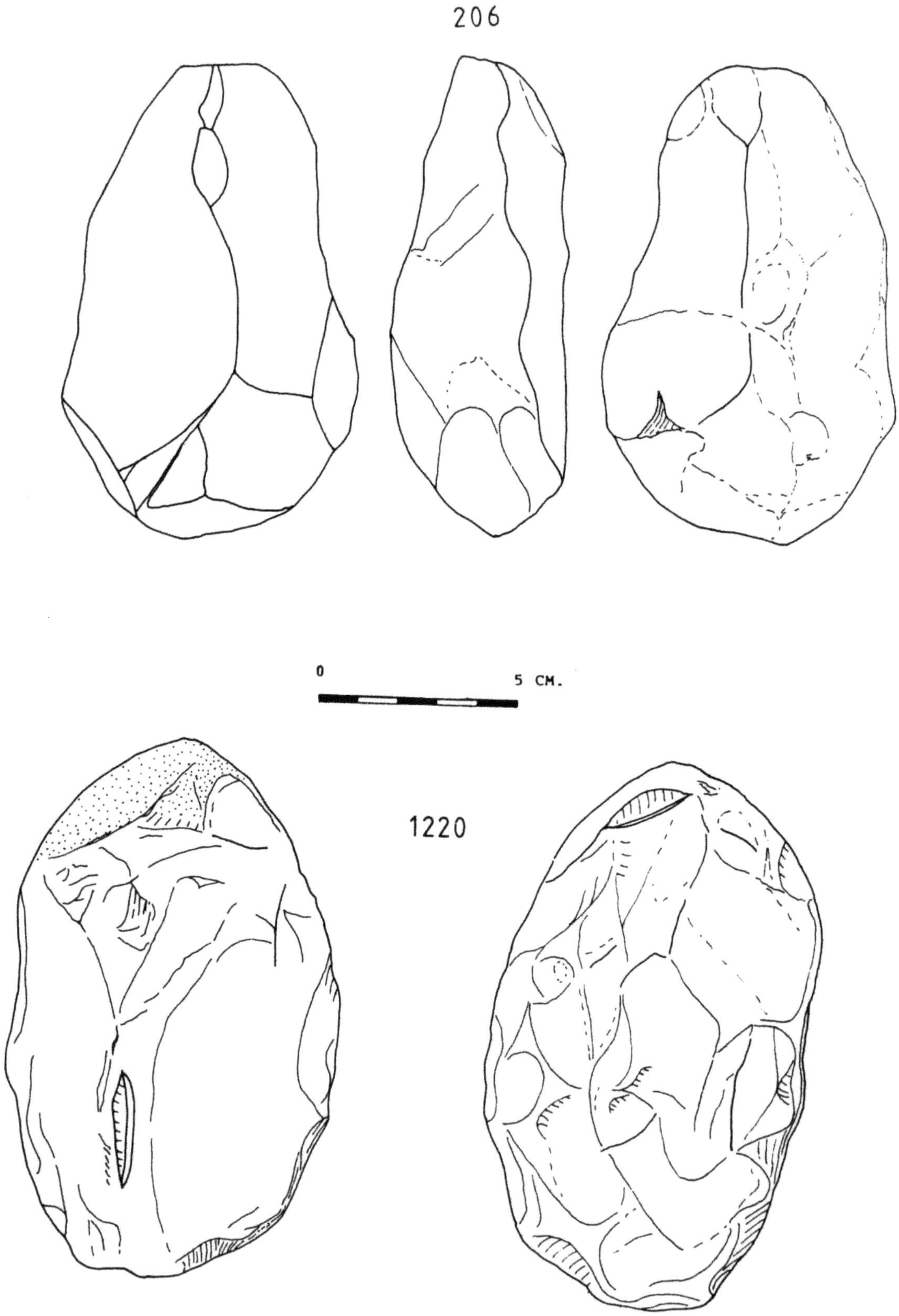

Fig. J.31 : Bifaces from site 55, Riwat

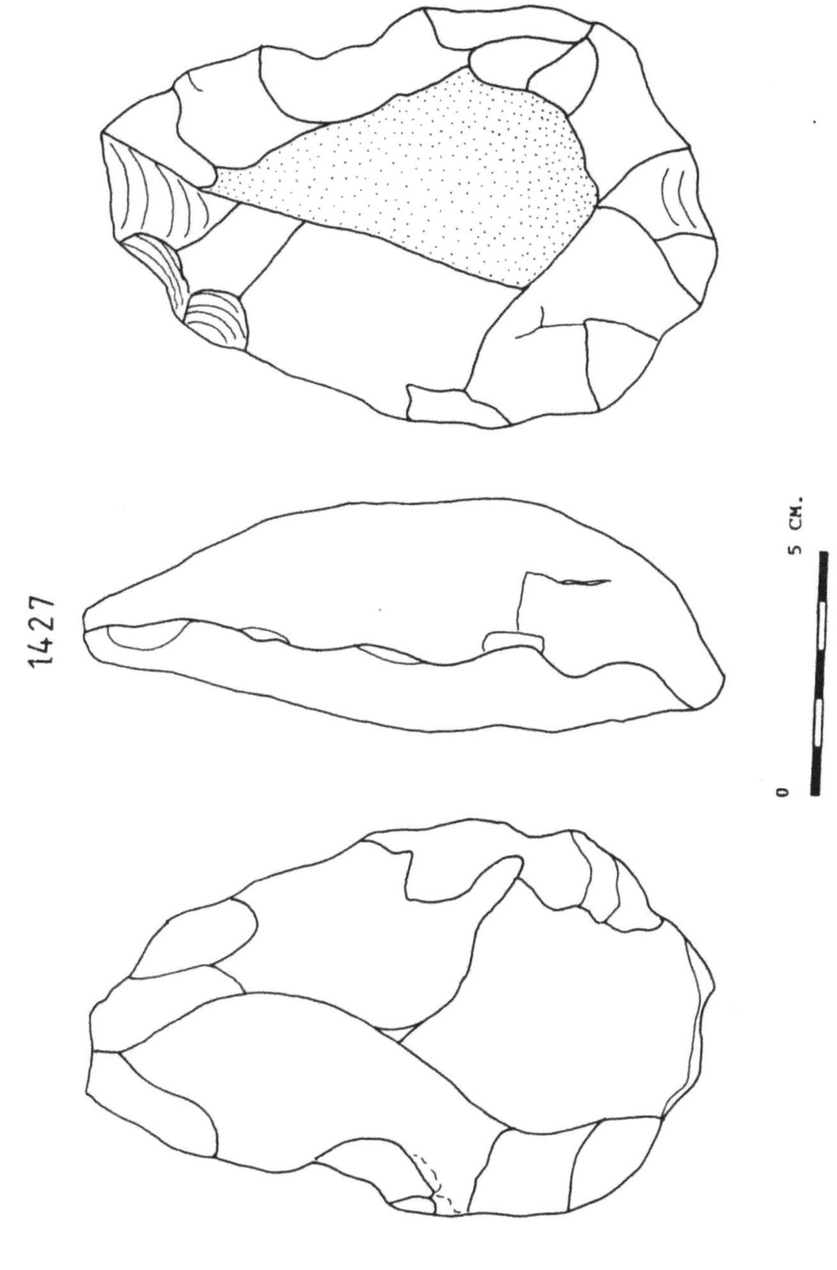

Appendix K

Catalogue of TL dates

sample	alpha counts /ks/cm^2	K_2O %	a value	annual dose Gy	equivalent dose Gy	TL age ka
PKXA5	1.01 ± 0.02	1.79 ± 0.05	0.075	0.00493	315.2 ± 31.5	63.95 ± 7.18
PKXB3	1.03 ± 0.02	1.74 ± 0.05	0.016	0.00580	342.8 ± 34.2	59.12 ± 6.22
PKXCB	1.03 ± 0.03	1.92 ± 0.06	0.088	0.00524	311.9 ± 31.2	59.52 ± 6.60
PKXCT	1.08 ± 0.03	1.92 ± 0.06	0.099	0.00562	319.4 ± 31.9	56.82 ± 6.19
PKS30	0.97 ± 0.03	1.89 ± 0.06	0.094	0.00520	241.9 ± 24.2	46.50 ± 5.10
PKS31	1.10 ± 0.03	1.91 ± 0.06	0.093	0.00520	231.9 ± 23.2	41.94 ± 4.52
PK553	0.97 ± 0.02	1.97 ± 0.06	0.100	0.00542	244.9 ± 24.9	45.11 ± 4.88
PKTRBT	0.93 ± 0.03	1.95 ± 0.06	0.093	0.00511	125.1 ± 12.5	24.47 ± 2.69
PK551	1.00 ± 0.03	2.03 ± 0.06	0.100	0.00559	138.2 ± 13.8	24.70 ± 2.68
PK552	1.03 ± 0.03	2.04 ± 0.06	0.160	0.00668	177.6 ± 17.8	26.57 ± 2.72
PKK3	0.95 ± 0.03	2.00 ± 0.02	0.095	0.00518	103.6 ± 10.0	19.95 ± 2.17
PKK2	0.92 ± 0.03	2.08 ± 0.02	0.095	0.00515	245.4 ± 24.0	47.77 ± 5.20
PKK1	0.82 ± 0.02	2.66 ± 0.03	0.095	0.00527	2220.0 ± 220.0	421.10 ± 56.90
PKRTUL	0.68 ± 0.02	2.00 ± 0.02	0.095	0.00422	313.9 ± 32.0	74.35 ± 8.36
PK60B	0.96 ± 0.02	2.20 ± 0.02	0.095	0.00543	394.0 ± 40.0	72.59 ± 7.99
PK57UL	0.83 ± 0.02	2.12 ± 0.02	0.095	0.00489	180.4 ± 18.0	36.85 ± 4.03
PK57LL	0.85 ± 0.02	2.41 ± 0.05	0.095	0.00520	193.5 ± 19.0	37.15 ± 4.10
PKRMG7	0.78 ± 0.02	2.76 ± 0.02	0.095	0.00517	878.7 ± 87.0	170.00 ± 20.40
PKL89UL	0.85 ± 0.02	2.46 ± 0.02	0.070	0.00497	202.2 ± 5.1	40.63 ± 4.60
PKL89LL	0.90 ± 0.02	2.30 ± 0.02	0.080	0.00514	256.4 ± 19.1	49.81 ± 4.90
PKL89B	0.90 ± 0.02	2.26 ± 0.02	0.080	0.00511	456.9 ± 45.0	89.35 ± 9.80
PKL58B	0.90 ± 0.02	2.20 ± 0.05	0.090	0.00521	261.9 ± 8.5	50.25 ± 4.90
PKL6	0.90 ± 0.02	1.90 ± 0.05	0.090	0.00496	130.0 ± 13.0	26.19 ± 2.85
PKR891	0.80 ± 0.02	2.50 ± 0.02	0.118	0.00536	173.8 ± 11.0	32.37 ± 3.20
PKR892	0.85 ± 0.02	2.50 ± 0.02	0.052	0.00469	187.2 ± 13.0	39.86 ± 4.25
PKR893	0.85 ± 0.02	2.40 ± 0.02	0.069	0.00475	217.4 ± 10.7	45.77 ± 4.10
PKR894	0.95 ± 0.02	2.30 ± 0.02	0.059	0.00482	226.0 ± 30.0	46.88 ± 4.20
PKR895	0.94 ± 0.02	2.20 ± 0.02	0.071	0.00501	292.3 ± 15.4	58.33 ± 5.30
PKR897	1.16 ± 0.02	2.20 ± 0.02	0.079	0.00573	340.4 ± 25.9	59.42 ± 5.90
PKR899	0.99 ± 0.02	2.20 ± 0.02	0.086	0.00537	402.6 ± 36.0	74.98 ± 7.90

sample	alpha counts /ks/cm^2	K_2O %	a value	annual dose Gy	equivalent dose Gy	TL age ka
PKR8911	0.88 ± 0.02	2.20 ± 0.02	0.094	0.00510	458.8 ± 34.5	89.94 ± 8.90
PKR8915	0.88 ± 0.02	2.10 ± 0.02	0.075	0.00476	457.5 ± 23.0	96.03 ± 8.70
PKR8917	0.90 ± 0.02	2.10 ± 0.02	0.080	0.00498	537.3 ± 16.9	107.96 ± 9.19
PKR8919	0.89 ± 0.02	2.10 ± 0.05	0.083	0.00501	680.3 ± 50.0	135.80 ± 12.60
PKC5	0.73 ± 0.02	2.32 ± 0.05	0.095	0.00472	113.6 ± 11.0	24.11 ± 2.67
PKC4	0.79 ± 0.02	2.28 ± 0.02	0.095	0.00483	267.5 ± 27.7	55.34 ± 6.20
PKC3	1.06 ± 0.05	2.08 ± 0.02	0.095	0.00565	333.5 ± 35.0	59.03 ± 6.73
PKC2	0.87 ± 0.02	1.74 ± 0.02	0.095	0.00472	288.4 ± 24.4	61.12 ± 6.22
PKRDSL	0.90 ± 0.02	2.21 ± 0.02	0.090	0.00522	93.9 ± 6.4	17.90 ± 1.68
PKRDUL	0.90 ± 0.02	2.36 ± 0.02	0.090	0.00534	196.4 ± 19.4	36.60 ± 4.00
PKRDML	0.90 ± 0.02	2.17 ± 0.02	0.090	0.00518	291.4 ± 13.8	56.20 ± 4.90

www.ingramcontent.com/pod-product-compliance
Lightning Source LLC
LaVergne TN
LVHW070045021026
849160LV00027B/1615